# Conservation and Adaptive Management of Seamount and Deep-Sea Coral Ecosystems

# Conservation and Adaptive Management of Seamount and Deep-Sea Coral Ecosystems

Robert Y. George and Stephen D. Cairns
Editors

Rosenstiel School of Marine and Atmospheric Science
University of Miami

*This volume may be referred to as:*

R. Y. George and S. D. Cairns, eds. 2007. Conservation and adaptive management of seamount and deep-sea coral ecosystems. Rosenstiel School of Marine and Atmospheric Science, University of Miami. Miami. 324 p.

*Conservation and adapted management of seamount and deep-sea coral ecosystems* was printed for the Rosenstiel School of Marine and Atmospheric Science by AllenPress, Inc., 800 East 10th Street, Lawrence, Kansas 66044, USA. © 2007 Rosenstiel School of Marine and Atmospheric Science, University of Miami.

ISBN 1–891276–58–1

Library of Congress Control Number: 2007942141

This volume is available from the Editorial Office of the *Bulletin of Marine Science,* Rosenstiel School of Marine and Atmospheric Science, University of Miami, 4600 Rickenbacker Causeway, Miami, Florida 33149-1031, USA. Telephone: (305) 421-4624, Fax: (305) 421-4600, E-mail: bmsaccounting@rsmas.miami.edu. The price is $50 ($53 international) including shipping and handling.

# Conservation and Adaptive Management of Seamount and Deep-Sea Coral Ecosystems

## Table of Contents

George, R. Y. and S. D. Cairns, eds. 2007. Conservation and adaptive
management of seamount and deep-sea coral ecosystems. Rosenstiel
School of Marine and Atmospheric Science, University of Miami.

# Preface

Research over the last two decades has revealed remarkable deep-sea coral ecosystems in the world's oceans. The 3rd International Symposium on Deep-Sea Corals was held at the University of Miami in December 2005. The symposium was attended by nearly 300 participants from 21 nations, and focused on the science and management of these spectacular but vulnerable ecosystems. The United States Department of Commerce's National Oceanic and Atmospheric Administration and the Department of the Interior's U. S. Geological Survey and Minerals Management Service are proud to have been major sponsors of this symposium.

We are pleased that proceedings of the Miami symposium are being published by the Rosenstiel School of Marine and Atmospheric Science, University of Miami, in this book dedicated to policy and management issues, *Conservation and Adaptive Management of Seamount and Deep-Sea Coral Ecosystems*. This book is the companion to an additional 18 scientific papers from the symposium published as "Deep-Sea Coral Ecosystems: Biology and Geology," a dedicated volume of *Bulletin of Marine Science* (Volume 81, Number 3), November 2007.

Since the First International Symposium on Deep-Sea Corals was held in Canada in 2000, there has been a rapid expansion of research on these unique ecosystems. Equally remarkable has been the speed with which discoveries of these biologically diverse and vulnerable ecosystems have been translated into conservation action. The 23 peer-reviewed papers in this volume represent the perspectives of resource managers, scientists, fishermen, and environmentalists, on a range of management and policy issues related to the conservation of deep-sea coral ecosystems. Integrated science, expanded research, and collaborative policymaking are central to ensuring the conservation of our ocean ecosystems and the sustainable use of their resources in the 21st century.

Since the Miami symposium, efforts to translate research into management action have continued to accelerate. Actions have been taken in the United Nations General Assembly, by Regional Fisheries Management Organizations (RFMOs), and by numerous countries. In 2006, the United States protected over 1.3 million km$^2$ of seafloor from bottom-trawling, including unique deep-sea coral habitats, and Congress included a new Deep-sea Coral Research and Technology Program in reauthorizing the nation's fisheries legislation. In 2007, the Joint Subcommittee on Ocean Science and Technology also established an Interagency Board on Deep-Sea Coral and other Vulnerable Ecosystems to assist in the coordination of U.S. research strategies.

We commend the editors of this volume, Dr. Robert Y. George and Dr. Stephen D. Cairns, for their life-long contributions to the understanding of deep-sea ecosystems. Dr. Cairns is Curator of Cnidaria at the Smithsonian Institution's National Museum of Natural History and has made extensive contributions to the biology, systematics and zoogeography of corals. Professor George (President, George Institute for Biodiversity and Sustainability) is a pioneer in deep-sea research and served as the co-organizer of

George, R. Y. and S. D. Cairns, eds. 2007. Conservation and adaptive management of seamount and deep-sea coral ecosystems. Rosenstiel School of Marine and Atmospheric Science, University of Miami.

1

the Miami deep-sea coral symposium. Dr. George also organized the "Harvard University Declaration Conference" in collaboration with the renowned conservation biologist Prof. Edward O. Wilson, for protecting biodiversity of deep-sea corals and associated fish and invertebrates in the continental margin within US Exclusive Economic Zone and on the seamounts.

*Timothy Keeney*
Deputy Assistant Secretary for Oceans and Atmosphere
U.S. Department of Commerce

*Kameran Onley*
Deputy Assistant Secretary
U.S. Department of the Interior

September 27, 2007
Washington DC

# Dedication

In solemn recognition of the late Dr. Robert M. Avent, the original discoverer of the *Oculina* cold coral reefs off the east coast of Florida, this book, *Conservation and Management of the Seamounts and Deep-Sea Coral Ecosystems,* from the Proceedings of the 3rd International Deep-Sea Coral Symposium in Miami is dedicated to his honor and memory. The accompanying photo shows Bob in his youth while entering the JOHNSON SEA-LINK submersible, which he used to study the *Oculina* coral reefs when he was Research Oceanographer with the Harbor Branch Oceanographic Institute in Fort Pierce, Florida.

Bob worked for the Mineral Management Service of the Department of Interior in New Orleans, Louisiana for 27 yrs and compiled the Outer Continental Shelf (OCS) Report on MMS "Biological Investigations in the Gulf of Mexico 1973–2000". Just a few months before he died prematurely after a lung cancer operation in December 2002, Bob Avent enthusiastically told me about the discovery of *Lophelia* coral lithoherms and chemoherms in the Gulf of Mexico.

Bob received his Ph.D. from the graduate department of Oceanography at Florida State University when I was on the faculty in the same department. Together we published papers on adaptations of marine animals to hydrostatic pressure. His comprehensive research paper, published in the *International Review of Hydrobiology* on the benthic communities associated with hard bottom off the Florida Atlantic coast, speaks to his depth of knowledge of species associated with cold coral ecosystems. I sailed with Bob on many cruises in the Gulf of Mexico and off the U.S. East Coast. His colleagues and friends will remember him for his charisma, enthusiasm, and his affable personality.

*Robert Y. George*

# Conservation and Adaptive Management of Seamount and Deep-Sea Coral Ecosystems

PROCEEDINGS OF THE THIRD INTERNATIONAL
SYMPOSIUM ON DEEP-SEA CORALS
ROSENSTIEL SCHOOL OF MARINE AND ATMOSPHERIC SCIENCE
UNIVERSITY OF MIAMI
MIAMI, FLORIDA, USA, NOV. 28–DEC. 2, 2005

## Introduction

In Halifax (Canada), Martin Willison of Dalhousie University and Susan Gass of the Ecology Action Center in collaboration with Mark Butler and others conducted the First International Symposium on Deep-sea Corals in the beginning of the 21st century (July 30–August 3, 2000). They wrote in the introduction of their book on the proceedings of the symposium (Willison et al., 2002): "Through this volume we hope to let a wide audience know that coral reefs are found in cold ocean waters, that corals are abundant in both Canada and the United States, and that threats to them are very real". Now we know that there are two kinds of coral reefs, the tropical coral reefs in the equatorial latitudes at temperature in excess of 21 °C and the cold deep-water coral reefs in temperate, boreal, and subarctic latitudes (in both northern and southern hemispheres). The azooxanthellate corals, lacking symbiotic zooxanthellate algae, occur mostly at bathyal depths (40–950 m). The Halifax deep-sea coral symposium also resulted in a special number of the International Journal of Limnology and Marine Sciences, *Hydrobiologia*, edited by Les Watling and Michael Risk (2002) with 18 peer-reviewed papers on deep-sea corals.

In Erlangen (Germany) Andre Freiwald and Murray Roberts (Scottish Marine Biological Association) conducted the Second International Symposium on Deep-sea Corals and they jointly edited the proceedings of the symposium as a book published by Springer Verlag (Freiwald and Roberts, 2005). It included 62 articles and 1210 pages. In Miami (Florida), Robert George of George Institute for Biodiversity and Sustainability (GIBS) and Robert Brock of National Oceanic and Atmospheric Administration (NOAA) conducted the Third International Deep-sea Coral Symposium in 2005 at the Rosenstiel School of Marine and Atmospheric Science (RSMAS) of the University of Miami. The deep-sea coral research community grew from Halifax to Miami, from 42 oral presentations (and 22 posters) to 102 oral presentations (and 88 posters).

We present 41 papers from participants of the Miami symposium for publication in the proceedings of the Third International Symposium on Deep-sea Corals to be published as an issue of the *Bulletin of Marine Science* as well as a companion book. Titled *Conservation and Management of Seamounts and Deep-Sea Coral Reef Ecosystems*, the book includes 23 peer-reviewed papers focused on policy and science related to

George, R. Y. and S. D. Cairns, eds. 2007. Conservation and adaptive management of seamount and deep-sea coral ecosystems. Rosenstiel School of Marine and Atmospheric Science, University of Miami.

5

the book includes 23 peer-reviewed papers focused on policy and science related to management of these ecosystems. The *Bulletin of Marine Science* issue consists of 18 peer-reviewed papers focused on scientific aspects of deep-sea corals.

We acknowledge the three main sponsors of the Miami deep-sea coral symposium in 2005: National Oceanic and Atmospheric Administration (NOAA), Mineral Management Service (MMS) of the U.S. Department of Interior, and United States Geological Survey (USGS) of the Department of Interior. The Miami symposium was also co-sponsored by the Smithsonian Institution, George Institute for Biodiversity and Sustainability (GIBS), Environmental Defense, PEW Institute for Ocean Sciences (PIOS), Marine Conservation Biology Institute (MCBI), and the International Council for the Exploration of the Seas (ICES). We also thank the board of directors of GIBS for providing travel funds for participants of the post-symposium "Harvard Declaration Conference" on October 24–25, 2006.

This companion book focuses on policies and management of deep-sea coral habitats. Here we highlight a few new trends and findings:

(1) Deep-sea coral reef ecosystems in the northeast Atlantic Ocean along the coast of the nations of the European Union have already experienced severe damages. Restoration and banning of bottom trawling must be soon implemented with the creation of seamount and coral mound MPAs within EEZs and in highseas.

(2) Conflict reduction between stakeholders is an important step to achieving deep-sea coral protection. Fishing industries that fish in the so-called "global commons" (100,000 seamounts inclusive) need to have meaningful dialogue in ICES-sponsored workshops, leading to a treaty under UN authorization involving ISA (International Seabed Authority) and FAO (Food and Agricultural Organization). Likewise "Science Priority Areas" (SPAs) are proposed for management by a committee of deep-ocean scientists with a genuine quest to establish areas reserved for repeated and long-term oceanographic studies and to protect scientific investments (Thiel).

(3) Ecosystem Based Fisheries Management (EBFM) should receive high priority (Morgan) for funding under RFMO (Regional Fisheries Management Organization) jurisdiction as well as UN agencies and the Census of Marine Life. A conceptual model of the foodweb in any chosen deep-sea coral ecosystem first must be developed (George et al.) and then followed by a precautionary model and establishment of EFH-HAPC (Essential Fish Habitats-Habitat Area of Particular Concern, as per Magnusson-Stevens Act reauthorization in 2007. The goal is to prevent further damage and biodiversity decline by over-fishing in these vulnerable habitats (Ahlfeld, Breeze, DeSanto). Seamounts and cold coral substrata are already known as oases, rest-areas, and feeding or spawning grounds for highly migratory species of tuna, swordfish and sharks: this companion book also adds substantial new information on fish associated with cold coral reefs (Auster, Sulak).

Chandra George helped in compiling the manuscripts in the form of CDs and hard copies, besides maintaining the files of all reviews. We particularly thank the reviewers who did the behind-the-scenes work to evaluate the scientific merit and accuracy of the papers. We list below the Associate Editors for this book along with the number of papers they handled in parentheses.

Editors: Robert Y. George, GIBS (9); Stephen Cairns, Smithsonian Institution (4); Associate Editors: John Reed, Harbor Branch Oceanographic Institute, Florida (2); Lance

Morgan, Marine Conservation Biology Institute, California (4); Kenneth Sulak, USGS, Univ. of Florida, Gainesville, FL (3); William Schroeder, University of Alabama (1)

Post Symposium Progress: George Institute for Biodiversity and Sustainability (GIBS), arranged a post-symposium event on October 24–25, 2006 at Harvard University. This "Harvard Declaration Meeting", was co-chaired by Edward O. Wilson of the Agassiz Museum of Harvard University, and Robert Y. George (GIBS). The recommendations of this conference led to the action of the Joint Subcommittee for Ocean Science and Technology (JSOST) to create the "Interagency Board on Deep-Sea Corals". Invited participants (22 in total), representing the United States Government (NOAA, MMS, and USGS—the three main sponsors of the Miami symposium and the Smithsonian Institution), several non-governmental organizations—NGOs (PEW Institute for Ocean Science, Environmental Defense, George Institute for Biodiversity and Sustainability, Marine Conservation Biology Institute, and World Water Watch Institute) and academia (Harvard University, Duke University, and University of Massachusetts) contributed to the unanimous recommendation to create the "Deep-Sea Coral Board".

Subsequently, the Harvard Declaration conference organizer (George) on December 4, 2006 reported the outcome of the Oct. 24–25 conference to the Department of Commerce in Washington, DC, with Assistant Deputy Secretary for Oceans and Atmosphere Hon. Timothy Keeney and Interior Department's Assistant Secretary Kameran Onley. A consensus was reached to create the "Deep-Sea Coral Board", which is within the jurisdiction of the Joint Subcommittee of Ocean Science and Technology (JSOST), under the COP (Committee for Ocean Policy) and CEQ (Council for Environmental Quality). This Interagency Deep-Sea Coral Board is now established and met for the first time on June 29, 2007. Robert George addressed this Deep-Sea Coral Board and representatives from several NGOs (Greenpeace, Oceana, MCBI, GIBS) by invitation on August 3 at NOAA headquarters in Silver Spring, Maryland, and offered advice to work toward protection of biodiversity in the vulnerable deep-water and deep-sea cold coral habitats on the shelf-slope environment, plateaus, canyons, seamounts, and adjacent seascapes including vents, cold seeps, trenches (hadal zone), and other abyssal zones within US EEZs and in the Areas Beyond National Jurisdiction (ABNJ).

It is also important to work with UN agencies such as FAO, UNEP, UNDP and World Bank, and in particular, with the Global Environmental Facility (GEF) in the efforts to assess and manage Large Marine Ecosystems (LMEs). Bob George recommended that the board give importance to corals on the summits and slopes of seamounts as habitats for fishes and invertebrates and organize two workshops: (1) the taxonomy of deep-sea corals to identify nomenclatural and distributional gaps, and (2) a 2008 expert workshop to address deep-sea coral and seamount ecosystems as integral subunits of the Large Marine Ecosystems within US EEZs, as organized by Kenneth Sherman (NOAA). Bob George also proposed this LME subunit concept for conservation and protection of deep-sea coral ecosystems in the 2nd Global LME Symposium in Qingdao, China (September 11–14, 2007).

We believe that the traditions established by the three international symposia on deep-sea corals will be pursued by the organizers of the December 1–5, 2008—4th International Deep-sea Coral Symposium, Di Tracey and Helen Neil of the NIWA, Wellington, New Zealand. We also believe that the 18 papers in *Bulletin of Marine Science*, volume 81, number 3, and the 23 papers in this book will lay the foundation for conservation and protection of deep-sea corals habitats in the depths of the world oceans. We hope that this book will inspire both scientists and managers alike to give importance to these that this book will inspire both scientists and managers alike to give importance to these unique deep-sea coral ecosystems that abound the summits of seamounts and the shelf-edge and slope environments within EEZs and in the high seas.

Editors

*Robert Y. George*
George Institute for Biodiversity and Sustainability (GIBS)
Wilmington, North Carolina, USA

and

*Stephen D. Cairns*
Smithsonian Institution, Washington DC, USA

**Literature Cited**

Freiwald, A. and M. Roberts, eds. 2005. Deep-water corals and Ecosystems. Proc. Second Int. Symp. on Deep-Sea Corals, Springer Verlag, Erlangen, Germany. 1210 p.
Watling, L. and M. Risk, eds. 2002. Biology of cold water corals. Hydrobiologia 471(Special issue): 166 p.
Willison, M., J. Hall, S. E. Gass, E. L. R. Kenchington, M. Butler, and P. Doherty, eds. 2002. Proc. First Int. Symp. on Deep-Sea Corals. Ecology Action Center, Halifax, Nova Scotia. 231 p.

ADDRESSES: (R.Y.G.) *George Institute for Biodiversity and Sustainability (GIBS), 305 Yorkshire Lane,Wilmington, North Carolina 28409.* (S.D.C.) *National Museum of Natural History, P.O. Box 37012, Smithsonian Institution, Washington DC 20560.*

# Ecosystem-based fisheries management of seamount and deep-sea coral reefs in U.S. waters: conceptual models for proactive decisions

Robert Y. George, Thomas A. Okey, John K. Reed, and Robert P. Stone

**Abstract**

Commercial fishing activities, primarily bottom trawling, have severely damaged vulnerable sea-floor communities such as undersea coral gardens and the summits of seamounts. Recreational fishing can also affect ecosystems adversely. The United States Ocean Commission (2004) recommended that fisheries be managed to protect marine ecosystems and their functions. The eight regional fisheries management councils in the United States under the jurisdiction of the National Marine Fisheries Service lack a sufficiently detailed understanding of ecosystem structure and function and of the target stocks and managed fisheries for making decisions that protect the stocks and ecosystems while allowing fisheries to proceed. Because the development of such detailed understanding is time consuming, we suggest that conceptual diagrammatic models can be used to express the generally known structures and functions of ecosystems so that precautionary management decisions can be made while more sophisticated models of marine ecosystems and fisheries are developed. This will protect resources while knowledge is gathered to enable exploitation that increases rather than degrades the overall value of the services provided by the ecosystem. Here we provide examples of such conceptual diagrammatic models for three US deep-sea coral ecosystems: (1) Aleutian gorgonian garden ecosystems, (2) Corner Rise Seamount, NW Atlantic, and (3) *Oculina* coral ecosystem off the Florida Atlantic coast, all of which have been established as Essential Fish Habitat and Habitat Areas of Particular Concern (EFH-HAPC). We also suggest how such models might be used by managers, scientists, and stakeholders.

Marine ecosystems have been severely degraded in recent decades by unsustainable and destructive fishing practices (Jackson et al., 2001; Myers and Worm, 2003), altered coastal water quality, and other pollution-related impacts (Anderson et al., 2002; Peterson et al., 2003), as well as global oceanographic changes such as intensification of the El Niño–Southern Oscillation and other long-term oceanographic changes (Anderson and Piatt, 1999; Fiedler, 2002). The resulting degradation of ocean ecosystems portends economic, cultural, and biological losses that are unprecedented and that have negative and uncertain consequences for Earth's life-support systems (Costanza et al., 1997; Vitousek et al., 1997, Costanza, 1999).

Trawling over vulnerable sea floor habitats such as undersea coral gardens or seamounts using gear with heavy metal chains, doors, or rollers stands out as a particularly severe and widespread agent of disturbance affecting marine communities (Watling and

George, R. Y. and S. D. Cairns, eds. 2007. Conservation and adaptive management of seamount and deep-sea coral ecosystems. Rosenstiel School of Marine and Atmospheric Science, University of Miami.

Norse, 1998; Freese et al., 1999; Koslow et al., 2001; NRC, 2002; Chuenpagdee et al., 2003; Morgan and Chuenpagdee, 2003).

Much of the reason human activities so severely affect the ecosystems that support humanity is that most economic, policy, and management systems lack common-sense frameworks that would allow assessment and accounting of the collateral effects of particular human activities (Hutton and Leader-Williams, 2003; Soderman, 2003; Okey and Wright, 2004). In addition, precedents for making protective decisions when information is lacking are still rare.

**Ecosystem-based fisheries management.**—We define *ecosystem-based management* (EBM) as the management of human activities so as to prevent net loss in the overall value of an ecosystem, considering all potential values and services associated with that ecosystem in the contexts of natural fluctuations and anthropogenic impacts. Managers consider the potential effects of their decisions on broad ecosystem components and values, as well as the effects of environmental and biological fluctuations on the component of interest. The recognition that ecosystems change and fluctuate naturally on multiple scales is a foundation of EBM, as is the recognition that ecosystems are defined by the integrative interactions of the physical, chemical, and biological aspects of the ecosystem. Ecosystem-based fisheries management (EBFM) considers the impacts of fisheries on both target and non-target stocks as well as cascading and interactive effects. Many instruments can be employed to achieve EBFM, but habitat protection combined with effort reductions is the simplest and most effective. Considerable lip service is given to the concept, but experts and casual observers agree that *operationalizing* ecosystem-based management is the central challenge (e.g., Babcock and Pikitch, 2004; Guerry, 2005).

Human activities in the oceans other than fisheries need to be similarly managed. For instance, the Puerto Rico trench was used as a deep-sea dump site for disposing of pharmaceutical wastes from the Arecibo Pziser penicillin factory on the assumption that this trench floor is anoxic and stagnant without any flow of the Antarctic Bottom Water. However, physical oceanographic studies and biological sampling found the opposite to be true and led to the closure of the dump-site (George and Herring, 2005).

There is a considerable amount of uncertainty about the structure, functions, biodiversity, and interactions of most marine ecosystems. The same is true about most fisheries and target stocks. It is now more possible than ever before to develop a sophisticated understanding of these ecological-social systems given investments of enough time and funds. It is likely, however, that vulnerable marine ecosystems such as many deep-water coral habitats would disappear during the time required to understand them in detail.

Our suggested alternative to fully understanding these systems is to use simple conceptual models of vulnerable ecosystems, here in the form of trophic interaction diagrams, to encapsulate known essential structures, membership, and functions such that a simple understanding of a given ecosystem and interactions can be achieved. This knowledge can be communicated to decision-makers and used as a communication tool among stakeholders, thereby enabling proactive and precautionary decisions for fisheries management in a holistic and broad-based approach that is not dependent on sophisticated and data-intensive models. As an example, we present three conceptual models of deep-water coral reef ecosystems: (1) Aleutian gorgonian garden ecosystems, (2) Corner Rise Seamount, NW Atlantic, and (3) *Oculina* Essential Fish Habitat–Habitat Area of Particular Concern (EFH-HAPC) ecosystem off the Florida Atlantic Coast (Fig. 1). These systems were cho-

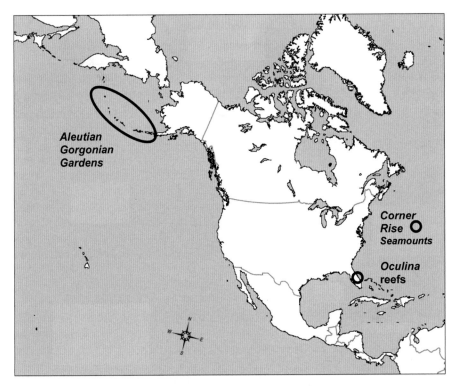

Figure 1. Locations of the three deep-sea coral and seamount ecosystems described in this paper (modified from basemap by Houghton Mifflin Company). The Aleutian gorgonian gardens occur between 150–350 m depths and spans 1800 km separating the North Pacific Ocean from the Bering Sea; the *Oculina* ecosystem occurs between 60 and 100 m depths off a limited area of central eastern Florida only; and coral habitats of the Corner Rise Seamount Cluster in the northwestern Atlantic occur generally between 1000 and 3000 m depths east of the New England Seamount chain.

sen as examples here because they represent vulnerable and somewhat poorly understood habitats and ecosystems.

We emphasize the need for a framework that accounts for interactions among species in the ecosystem as well as between that biota and the changing physical and chemical environment. We also emphasize that such a framework should employ current descriptive and general knowledge about each system while a more thorough and quantitatively precise understanding of the system emerges over time. This sort of approach was suggested by Botsford et al. (1997). We further suggest that potentially harmful effects of human activities can be evaluated in a useful way only by employing comprehensive and integrated frameworks, whether qualitative or quantitative, or both, simply because of the "comprehensive" nature of ecosystems. Although many types of conceptual or integrated approaches are possible, we suggest that this general approach of employing rapid preliminary conceptual models is an effective proactive and precautionary solution for operationalizing EBM while more rigorous models are generated. Immediate management decisions can, and should, be made in poorly understood ecosystems (most marine ecosystems) based on such descriptions and simple models, which distill and represent the essential current knowledge about a given ecosystem's structure and functions. For example, the following descriptions and simple models can be used by decision makers and stakeholders to protect the value of ecosystems by using them to hedge against uncertainty. Such simple approaches would, moreover, allow all stakeholders to see the linkages in the ecosystem and the ways in which fishing and other stressors would likely impact the system.

*1. Aleutian Gorgonian Garden Ecosystem.*—The Aleutian Islands Archipelago, which separates the deep North Pacific Ocean from shallow shelf waters of the Bering Sea, spans nearly 1800 km—approximately as long as Australia's Great Barrier Reef. The RV ALBATROSS expedition documented the existence of diverse benthic communities in the Aleutian Islands more than a century ago but little attention was given to this remote area of the world until recent examination of deep-sea corals incidentally caught in fisheries there prompted major taxonomic revisions. Fisheries bycatch records indicated that the Aleutian Islands may harbor the highest abundance of cold-water corals in the world (Heifetz et al., 2005) and submersible expeditions quickly launched in response have confirmed that the benthic ridge of the Aleutian Archipelago supports incredibly rich benthic ecosystems (Stone, 2006). In response, the U.S. National Marine Fisheries Service recently closed vast areas of this remote undersea area to all bottom-trawling.

Submersible studies along the Aleutian Archipelago have revealed extensive octocoral gardens at depths between 150 and 350 m, harboring a diverse array of sessile and sedentary invertebrate fauna and an abundance of fishes (Stone, 2006). The sedentary suspension feeders that form the physical structure of this ecosystem are dominated by gorgonian octocorals (likely included several undescribed species), but they also include hydrocorals, sponges, hydroids, and bryozoans. Scleractinian corals, however, with their skeletons composed principally of aragonite, are relatively poorly represented in terms of species diversity, unlike the southern Blake-Plateau ecosystems off the southeastern United States and the Caribbean deep-water lithoherm ecosystems (Cairns, 2007). The paucity of scleratinian fauna in the North Pacific Ocean may be due to the shallow depth of the aragonite saturation horizon found at some high latitude regions (Guinotte et al., 2006).

This discovery raises several important questions: (1) Is the Aleutian cold coral ecosystem comparable to other similar deep-water coral ecosystems at lower latitudes in terms of species diversity and food-chain complexity? (2) How important are these octocoral gardens to local and regional populations of mobile fauna and broader biological communities and mankind? (3) What are the effects of fishing activities on these octocoral garden ecosystems? (4) What are the influences of oceanographic and climate changes on these octocoral ecosystems? Answers to these questions will be the critical ingredients of refined ecosystem-based fishery management strategies for the region.

Gorgonian gardens are structurally complex, resembling low latitude tropical coral reefs (Fagerstrom, 1987), and are supported by a rigid framework of living organisms. The complex vertical relief provides refuge and food to myriad species of fish and invertebrates. The conventional theory of high species diversity found only in tropical ecosystems calls for revaluation in the light of the present finding of high species diversity in this Aleutian marine ecosystem (Heifetz et al., 2005). The biocomplexity of the Aleutian coral gardens is unclear due to our lack of understanding of the intricate foodwebs in this ecosystem (Livingston and Tjelmeland, 2000). Octocoral gardens are clearly unique and vulnerable and, therefore, a necessarily simplified understanding of the ecosystem is needed immediately to further develop precautionary ecosystem-based policies.

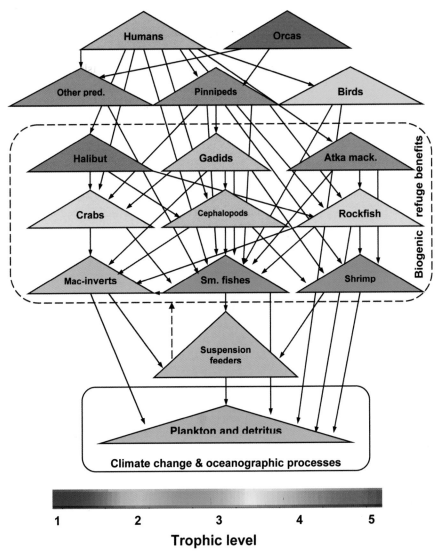

Figure 2. Conceptual model depicting trophic structure and energy flow in the Aleutian Islands Gorgonian Coral Gardens in the North Pacific Ocean. Solid arrows denote directly imposed mortalities (current or historical); triangles denote biotic components in flux (size of triangles has no meaning); dashed lines denote biogenic refuge benefits; solid box represents physical oceanography; colors indicate trophic level. In this figure, suspension feeders include the octocorals *Paragorgia* and *Primnoa*, gorgonians, and other sessile invertebrates such as sponges, tunicates, sea anemones, bryozoans, hydrozoans, tube-worms, and crinoids. Gadids include walleye pollock (*Theragra chalcogramma*) and Pacific cod (*Gadus macrocephalus*). Atka mackeral (*Pleurogrammus monopterygius*) is a greenling. Other predators include sharks and marine mammals.

Typical of boreal shelf ecosystems worldwide (Livingston and Tjelmeland, 2000), the Aleutian marine ecosystem is a complex web of trophic levels with strong interactions between biotic and abiotic features (Fig. 2). Cyclic changes in atmospheric and oceanographic conditions may cause changes in community structure that cascade through the ecosystem (Anderson and Piatt, 1999). Fisheries affect marine ecosystems directly through the over-fishing of target species, fishing mortality on non-target species, and

Table 1. Examples of target species in the benthic and demersal fisheries of the three benthic ecosystems examined.

| Ecosystem | Gear type | Target species | |
|---|---|---|---|
| | | Common name | Species |
| Aleutian gorgonian garden | Bottom trawling | Atka mackerel | *Pleurogrammus monopterygius* |
| | | Pacific ocean perch | *Sebastes alutus* |
| | | Northern rockfish | *Sebastes polyspinis* |
| | Longlining | Pacific cod | *Gadus macrocephalus* |
| | | Pacific halibut | *Hippoglossus stenolepis* |
| | | Sablefish | *Anoplopoma fimbria* |
| | Single pot | Pacific cod | *Gadus macrocephalus* |
| | | Sablefish | *Anoplopoma fimbria* |
| Corner Rise Seamount | Longline pot | Golden king crab | *Lithodes aequispina* |
| | Bottom trawling | Wreckfish | *Polyprion americanus* |
| | | Rat tail species | *Coryphaenoides guentheri* |
| | | | *Coryphaenoides armatus* |
| | | | *Coryphaenoides rupestris* |
| | | Cutthroat eel | *Diastobranchus capensis* |
| | Longlining | Blue hake | *Antimora rostrata* |
| | | False boarfish | *Neocyttus helgae* |
| | | Various shark species | |
| | | Tuna and billfish | |
| Florida *Oculina* reefs | Trapping | Red crabs | *Chaceon* sp. |
| | Dredge | Calico scallop | *Astropecten gibbus* |
| | Bottom trawling | Rock shrimp | *Sicyonia brevirostris* |
| | | Penaeid shrimps | *Penaeus* spp. |
| | Hook-and-line | Scamp grouper | *Mycteroperca phenax* |
| | | Gag grouper | *Mycteroperca microlepis* |
| | | Northern red snapper | *Lutjanus campechanus* |
| | | Common seabream | *Pagrus pagrus* |
| | | Various shark species | Carcharinidae, Sphyrnidae |

damage to the seafloor and the benthos including those providing biogenic structure (Goni, 1998). Other effects include mediation of biological interactions, altering community structure, and effects on apex predators (Goni, 1998).

Aleutian coral garden habitat appears to be particularly sensitive to bottom disturbance from fishing gear (Stone, 2006). Recent examination of fisheries bycatch records and archived specimens indicated that corals are widely distributed through the Aleutian Island Archipelago and that there is substantial interaction between gear and coral habitat for all major fisheries. Four benthic gear types—bottom trawl, longline, single pot, and longline pot—are currently used throughout the archipelago and along the slope to depths of at least 1000 m (Table 1). These fisheries are managed under both state and federal management, with an observer program for each fishery. Alaskan fisheries have a history of sequential (serial) over fishing (Berger et al., 1986), especially in the Bering Sea (Loher et al., 1998), and new areas of seafloor are potentially disturbed as new species and stocks are exploited.

During the 1990s, fishery managers implemented several measures to curtail fish stock depletion in the region, effectively redistributing fishing effort in the Aleutians. For example, in 1994 the harvest limit for Atka mackerel was reapportioned among smaller ample, in 1994 the harvest limit for Atka mackerel was reapportioned among smaller sub-areas to prevent over-fishing of localized stocks. In 1998, fishing was restricted near critical habitats of endangered Steller sea lions. This conservation strategy is, in part, a reaction to public concern to protect an endangered and charismatic marine mammal, similar to the measure implemented to protect the endangered monk seal in the cold coral reefs off northwest Hawaiian Islands. In particular, fisheries for walleye pollock, Pacific cod, and Atka mackerel were excluded from areas surrounding rookeries and haulouts to minimize modification to prey fields and competition for prey between fisheries and Steller sea lions. While these actions may have curtailed over-fishing of local fish stocks, they also may have pushed fisheries into new regions and deeper waters where pristine (fragile) deep-water coral ecosystems occur. Unfortunately, knowledge of these habitats is meager at best. Despite the increased economic interest in the deep-water resources in the Aleutians, the underwater seascape in the Aleutian Island chain remains one of the least-explored areas of the US Exclusive Economic Zone (EEZ). The logical step at this juncture is to highlight the need for precautionary protections and research initiatives including mapping and monitoring so that these ecosystems can be managed on the basis of sound science as envisioned in the 2004 Ocean Commission Report.

A comprehensive ecosystem-based approach to management strategies for Alaska groundfish fisheries exists (Witherell et al., 2000), but fully implementing the proposed policies in the Aleutian Islands remains a challenge given the complexity and interdependence of the components of this cold coral ecosystems. One strategy is to define major components and interactions (Fig. 2). The obvious structure-forming biota in this ecosystem are suspension feeders dominated by octocorals (Fig. 3), especially the gorgonians *Paragorgia* and *Primnoa*, which occur with other sessile invertebrates including sponges, tunicates, sea-anemones, bryozoans, hydrozoans, tube-dwelling worms, and crinoids. Fauna inhabiting biogenic refugia include an array of species that are mostly motile and that live on the "bio-buildups," either using it as refuge (a habitat or home) or as an opportunistic site for foraging on the assemblage of species inhabiting the substratum. The upper triangles in Figure 2 represent top predators such as the gadids, walleye pollock *Theragra chalcogramma* (Pallas, 1811) and Pacific cod *Gadus macrocephalus* Tilesius, 1810, and Atka mackerel *Pleurogrammus monopterygius* (Pallas, 1810). At the

Figure 3. In situ photograph of the seascape of the Aleutian cold coral undersea-garden, showing abundance of octocorals and fishes. Note a mixed school of Pacific cod *Gadus macrocephalus* and Atka mackerel *Pleurogrammus monopterygius* swimming over a coral garden at 120 m depth in the Aleutian Islands, Alaska.

summit of this food web are larger predators such as humans and marine mammals (Estes and Palmisano, 1974).

In June 2006, the National Marine Fisheries Service took a major precautionary step to protect coral habitat in the Aleutian Islands when they implemented the closure of about 100,000 km$^2$ of the fishing grounds to bottom trawling. Additionally, the measure includes closure of six known gorgonian gardens to all bottom contact gear. These are the first MPAs in the Aleutian Islands designed specifically to protect cold coral habitats and their functions in these unique benthic ecosystems.

*2. Corner Rise Seamount Cluster, Northwest Atlantic Ocean.*—As early as 1959 seamount ecosystems were recognized as unique among marine ecosystems (Hubbs, 1959). It is now known that there may be 100,000 seamounts with elevations ≥ 1000 m in the world oceans both within EEZs and in the high-seas (Wilson and Kaufman, 1987; Rogers, 1994). Seamounts are most common in the Pacific Ocean (more than 50%), but the New England seamount chain in the northwest Atlantic is the most conspicuous and ecologically important seamount habitat west of the Mid-Atlantic ridge (Moore et al., 2004). Fewer than 300 seamounts thus far have been sampled globally, despite the growing awareness of perturbations to these fragile deep-sea habitats by orange roughy *Hoplostethus atlanticus* Collett, 1889 fisheries (Koslow, 1996; Koslow et al., 2000). Nevertheless, preliminary surveys have revealed that seamount ecosystems are "forgotten oases" in the deep ocean in spite of the presence of aggregations of commercially important fish species in and around the seamount summits. These include highly migratory fish species such as billfish and Atlantic bluefin tuna, which are found at the Corner Rise Seamount cluster in the northwest Atlantic (Vinnichenko, 1997), and commercially important deep-sea mesopelagic fish such as orange roughy found at the New Zealand and Tasmania Seamounts (Clark et al., 2000; Koslow et al., 2001).

Seamounts can occur nearshore within EEZs and also offshore in the high-seas as in the case of the New England seamount chain, which extends perpendicularly from northeast U.S. coast to the Mid-Atlantic ridge. The fauna of the New England seamount chain (e.g., Physalia, Bear, Balanus, and Manning Seamounts) are all similar with reduced endemism in any one of the seamounts (Moore et al., 2004). This is possibly due to high gene flow aided by eastward flowing currents that transport broadcast larvae (with exceptions of some octocoral species; A. Baco, Woods Hole Oceanographic Institution, pers. comm.). In contrast, seamounts farther to the east such as the West and East Corner Rise Seamounts appear to harbor a more endemic fauna possibly not genetically linked to nearshore seamounts (Moore et al., 2004). Seamounts in this region rise steeply to 400–100 m above continental rise or abyssal plain, unlike mid-ocean islands.

The summits (top and flanks) of seamounts are generally composed of volcanic rocks of various geological ages. This is particularly true of the Corner Rise Seamount, which is adjacent to the Mid-Atlantic Ridge and younger than the Indo-Pacific and New Zealand seamounts. The impact of the Gulf Stream on the seamount biota on these mid-ocean seamounts began < 3–6 million yrs ago with the formation of the Isthmus of Panama (George, 2003). Isotopic studies of deep-sea cold corals indicate that the Gulf Stream has extended northward and northeastward in recent decades (M. Risk, pers. comm.). The northward movement of the Gulf Stream northward along the eastern continental margin transports larvae of broadcast-spawning sessile species such as the soft and hard corals to the Corner Rise Seamounts.

Partly for the purpose of protecting the northeast Atlantic seamount ecosystems, the European Commission initiated research to promote the implementation of "Sustainable Marine Ecosystem" under their "Fifth Framework Programme" (Thiel and George, 2003). A first step in such a research program is to implement precautionary ecosystem-based management to develop a conceptual model of regional seamount ecosystems. It is essential that seamount ecosystem modeling consider the following five components:

A. Knowledge of physical forcing mechanisms affecting the functioning of the seamount ecosystems, including an understanding of the fine-scale bathymetry and hydrography, current field and vortices around the seamount under study, and accurate characterization of the benthic mixed layer (BML).

B. In formation on surface sediments and dynamics of organic particulate material on the surface of seamount, which implies quantitative knowledge of the flux of carbon from surface primary and secondary production (zooplankton carcasses and fecal vertical rain) and benthopelagic coupling.

C. A quantitative synthesis of the biodiversity of seamount biota and fisheries involving both resident (endemic) and transitional migratory large pelagic fauna.

D. Development of a preliminary trophic model (mass-balance) of the seamount ecosystem by synthesizing existing knowledge of the ecology of the seamount chosen for ecosystem-based fisheries modeling.

E. Integration of scientific models within a practical decision-making process of international management with the UN infrastructure for high-seas and national or regional management (e.g., Oslo and Paris Convention region [OSPAR-region]) for seamounts within EEZs. For the northwestern Atlantic Ocean, which includes the New England Seamount, a similar governance structure needs to be founded within the advisory fisheries management infrastructure of the International Council for Exploration of the Seas (ICES). This may entail a formation of a working group for northwestern Atlantic within ICES jurisdiction. The working group should involve the Atlantic Fisheries Management Commission that oversees highly migratory fish and marine mammals, including the seamounts and cold coral reefs in the Blake Plateau and in the Gulf and Caribbean region. New England Fisheries Management Council should also be represented in the ICES WG.

We recommend that immediate steps be taken to reduce the rapid decline of the world's most accessible fish stocks due to overfishing (Myers and Worm, 2003) which also leads to enormous fuel consumption of fishing fleet (Tyedmers et al., 2005). We also recommend that commercial fishing vessels from the EU (Spain, in particular), Russia, New Zealand, Canada, and the United States refrain from using destructive fishing gear that severely damage cold coral-beds on summits and top flanks of seamounts. These destructive bottom-trawling practices continue both legally and illegally with enormous bycatch mortalities. George (2006) recommended selecting, on the basis of The World Conservation Union (IUCN) guidelines, both southern and northern hemisphere seamount chains as MPAs. We also recommend giving endangered species status to species such as onion-eye grenadier *Macrourus berglax* Lacépède, 1801, roundnose grenadier *Coryphaenoides rupestris* Gunnerus, 1765, and blue hake *Antimora rostrata* (Günther, 1878), which are now victims of over-fishing with dwindling populations (Moore et al., 2004).

There is a compelling reason to save the Corner Rise Seamount from similar processes that lead to the extinction of orange roughy (Koslow et al., 2001) in seamounts close to New Zealand and Australia (Tasmania). Designation of high-seas MPAs is the logical next step as implied in the World Wildlife Fund (WWF) seamount report (2003) and IUCN classification of Marine Protected Areas (MPAs) (Probert et al., 2006). Thiel and George (2003) appealed to the UN for immediate action by submitting a statement with signatures of 130 deep-sea biologists.

Our knowledge about macrofauna and invertebrate megafauna on which many higher trophic level fish of seamounts feed is incomplete. High proportions of the species found at seamounts are new to science and still undescribed, as revealed in the recent study of fauna of Lord Howe Seamounts and Norfolk Ridge Seamounts off Tasmania (de Forges et al., 2000). For example, Tasmanian seamounts have shown 16%–33% endemism. Approximately 600 invertebrate species are thus known from seamounts, but this number is miniscule if we take into account small infauna and minute predatory crustaceans (0.5–10 mm), such as cumaceans, mysids, isopods, and amphipods. These species are abundant near seamount corals as commensals or predatory crustaceans (George, 2002). The unique biological complexity, diversity, and endemicity of seamount biota lends support for immediate and strict conservation measures to save seamounts from destructive commercial exploitation, as witnessed in orange roughy fisheries.

The Russian seamount ecologist Vinnichenko (1997), who has studied the commercial fish and fisheries in the Corner Rise Seamount, postulated that many species of fish [*Coryphaenoides guntherii* (Vaillant, 1888), *Hydrolagus pallidus* Hardy and Stehmann, 1990, *Epigonus telescopus* (Risso, 1810), and *Brotulotaenia crassa* Parr, 1934] originated from Bear Seamount of the northeastern Atlantic via Mid-Atlantic Ridge. It is also evident from trawling studies in the vicinity of *Lophelia* Reefs (northernmost Agassiz Coral Hill) off North Carolina (George, 2002, 2004c, 2005) that the commercially important rattail *Coryphaenoides armatus* (Hector, 1875) is abundant from 350 to 900 m whereas its congers *C. rupestris* and *C. guenthteri* are abundant in the Corner Rise Sea Mounts. Establishment of MPAs in the Corner Rise Seamounts will protect these latter deep-sea fish species which should be given endangered species status.

A preliminary conceptual model that can be used for precautionary decision-making (Fig. 4; see Fig. 5 for associated seascape) first step in the conservation and protection of biodiversity of the Corner Rise Seamount. The model includes climate-induced changes in the northeast oscillation of the Gulf Stream as indicated in the bottom box of blue triangles. The turquoise triangle just above this bottom box represents myriad species of sessile invertebrates such as scleractinian corals, soft octocorals, sponges, anemones, bryozoans, and crinoids. Together this mat of sessile fauna constitute a food-base and refugium for many motile species in the mid-ocean: Biogenic habitat benefits in the Corner Rise Seamount are attributed to several known and unknown fish and invertebrate species that are depicted in the middle box comprising scavengers, small fish, and carnivorous fish.

What makes this seamount ecosystem unique is really the top orange and red triangles. This includes marine bird, marine mammals, and giant squids, and more importantly, the commercially important rattail, alfonsino, billfish, and tuna species that are illustrated as top predators with a link to humans (fishers). The large migratory fishes use the "ocean highways" of transatlantic currents for movement in and out of this pelagic ecosystem above the Corner Rise Seamount. Pauly et al. (1998) and Myers and Worms (2003) cautioned that fishing at the top trophic level of seamounts can alter the ecosystem structure and function in a irrecoverable manner. Immediate precautionary management action is therefore crucial to protect biodiversity on the seamounts.

*3.* **Oculina *EFH-HAPC ecosystem off eastern Florida*.**—The term *essential fish habitat* (EFH) was introduced legally in the 1996 re-authorization of the 1976 Magnuson Fishery Conservation and Management Act to protect marine habitats that are es-

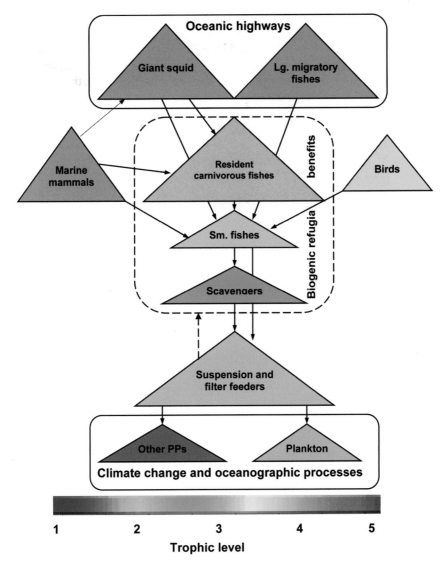

Figure 4. Conceptual model depicting trophic structure and energy flow in the Corner Rise Seamount in the NW Atlantic Ocean. Box shapes and colors per Figure 2.

sential for the sustainability of fish and other targeted organisms, recognizing the nature and wide geographic distribution of species under protection. Likewise, a Habitat Area of Particular Concern (HAPC) is legally defined as an area within EFH that may need additional protection. The *Oculina* coral reserve on the Atlantic Coast of Florida was the first in the United States to receive the EFH and HAPC status.

Two types of cold deep-water coral reefs occur off the southeastern United States: *Oculina* reefs and *Lophelia* reefs. *Oculina varicose* Lesueur, 1820, grows at depths of 60–100 m, and *Oculina* reefs are known only off central eastern Florida. *Lophelia pertusa* (Linnaeus, 1758) reefs in this region occur at depths of 490–870 m from North Carolina to Florida on the Blake Plateau to the Straits of Florida and Bahama Archipelago (Avent et al., 1977; Reed, 1980, 2002a). Both of these types of cold coral reefs are

Figure 5. A false boarfish *Neocyttus heigae* swims above coral rubble at approximately 1100 m depth near the summit of Verril Peak, Calcoosahatchee Seamount, one of the Corner Rise seamount group in the North Atlantic. Image courtesy of Deep Atlantic Stepping Stones Science Party, IFE, URI-IAO, and NOAA.

strikingly different from typical tropical warm coral reefs in that they consist primarily of a single species of azooxanthellate coral, they occur in areas of strong currents or upwelling zones, and they form high-relief mounds or pinnacles. Both cold coral reefs provide EFH for diverse communities of fish and invertebrates (Reed 2002a,b; Reed at al., 2006).

*Oculina* reefs are known as ideal habitats for commercially important grouper, snapper, and mackerel species (Koenig et al., 2000). They are formed by single species of a branching scleractinian coral, *O. varicosa* and they consist of hundreds of individual coral pinnacles, mounds, and ridges that are high relief structures ranging from 3 to 35 m in height and up to 100–300 m in width. Each pinnacle is actually a veneer of living coral overlying a mound of sand and mud sediment, coral debris, and oolite base formed during the Holocene transgression (Macintyre and Milliman, 1970: Reed, 1980). A large 1.5 m colony may be nearly a century old, and the reefs are estimated to exceed 1000 yrs (Reed, 1981; Hoskin et al., 1987).

In 1984, the uniqueness, productivity, and vulnerability of the *Oculina* habitat prompted the South Atlantic Fisheries Management Council (SAFMC) to declare a 315 km$^2$ (92 nmi$^2$) portion of the reefs as "EFH-HAPC". This legal action was taken in order to protect the coral reef and associated fauna from bottom trawling, dredging, longlines, traps, and other destructive bottom gears which are known to obliterate of live bottom cold coral beds (Watling and Norse, 1998). Impacts of overfishing on grouper spawning aggregations further stimulated the SAFMC to close the original HAPC for bottom hook and line fishing in 1994 as a precautionary action to enhance the reproductive behavior of commercially and recreationally important grouper and snapper species. These fish species, such as the snowy and gag groupers and red snappers (more than 70 species present), are the most dominant fishes on hard bottom areas in the continental shelf off the southeastern United States. This fish assemblage is in contrast to rockfish species (*Sebastes* spp.) in the northeast Pacific shelf region off the Oregon-Washington coast.

*Oculina* reserve became the first MPA designated for habitat protection off the southeastern United States and it serves as a model for other MPAs for fisheries management. In 2000, the *Oculina* MPA was further expanded to 1029 km$^2$ (300 nmi$^2$). Nevertheless, illegal fishing activities continued, resulting in severe damage to *Oculina* coral habitat. Bottom trawling for rock shrimp *Sicyonia brevirostris* Stimpson, 1871 and panaeid shrimps are the primary culprit for extensive destruction of the physical framework of the *Oculina* coral matrix off Florida's cast coast. This kind of cold deep-water coral reef destruction is comparable to the irreversible damages done to *Lophelia* reefs in the European coast, including at the Koster Fjord on the west coast of Sweden (George, 2004a,b), which also suffered severe coral mortality due to commercial trawling activities for *Pandalus borealis* Krøyer, 1838. Hurricanes and bioerosion also impact *Oculina* reefs, but the "coral graveyards" now present in the form of vast stretches of dead coral rubble denote the failure of managers to restrict or ban shrimp trawling over fragile ecosystems such as the *Oculina* reef complex. Our knowledge about the impact of thermal changes related to climate change (global warming) on *Oculina* and *Lophelia* reefs is minimal.

The deep-water *Oculina* reefs are an excellent pilot setting to operationalize EBFM. The rainbow gradient conceptual model of the foodweb (Fig. 6) presented here represents the basic known interactions within this ecosystem. In the bottom blue triangle we have emphasized the potential impact upon *Oculina* reef of four physical events. Gulf Stream seasonal influences, upwelling episodes, climate change related to global warming, and coastal eutrophication. Many would argue that our knowledge is currently too meager to provide a rigorous characterization of trophic connections among species inhabiting this *Oculina* reef ecosystem. We have limited knowledge of benthopelagic couplings and the precise influences of climate-induced changes in Gulf Stream on a decadal scale. We also have not yet documented the recovery of the grouper-snapper fish populations since MPA establishment. For example, the seasonal spawning aggregation-behavior of scamp grouper (Fig. 7) in the *Oculina* reefs is poorly known and calls for careful studies.

It is evident that physical factors are a critical component for these reefs. The primary factors controlling these reefs are: (1) Available substrate for the coral to grow on, i.e., Pleistocene bedrock, (2) cold-water upwelling events providing increased nutrients, and (3) overlying Florida Current (Gulf Stream) waters providing plankton and particulate organic matter (POM). Since the light levels on the 80 m deep *Oculina* reefs are low (0.33% transmitted surface light), corals lack zooxanthellae, and macroalgal blooms are not possible in spite of high nutrients provided through upwelling or coastal eutrophication. Upwelling from the Straits of Florida is, however, a critical factor as this produces episodic intrusions of cold water (< 10°C) throughout the year at the shelf edge (Smith, 1981; Reed, 1983). Upwelled nutrients enhance *Oculina* growth through increases in overlying phytoplankton and zooplankton (Reed, 1981).

The *Oculina* triangle in Figure 6 includes a vast array of suspension feeders, including colonies of *Oculina* corals. Very dense populations of associated invertebrates that are sessile, sedentary, and benthopelagic (Reed et al., 1982; Reed and Mickelsen, 1987) are essentially a mixture of detritivores, carnivores, and corallivores. The overall biodiversity of this cold coral ecosystem is thought to be comparable to shallow tropical reefs. Over 20,000 invertebrates have been found living among and within the branches of 42 small *Oculina* colonies, yielding 230 species of molluscs, 50 species of decapods, 47 species of amphipods, 21 species of echinoderms, 15 species of pycnogonids, numerous polychaete species belonging to 23 families, and a variety of other taxa (not yet fully identified) e.g., sipunculids, nemertines, isopods, tanaids, cumaceans, ostracods, and herpacticoid copepods (Miller and Pawson, 1979; Reed et al., 1982; Reed and Hoskin, 1987; Reed and Mikkelsen, 1987; Child, 1998).

In comparison, quantitative analysis of *L. pertusa* colonies from the Faeroe Shelf in the northeastern Atlantic (Jensen and Fredricksen, 1992) revealed 298 species of inverte-

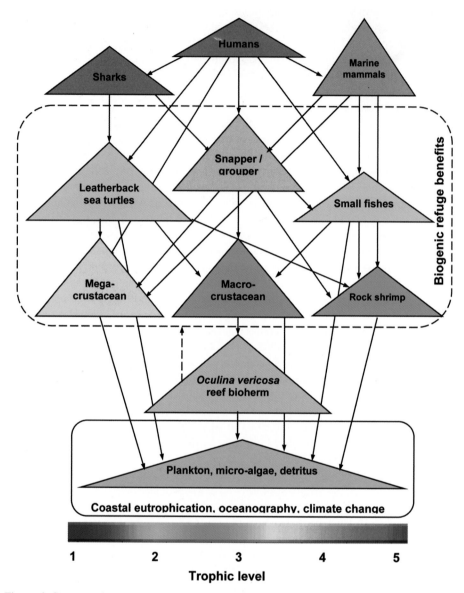

Figure 6. Conceptual model depicting trophic structure and energy flow in the *Oculina* HAPC ecosystem off Florida Atlantic Coast in the Northwest Atlantic Ocean. Box shapes and colors per Figure 2.

brate taxa, mostly dominated by polychaetes (67 spp.), bryozoa (45 spp.), and sponges (29 spp.). Although the molluscan community was numerically dominant in that study, only 31 species were recorded compared to 230 species associated with *Oculina* coral. The high molluscan diversity of the *Oculina* reefs off Florida vs apparently lower diversity in *Lophelia* reefs is consistent with other observed latitudinal patterns of diversity.

Reed and Mikkelsen (1987) reported on the feeding habits of *Oculina* molluscan assemblage and demonstrated that 29% of the species were filter feeders (including suspension feeders and mucoid entrappers), 24% parasitic carnivores, 17% non-parasitic

Figure 7. Scamp grouper *Mycteroperca phenax* use the deep-water *Oculina* reefs (80 m) as spawning and feeding grounds. Spawning aggregations consist of large males with distinct coloration and hundreds of females which swarm over the reef releasing sperm and eggs.

carnivores, 16% herbivores, 7% detritovores, 4% scavengers, and 4% corallivores. These associated invertebrates are represented in the two dark green triangles in Figure 6.

Small fish associated with *Oculina* reefs such as angelfishes, anthiids, and butterflyfishes, are represented in the light green triangle on the right. At the very top of the "*Oculina* biogenic refuge" box is the light orange triangle that accomodates the different species of groupers, snappers, and sharks. Overfishing of grouper and snapper species and bycatch removal of non-target species by the 1990s severely degraded the *Oculina* reef ecosystem and restoration projects are in progress (Koenig, 2001).

Surveys of fish populations and grouper behavior were conducted concurrently with the above habitat and coral studies from 1976 to 1983. The deep-water *Oculina* reefs once formed impressive breeding grounds for commercially important populations of snapper (*Lutjanus* spp.) and grouper (Serranidae), and dense populations of gag grouper *Mycteroperca microlepis* (Goode and Bean, 1879) and scamp grouper *Mycteroperca phenax* Jordan and Swain, 1884 were common (G. Gilmore, pers. comm.; Reed, 1985; Reed and Hoskin, 1987; Gilmore and Jones, 1992; Reed, 2002b). Scamp groupers were seasonally abundant, forming dense spawning aggregations of several hundred individuals per hectare. Population densities for dominant fish species correlated highly with habitat type (Koenig et al., 2000). Speckled hind (*Epinephelus drummondhayi* Goode and Bean, 1878), which might be designated in the near future as threatened species, and gag and scamp groupers are associated predominantly with the intact coral habitat. Few grouper or snapper are present in areas where the *Oculina* coral has been reduced to rubble (Koenig et al., 2000). Transplantation of live coral and reef ball modules, as conducted in tropical coral reef restoration, might allow these *Oculina* reefs to recover, but the rate of recovery is difficult to predict. Control of illegal fishing (recreational and commercial) is the greatest challenge and obstacle to such a restoration program. Strict implementation of a vessel monitoring system (VMS), as done in the George Bank fishing grounds, may help in regulatory measures enforced by the SAFMC. The benthic fisheries that have operated in the region of these *Oculina* reefs are listed in Table 1 and discussed in Koenig et al. (2000). In addition, several species of large migratory fish are present due to the proximity of the *Oculina* reefs to the western boundary of the Gulf

Stream. The pelagic fisheries of this region, which are focused on midwater, migratory, and nocturnal species (Fig. 6, central orange triangle), have been intense in *Oculina* reef areas despite quotas, closures, and size class restrictions imposed by the National Marine Fisheries Service (NMFS). These surface water fisheries may not cause direct physical damage to *Oculina* reefs, but dumping of wastes and illegal bottom fishing can. In 1994, a moratorium on bottom hook-and-line fishing was enacted in an *Oculina* MPA on the basis of recommendations from SAFMC. In addition, quotas for both benthic and pelagic species, and closure and size restrictions were imposed by fishery regulators (NMFS). Subsequent submersible and remotely-operated vehicle (ROV) surveys (in 2001 and 2003) suggested an increase in grouper number and size since 1995 survey (Reed et al., 2006). There was a clear increase also in the reproductive activities of the gag and scamp groupers, suggesting the possible recurrence of spawning aggregations of both species (Shepard and Reed, 2003).

Despite the improvement in fishery management, recent ROV and submersible surveys have confirmed poaching by shrimp trawlers and fishers within the *Oculina* MPA. Trawl nets, bottom longlines, and fishing lines are evident on the bottom reef-ball modules destroyed from apparent mechanical damage. Evidence of habitat damage, during the interval between the discovery of the *Oculina* reefs and the present, was especially evident from a submersible dive made in 2001 on a 20-m high *Oculina* pinnacle off Cape Canaveral (Reed et al., 2007). In 1976, this reef was described from a submersible dive (JOHNSON-SEA-LINK II-63) as having an estimated 25%–100% cover of live *Oculina varicosa* coral thickets on the slopes and crest of the reef (Reed et al., 2007). This region, however, had been open to trawling until 2000 when *Oculina* MPA was expanded. The 2001 submersible survey (CLELIA 616) found the coral thickets on the mound had been reduced to 100% rubble.

A multibeam survey in 2002 and ROV video-transects in 2003 discovered a series of high-relief *Oculina* bioherms that extend just outside the western boundary of the current *Oculina* MPA. It is now estimated that over 100 bioherms may exist in this unprotected zone. In addition, based on a few submersible dives in 1982, *Oculina* pinnacles extend at least 55 km north of the current *Oculina* MPA, but no recent studies have been made in this region. It is therefore easy to argue that it is imperative that the protected areas be expanded. Such decisions rely on the consent of broad stakeholder interests, which can be achieved only if at least a basic understanding of the structure and functions of this ecosystem can be expressed transparently (e.g., Fig. 6).

## Discussion

The three "conceptual models" presented in this paper are simplified and incomplete caricatures of the ecosystems they represent. Still, three main features of these ecosystems become apparent when using this first step conceptual approach toward ecosystem-based management of these vulnerable habitats: (1) the biotic components are interconnected; (2) certain "foundation species" provide biogenic structure and refuge benefits; (3) physical aspects of each environment influence the overall biological communities, i.e., biological and physical forces are integrated. These models are presented here as conceptual frameworks and catalysts for developing ecosystem-based fisheries management approaches for vulnerable ecosystems and habitats. They indicate to government managers and scientists alike the key interdependencies of these ecosystems. They show, for example, that degradation of foundation groups such as corals and sponges, which provide biogenic habitat refuge (and energy), is likely to have a negative effect on many of the target fish species that are supported by these interconnections.

These color-coded energy-flow models emphasize the basic structure and functions of complex biological communities, and they do so without the need to wait for detailed information. These preliminary models should provide a broad overview to regulate human interactions with each ecosystem by considering the potential direct and indirect effects on each ecosystem component in the context of the natural environmental fluctuations. Higher-level approaches to rapid conceptualization of system structure and function that can also be used for proactive decision making include the qualitative modeling employed by Dambacher et al. (2002). Fully quantitative approaches to understanding ecosystem structure, functions, and dynamics for ecosystem-based management decisions include ecosystem models such as Ecopath with Ecosim (e.g., Christensen and Walters, 2004; Okey and Wright, 2004). Managers should consider combining these approaches in sequence. Such approaches will enable "Ecosystem based" economics, policies, and management to improve the sustainability of ecosystems, cultures, and societies through explicit consideration of human impacts on all potential values now and in the future (Costanza et al., 1997; Sumalia and Walters, 2005).

Ecosystem-based Fisheries Management is the current slow revolution in fisheries management. Lubchenco (1998) perceptively suggested that this would be a challenge for the 21st Century. EBFM will manifest as systems and decisions that consider the impacts of fisheries on non-target components of ecosystems and on whole biological communities—decisions that protect and preserve broad values that these ecosystems represent and provide. These systems can be formalized rules and guidelines about how to make decisions in the context of varying degrees of uncertainty, or they can simply be MPAs that hedge against uncertainty—a very effective and proven tool despite common arguments that confound detectability of effects with effectiveness. We consider the tool presented here as the ultimate and critical starting point for decisions making in situations with limited information. The simple approach is necessary because, as in the case of deep-sea and cold-water corals, a lack of protective decisions is a decision. An EBFM system is not an integrated software; such tools can help considerably, but they often confuse and delay unnecessarily. The most effective EBFM systems will consist of simple logic and simple tools that communicate basic understandings of ecosystems to decision makers and stakeholders.

Our discussion is focused on fisheries because of their immediate potential for impacts on deep-sea corals and seamounts. However, we emphasize that ecosystem-based management of biogenic marine habitats also involves consideration of all physical and biological agents of change, such as climate change, water quality change, oil and gas explorations and production, and shifts in traditional uses of living marine resources. Given these multi-faceted considerations, implementation of EBFM schemes will depend on uncharacteristic transcendence of institutional and disciplinary boundaries by scientists and managers as argued persuasively by Wilson (1998) and it ought to be facilitated by enormous fiscal resources where necessary (George, 2006).

Detailed examination of the patterns of degradation and recovery from anthropogenic impacts such as bottom trawling, thermal stress (as induced by global warming), acidification, and pollution will help generate an understanding of the resilience of deep-water coral and seamount ecosystems and the synergistic effects on them. There is also a need to understand the effects of changing oceanographic cycles, recent decadal modulations in the north-northeast extension, and eddies of the Gulf Stream. This is evident from

2002). A shift of 1 °C in a stenothermal ecosystem such as the Antarctic (George, 2004a) or tropical coral reefs (Glynn et al., 2001) might by itself bring about major shifts in the ecosystem form and function.

It is also common sense that fishing community participation (both large and small enterprises as well as commercial and sports-fishing entities) and the participation of other extractive user groups is a crucial ingredient for the success of the EBFM implementation programs (Dyer, 1994). US Congressional support of these fiscal responsibilities is a pivotal factor in supporting the recommendations of the 2005 US Ocean Commission recommendations for ecosystem-based management. Okey (2003) and George (2006) emphasized that fisheries management must involve input from various human interests and representation of the general public rather than just interested stakeholders. Therefore, EBFM schemes cannot optimize the public interest without deliberately-designed collaboration involving academia, NGOs, government policy makers, and the general public.

Scientists and managers seeking to develop ecosystem-based management plans for deep-water corals and seamount ecosystems could benefit from lessons learned from managers of tropical coral reefs (Hall-Spencer, 2007). The most effective and value-protective type of management for deep-sea coral reefs and seamounts will be proactive decisions and protection and closure of vulnerable biogenic seafloor habitats and seamount summits before and until adequate scientific information becomes available to determine which types and levels of fishing are appropriate for these ecosystems. Conceptual models in the form of schematic diagrams such as those presented herein might prove useful for developing ecosystem-based fisheries management approaches for poorly known ecosystems that are currently threatened by human activities, if only because they communicate simple concepts and logic to and along a side range of stakeholders. A main conclusion that can be gleaned from the conceptual models of these three systems and the ecosystem descriptions herein is that protection of the corals themselves would enhance the biogenic habitat refugia value of these ecosystems. This, in turn, would enhance fish abundance and production as well as high biodiversity.

## Acknowledgments

We acknowledge the help provided at the Friday Harbor Laboratories of the University of Washington where both T.O. and R.G. spent time together to develop the concept underlying this paper and the rainbow color-gradient conceptual models. J.R. acknowledges the help of all those who dived with him on JOHNSON-SEA-LINK submersibles to obtain valuable information on deep-sea corals. R. George is grateful for help of the captains and crew of research vessels (R/V EASTWARD, R/V CAPE HATERRAS, R/V GILLISS, ALBATROSS IV) that carried him to both *Oculina* and *Lophelia* reefs along the southeast US and to Georges Bank and Corner Rise Seamount. R. Stone is thankful all those who helped him in his survey of the Aleutian cold coral reefs. T. Okey is grateful for the help he received at the Fisheries Center of the University of British Columbia and at CSIRO Marine and Atmospheric Research in Australia. We thank S. Keys for her editorial assistance. We also thank the funding agencies that made the research possible. These include the National Science Foundation, the National Marine Fisheries Service (NOAA) and the National Undersea Research Program at the University of North Carolina at Wilmington.

# Literature Cited

Anderson, D. M., P. M. Glibert, and J. M. Burkholder. 2002. Harmful algal blooms and eutrophication: Nutrient sources, composition, and consequences. Estuaries 25: 704–726.

Anderson, P. J. and J. F. Piatt. 1999. Community reorganization in the Gulf of Alaska following ocean climate regime shift. Mar. Ecol. Prog. Ser. 189: 117–123.

Avent, R. M., M. E. King, and R. H. Gore. 1977. Topographic and faunal studies of shelf-edge prominences off central eastern Florida coast. Int. Rev. Gesamt. Hydrobiol. 62: 185–208.

Babcock, E. A. and E. K. Pikitch. 2004. Can we reach agreement on a standardized approach to ecosystem-based fishery management? Bull. Mar. Sci. 74: 685–692.

Berger, J. B., J. E. Smoker, and K. A. King. 1986. Foreign and joint-venture catches and allocations in the Pacific Northwest and Alaska fishing area under the Magnuson Fishery and Conservation Act, 1977-84. U.S. Department of Commerce, NOAA Tech Memorandum. NMFS F/NWC-99.

Botsford, L. W., J. C. Castilla, and C. H. Peterson. 1997. The management of fisheries and marine ecosystems. Science 277: 509–515.

Cairns, S. 2007. Deep-Sea Corals: An overview with special reference to diversity and distribution of deep-water scleractinian corals. Bull. Mar. Sci. 81: 311–322.

Child, C. A. 1998. *Nymphon torulum*, new species and other pycnogonida associated with the coral *Oculina varicosa* on the east coast of Florida. Bull. Mar. Sci. 63: 595–604.

Christensen, V. and C. J. Walters. 2004. Ecopath with Ecosim: methods, capabilities and limitations. Ecol. Model. 172: 109–139.

Chuenpagdee, R., L. E. Morgan, S. M. Maxwell, E. A. Norse, and D. Pauly. 2003. Shifting gears: Assessing collateral impacts of fishing methods in US waters. Front. Ecol. Environ. 1: 517–524.

Clark, M. R., O. F. Anderson, R. Francis, and D. M. Tracey. 2000. The effects of commercial exploitation on orange roughy (*Hoplostethus atlanticus*) from the continental slope of the Chatham Rise, New Zealand, from 1979 to 1997. Fish. Res. 45: 217–238.

Costanza, R. 1999. The ecological, economic, and social importance of the oceans. Ecol. Econ. 31: 199–213.

_____, R. dArge, R. deGroot, S. Farber, M. Grasso, B. Hannon, K. Limburg, S. Naeem, R. V. Oneill, J. Paruelo, R. G. Raskin, P. Sutton, and M. vandenBelt. 1997. The value of the world's ecosystem services and natural capital. Nature 387: 253–260.

Dambacher, J. M., H. W. Li, and P. A. Rossignol. 2002. Relevance of community structure in assessing indeterminacy of ecological predictions. Ecology 83: 1372–1385.

de Forges, B. R., J. A. Koslow, and G. C. B. Poore. 2000. Diversity and endemism of the benthic seamount fauna in the southwest Pacific. Nature 405: 944–947.

Dyer, C. L. 1994. Proaction versus reaction: Integrating applied anthropology into fisheries management. Hum. Org. 53: 83–88.

Estes, J. A. and J. F. Palmisano. 1974. Sea otters: their role in structuring nearshore communities. Science 185: 1058–1060.

Fagerstrom, J. A. 1987. The evolution of reef communities. Wiley, New York.

Fiedler, P. 2002. Environmental change in the eastern Tropical Pacific Ocean: Review of ENSO and decadal variability. Mar. Ecol. Prog. Ser. 244: 265–283.

Freese, L., P. J. Auster, J. Heifetz, and B. L. Wing. 1999. Effects of trawling on seafloor habitat and associated invertebrate taxa in the Gulf of Alaska. Mar. Ecol. Prog. Ser. 182: 119–126.

George, R. Y. 2002. Ben Franklin temperate reef and deep sea 'Agassiz Coral Hills' in the Blake Plateau off North Carolina. Hydrobiologia 471: 71–81.

_____. 2003. *Bermudasignum frakenbergi n.* gen., n. sp (Crustacea : Isopoda) off Bermuda and evolution within the deep-sea family Mesosignidae. Bull. Mar. Sci. 73: 699–712.

_____. 2004a. Conservation and management of deep-sea coral reefs: Need for new definition, an agenda and action plan for protection. Pages 26–29 *in* T. Wolff, ed. Deep-Sea Newsletter.

_____. 2004b. Deep-sea asellote isopods (Crustacea, Eumalacostraca) of the north-west Atlantic: the family Haploniscidae. J. Nat. Hist. 38: 337–373.

_____. 2004c. *Lophelia* bioherms and lithoherms as fish habitats on the Blake Plateau: Biodiversity and Sustainability. Proc. ICES Annual Science Conference (CD) Vigo, Spain.

_____. 2005. Ecosystem-based fisheries management for *Lophelia* reef HAPC at the northern-most 'Agassiz Coral Hills' at bathyal depths off Cape Lookout, North Carolina. Proc. 2005 ICES Annual Science Conference; Aberdeen, Scotland.

_____. 2006. Management of seamounts and deep-sea coral habitats: Primer from the Miami 2005 symposium. 11th International Deep-sea Biology Symp., Southampton, UK (Abstract).

_____ and J. K. Berger. 2005. Concept of biobank and regional species inventory: Priorities in the Southeastern United States. Association for Southeastern Biologists (annual mtg.) Gatlinburg, Tennessee. [Abstract]

_____ and D. Herring, 2005. Multifaceted sampling strategy for EBFM-modeling: How can 'AUV' help? Proc. 14th Int. Symp. on Unmanned Untethered Submersible Technology (CD), University of New Hampshire, Durham, NH.

Gilmore, R. G. and R. S. Jones. 1992. Color variation and associated behavior in the epinepheline groupers, *Mycteroperca microlepis* (Goode and Bean) and *M. phenax* Jordan and Swain. Bull. Mar. Sci. 51: 83–103.

Glynn, P. W., J. L. Mate, A. C. Baker, and M. O. Calderon. 2001. Coral bleaching and mortality in Panama and Ecuador during the 1997–1998 El Nino-Southern oscillation event: Spatial/temporal patterns and comparisons with the 1982–1983 event. Bull. Mar. Sci. 69: 79–109.

Goni, R. 1998. Ecosystem effects of marine fisheries: An overview. Ocean. Coast. Manage. 40: 37–64.

Guerry, A. D. 2005. Icarus and Daedalus: conceptual and tactical lessons for marine ecosystem-based management. Front. Ecol. Environ. 3: 202–211.

Guinotte, J. M., J. Orr, S. Cairns, A. Freiwald, L. Morgan, and R. George. 2006. Will human-induced changes in seawater chemistry alter the distribution of deep-sea scleractinian corals? Front. Ecol. Environ. 4: 141–146.

Hall-Spencer, J. 2007. Historical deep-sea coral distribution on seamount oceanic island and continental shelf-slope habitats in the NE Atlantic. Pages 135–146 *in* R. Y. George and S. D. Cairns, eds. Conservation and adaptive management of seamount and deep-sea coral ecosystems. Rosenstiel School of Marine and Atmospheric Science, University of Miami. Miami. 324 p.

Heifetz, J., B. L. Wing, R. P. Stone, P. W. Malecha, and D. L. Courtney. 2005. Corals of the Aleutian Islands. Fish. Oceanogr. 14 (Suppl. 1): 131–138.

Hoskin, C. M., J. K. Reed, and D. M. Mook. 1987. Sediments from a living shelf-edge reef and adjacent area off central eastern Florida. Proc. Symp. South Florida Geol., Miami Geol. Soc. Mem. 3: 42–57.

Hubbs, C. 1959. Initial discoveries of fish fauna on the seamounts and offshore banks in the eastern Pacific. Pac. Sci. 13: 311–316.

Hutton, J. M. and N. Leader-Williams. 2003. Sustainable use and incentive-driven conservation: realigning human and conservation interests. Oryx 37: 215–226.

Jackson, J. B. C., M. X. Kirby, W. H. Berger, K. A. Bjorndal, L. W. Botsford, B. J. Bourque, R. H. Bradbury, R. Cooke, J. Erlandson, J. A. Estes, T. P. Hughes, S. Kidwell, C. B. Lange, H. S. Lenihan, J. M. Pandolfi, C. H. Peterson, R. S. Steneck, M. J. Tegner, and R. R. Warner. 2001. Historical overfishing and the recent collapse of coastal ecosystems. Science 293: 629–638.

Jensen, A. and R. Frederiksen. 1992. The fauna associated with the bank-forming deep-water coral *Lophelia pertusa* (Scleractinia) on the Faroe shelf. Sarsia 77: 53–69.

Koenig, C. C. 2001. Oculina Banks: Habitat, fish populations, restoration, and enforcement. South Atlantic Fishery Management Council, Charleston, SC. Special Report 2001. 24 p.

_____, F. C. Coleman, C. B. Grimes, G. R. Fitzhugh, K. M. Scanlon, C. T. Gledhill, and M. Grace. 2000. Protection of fish spawning habitat for the conservation of warm-temperate reef-fish fisheries of shelf-edge reefs of Florida. Bull. Mar. Sci. 66: 593–616.

Koslow, J. A. 1996. Energetic and life-history patterns of deep-sea benthic, benthopelagic and seamount-associated fish. J. Fish. Biol. 49 (Suppl A): 54–74.

_____, G. W. Boehlert, J. D. M. Gordon, R. L. Haedrich, P. Lorance, and N. Parin. 2000. Continental slope and deep-sea fisheries: Implications for a fragile ecosystem. ICES J. Mar. Sci. 57: 548–557.

_____, K. Gowlett-Holmes, J. K. Lowry, T. O'Hara, G. C. B. Poore, and A. Williams. 2001. Seamount benthic macrofauna off southern Tasmania: Community structure and impacts of trawling. Mar. Ecol. Prog. Ser. 213: 111–125.

Livingston, P. A. and S. Tjelmeland. 2000. Fisheries in boreal ecosystems. ICES J. Mar. Sci. 57: 619–627.

Loher, T., P. S. Hill, G. Harrington, and E. Cassano. 1998. Management of Bristol Bay red king crab: a critical intersections approach to fisheries management. Rev. Fish. Sci. 6: 169–251.

Lubchenco, J. 1998. Entering the century of the environment: A new social contract for science. Science 279: 491–497.

Macintyre, L. G. and J. D. Milliman. 1970. Physiographic features on outer shelf and upper slope, Atlantic continental margin, southeastern United States. Bull. Geol. Soc. Am. 81: 2577–2588.

Miller, J. E. and D. L. Pawson. 1979. A new subspecies of *Holothuria lentiginosa* Marenzeller from the western Atlantic Ocean. Proc. Biol. Soc. Wash. 91: 912–922.

Moore, J. A., M. Vecchionne, B. B. Collottte, R. Gibbons, and K. Hartel. 2004. Selected fauna of the Bear Seamount (New England Seamount chain), and the presence of "natural invader" species. Arch. Fish. Mar. Res. 51: 241–250.

Morgan, L. E. and R. Chuenpagdee. 2003. Shifting gears: addressing the collateral impacts of fishing methods in U.S. waters. Island Press, Washington, D.C. 42 p.

Myers, R. A. and B. Worm. 2003. Rapid worldwide depletion of predatory fish communities. Nature 423: 280–283.

NRC. 2002. Effects of trawling and dredging on seafloor habitat. Committee on Ecosystem Effects of Fishing: Phase 1 -- Effects of Bottom Trawling on Seafloor Habitats, National Research Council, National Academy Press, Washington, D.C.

Okey, T. A. 2003. Membership of the eight Regional Fishery Management Councils in the United States: Are special interests over-represented? Mar. Policy 27: 193–206.

_____ and B. A. Wright. 2004. Toward ecosystem-based extraction policies for Prince William Sound, Alaska: integrating conflicting objectives and rebuilding pinnipeds. Bull. Mar. Sci. 74: 727–747.

Pauly, D., V. Christensen, J. Dalsgaard, R. Froese, and F. Torres. 1998. Fishing down marine food webs. Science 279: 860–863.

Peterson, C. H., S. D. Rice, J. W. Short, D. Esler, J. L. Bodkin, B. E. Ballachey, and D. B. Irons. 2003. Long-term ecosystem response to the *Exxon Valdez* oil spill. Science 302: 2082–2086.

Probert, P. K., S. Christensen, K. M. Gjerde, S. Gubbay, and R. S. Santos, 2006. Management and conservation of seamounts. *In* C. Hollingworth, ed. Seamounts: Ecology, Fisheries and Conservation, Blackwell Publishers. New York.

Reed, J. K. 1980. Distribution and structure of deep-water *Oculina varicosa* coral reefs off central eastern Florida. Bull. Mar. Sci. 30: 667–677.

_____. 1981. In situ growth rates of the scleractinian coral *Oculina varicosa* occurring with zooxanthellae on 6-m reefs and without on 80-m banks. Proc. 4th Int. Coral Reef Symp. 2: 201–206.

_____. 1983. Nearshore and shelf-edge *Oculina* coral reefs: the effects of upwelling on coral growth and on the associated faunal communities. National Oceanic Atmospheric Administration, Symp. Ser. Undersea Res. 1: 119–124.

_____. 1985. Shelf edge *Oculina* reefs. Pages 466–468 *in* W. Seaman, Jr., ed. Florida aquatic habitat and fishery resources. Florida Chapter of American Fisheries Society, Kissimmee, Florida.

_____. 2002a. Comparison of deep-water coral reefs and lithoherms off southeastern USA. Hydrobiologia 471: 57–69.

_____. 2002b. Deep-water *Oculina* coral reefs of Florida: biology, impacts, and management. Hydrobiologia 471: 43–55.

_____ and C. M. Hoskin. 1987. Biological and geological processes at the shelf edge investigated with submersibles. National Oceanic Atmospheric Administration, Symp. Ser. Undersea Res. 2: 191–199.

_____ and P. M. Mikkelsen. 1987. The molluscan community associated with the scleractinian coral *Oculina varicosa*. Bull. Mar. Sci. 40: 99–131.

_____, C. C. Koenig, and A. N. Shepard. 2007. Impacts of bottom trawling on a deep-water *Oculina* coral ecosystem off Florida. Bull. Mar. Sci. 81: 481–496.

_____, D. Weaver, and S. A. Pomponi. 2006. Habitat and fauna of deep-water *Lophelia pertusa* coral reffs of southeastern USA: Blake Plateau, Straits of Florida, and Gulf of Mexico. Bull. Mar. Sci. 78: 343–375.

_____, R. H. Gore, L. E. Scotto, and K. A. Wilson. 1982. Community composition, structure, areal and trophic relationships of decapods associated with shallow-water and deep-water *Oculina varicosa* coral reefs - studies on decapod Crustacea from the Indian River region of Florida. Bull. Mar. Sci. 32: 761–786.

Rogers, A. D. 1994. The biology of seamounts. Adv. Mar. Biol. 30: 305–350.

Shepard, A. and J. K. Reed. 2003. *Oculina* Banks 2003: characterization of benthic habitat and fish populations in the *Oculina* Habitat Area of Particular Concern (OHAPC), Mission Summary Report. NOAA/ NURC and South Atlantic Fishery Management Council, Charleston, South Carolina.

Smith, S. V. 1981. Marine macrophytes as a global carbon sink. Science 211: 838–840.

Soderman, M. L. 2003. Including indirect environmental impacts in waste management planning. Resour. Conserv. Rec. 38: 213–241.

Stone, R. P. 2006. Coral habitat in the Aleutian Islands of Alaska: depth distribution, fine-scale species associations, and fisheries interactions. Coral Reefs 25: 229–238.

Sumaila, U. R. and C. Walters. 2005. Intergenerational discounting: A new intuitive approach. Ecol. Econ. 52: 135–142.

Thiel, H. and R. Y. George. 2003. "Recommendations for highseas marine protected areas", Letter to Hon. Koffi Annan, General Secretary, United Nations (with signatures of 130 deep-sea biologists).

Tyedmers, P. H., R. Watson, and D. Pauly. 2005. Fueling global fishing fleets. Ambio 34: 619–622.

Vinnichenko, V. I. 1997. Russian investigations and deep water fishery on the Corner Rise Seamount in Subarea 6. NAFO Sci. Count. Studies 30: 41–49.

Vitousek, P. M., H. A. Mooney, J. Lubchenco, and J. M. Melillo. 1997. Human domination of Earth's ecosystems. Science 277: 494–499.

Watling, L. and E. A. Norse. 1998. Disturbance of the seabed by mobile fishing gear: a comparison to forest clearcutting. Conserv. Biol. 12: 1180–1197.

Wilson, E. O. 1998. Consilience: The unity of knowledge, 1st edition. Knopf : Distributed by Random House, New York.

Wilson, R. R. and R. S. Kaufman, 1987. Seamount biota and biogeography. Pages 355–377 *in* B. H. Keating, P. Fryer, R. Batiza, and G. W. Boehlert, eds. Seamounts, islands, and atolls. Geophys. Mono. 43, American Geophysical Union, Washington, DC.

Witherell, D., C. Pautzke, and D. Fluharty. 2000. An ecosystem-based approach for Alaska groundfish fisheries. ICES J. Mar. Sci. 57: 771–777.

WWF-Germany. 2003. Seamounts of the North-East Atlantic. Frankfurt.

ADDRESSES: (R.Y.G.) *George Institute for Biodiversity and Sustainability (GIBS), 305 Yorkshire Lane, Wilmington, North Carolina 28409.* (T.O.) *CSIRO Marine and Atmospheric Research, PO Box 120, Cleveland Old 4163, Australia.* (J.K.R.) *Harbor Branch Oceanographic Institute, Fort Pierce, Florida 34946.* (R.P.S.) *National Marine Fisheries Service, Auke Bay Laboratory, 11305 Glacier Highway, Juneau, Alaska 99801-8626.* CORRESPONDING AUTHOR: (R.Y.G.) *E-mail: <georgeryt@cs.com>.*

# Science Priority Areas on the high seas

Hjalmar Thiel

## Abstract

The establishment of Science Priority Areas (SPAs) on the high seas is proposed to avoid any disturbance particularly of long-term research activities. Such areas should become management units independent from Marine Protected Areas (MPAs). SPAs are consistent with the United Nations Law of the Sea Convention and should not be subordinated within the framework of The World Conservation Union's (IUCN) categories for MPAs. Scientists need to become stakeholders in their own interest and states should establish SPAs through international cooperation.

Science Priority Areas (SPAs)—also previously termed Unique Science Priority Areas (USPAs)[1]—are large regions of seafloor and/or water column, in which long-term research projects have been established, or might be created in the future, to address issues relating to climate change and man's impact on the environment, as well as basic research (Thiel, 2001, 2002, 2003a,b). SPAs particularly aim to protect long-term scientific studies from being disturbed by other activities on the high seas. SPAs do not necessarily exclude other activities, but science would have absolute priority above other uses.

The concept of SPAs developed during the organization of the 2001 Vilm Workshop on Marine Protected Areas on the High Seas. The workshop was stimulated by the development of fisheries on the high seas, particularly on and above seamounts, and by concerns expressed by the World Conservation Union (IUCN). It was the first common conference of jurists, conservationists, and scientists (Thiel and Koslow, 2001). The workshop identified the major current and potential commercial uses of the high seas (e.g., fisheries, manganese nodule mining, waste disposal) and considered the practical and legal steps that would be needed to create high seas marine protected areas (HS-MPAs). It became evident that there had been significant intrusions into the high seas in the preceding 50 yrs, and that a number of new schemes for the utilization of living and non-living resources were at various stages of planning and exploration.

Until the end of the last century, deep-sea scientists conducted research in an isolated and independent fashion based on the principle of the freedom of the high seas, as expressed by the United Nations Convention on the Law of the Sea [UNCLOS, Article 87 (1)]. At the time, there was little need for extra provisions to secure research activities in the deep-sea as competition for the vast spaces of the open ocean did not exist. But the growing exploitation of deep-sea resources by various stakeholders has interfered increasingly with the interests of scientists (e.g., the impact of tuna fisheries on long-term moorings and observatories placed in the oceans to monitor climate change). Therefore, scientists should now look to secure their activities by developing stakeholder rights and demands.

---

[1] The term SPA was modified to PASR (Priority Areas for Scientific Research). *See* Thiel, H. 2007. Priority Areas for Scientific Research: protecting scientific investments. Ocean Challenge 15: 6–7.

## Earlier approaches for the separation of science areas in the deep sea

Based on IUCN discussions in 1978, and further consultations by the National Research Council (1984) of the United States National Academy of Sciences, specific long-term research areas were proposed for the abyss in the context of polymetallic nodule mining. As the environmental impacts of mining and the re-establishment of deep-sea communities could not be predicted, the concept of Stable Reference Areas (SRAs) was developed. Two main categories were defined:

(1) the Impact Reference Area (IRA) to study the natural development of the community during the years after mining, and

(2) the Preservation Reference Area (PRA) to study community development in an undisturbed area for comparison with the situation in the IRA.

These specific research areas have never been established, because mining has not yet passed beyond the exploration phase. However, the concept remains of interest to the members of the International Seabed Authority and should be considered again if mining is developed in the future. SRAs would serve as specific sites for monitoring mining impacts, but they are not defined as general areas for all types of research activities as is the case with SPAs.

Under the 1959 Antarctic Treaty [402 UNTS 71, Article IX (1)], the 1991 Madrid Protocol on Environmental Protection (30 IML 1461) provides for environmental regulations. Its Annex V, Articles 2 and 3(1) allow the designation of Antarctic Specially Protected Areas (ASPAs) which may protect among other things areas of outstanding scientific value and ongoing or planned research. This may have been formulated with particular regard to Antarctic land regions as many such regions have since been established, and marine areas are explicitly included in the ASPA framework (Antarctic Treaty, 2006).

Separation and restriction of research areas was also discussed by the scientists of the InterRidge community after they discovered their own research impacts on hydrothermal vent communities (Tunnicliffe, 1990; Mullineaux et al., 1998). The goal of these agreements among scientists was to minimize or eliminate adverse environmental impacts on hydrothermal vent communities and to optimize research access for scientists working on hydrothermal vent habitats (Juniper, 2003). In this respect, the competing stakeholders were the cooperating scientists of various scientific subjects and nations. InterRidge scientists also discussed a code of conduct for their work, including a number of self-restrictions for the conservation of species and habitats (Juniper, 2001). The "InterRidge statement of commitment to responsible research practices" demonstrates the need for "gathering information …, both to understand the systems and to form a basis for sustainable use strategies", and it states the "interest in environmental stewardship" and responsibility of scientists in conserving habitats and their fauna (Devey, 2006a,b). Such self-regulatory restrictions are frequently regarded as ineffective because they are not binding for the individual scientist or research group. For example, Leary (2005a) classifies this environmental stewardship as "no more than an aspirational or rhetorical statement," not taking into account the serious willingness and responsibility of scientists. National research councils should develop codes of conduct and exert control of its compliance with the approval of research funds.

## The need for Science Priority Areas

The importance of conducting basic research on and in the oceanic commons is generally accepted. The vast scale of the oceans and the interconnections of its waters by horizontal and vertical mixing processes mean that we have to study the ocean as a whole, including the deep-sea. The various uses of the deep-sea and climate change scenarios force us, in the interest of society, to evaluate oceanic ecosystems and especially regions of the high seas beyond national jurisdiction. Long-term studies are one approach to this end, but disturbance of a site for long-term studies would likely create irretrievable loss of resources. SPAs are the solution for their safeguarding.

Originally, the establishment of SPAs was proposed on purely theoretical considerations: the possibility that mining of the abyssal seafloor or the disposal of wastes in the deep-sea beyond the bounds of national jurisdiction may include an area where long-term scientific research was established. A few such research stations already exist in the northeast Atlantic Ocean, the Norwegian Sea, and the Pacific Ocean, and disturbance by other stakeholders may interrupt and terminate long-term research in these areas, compromising a large investment in research activities built up over many years. Such intervention may render formerly collected data less valuable and destroy the basis for comparison with future results. Long-term studies are of high importance to understanding natural variability and effects of industrial impacts or climate change. This is important not only in shallow marine seas but also in the deep-sea, where the ocean's remote, cold, and energy-poor realm exhibits a slow pace of life (see also Soltwedel et al., 2005).

Why ask for the establishment of SPAs at the present time? Mining has not developed beyond the exploration phase and the exploitation of polymetallic nodules continues (since the 1970s) to be projected to commence one or two decades ahead. Marine sulphide mining may develop in exclusive economic zones under national jurisdiction, but exploration and exploitation of oil and gas are intensified at bathyal depths. The deep-sea disposal of waste products (muds, nuclear waste products, redundant munitions) has been terminated. However, the increasing environmental pressures on terrestrial and shallow-water marine environments may lead to greater utilization of deep-sea resources. While this may seem rather unlikely, it cannot be discounted and used as an argument against the establishment of SPAs, nor against the recognition of marine scientists as major stakeholders in the management of the oceans. The costs to establish SPAs remain low when arrangements are settled before competitive situations emerge.

Ironically, only months after the publication of a paper on the need to designate SPAs in the northeast Atlantic Ocean (Thiel, 2002), a British expedition on RRS DISCOVERY returned to its long-term research station off Ireland and found that it was no longer possible to continue their research there because a cable company had positioned a communication cable through the study area. This problem could have been easily avoided if scientists had been respected stakeholders. The cable company would have placed the cable in a more distant position had they known about this long-term research of international importance, as confirmed by later negotiations with the cable company (Lampitt, 2004, pers. comm. 2005).

A second example involved several participants at the present symposium. During the Vilm Workshop in 2001, World Wide Fund for Nature (WWF) presented a number of proposals for HSMPAs, including an area in the northeast Atlantic where the BI-OTRANS research project was located (Cripps and Christiansen, 2001). A proposal for a much larger SPA was simultaneously submitted that also included the BIOTRANS area (Thiel, 2001, 2002, 2003a,b). The BIOTRANS area, located at ~ 47°N and 20°W was the first central oceanic region intensively studied by German and European scientists in the late 1980s and early 1990s (e.g., Thiel et al., 1988/89; Christiansen and Thiel, 1992; Lochte, 1992; Koppelmann and Weikert, 1993; Pfannkuche, 1993; Boetius and Lochte, 1994; Pfannkuche et al., 1999), with plans to return to this site. The double nomination as MPA and as SPA did not lead to a conflict, but demonstrated the interest of different stakeholders in the same area. A list of potential areas or regions for HSMPA nomination was presented by Gubbay (2003) and included scientifically most interesting spots.

Such potential conflicts may occur more often in the future. NGOs and governmental agencies have assumed a role in proposing HSMPAs. NGOs have become important driving forces in the political scene by stimulating the adoption of regulations for HSMPAs by the United Nations. In order to make the case for HSMPAs, information is needed on topography, geology, hydrography, and ecology of the region. Because this knowledge must be obtained from the scientific literature, well known areas are particularly suitable for formulating good arguments, increasing the chances that environmental protection plans will be accepted. However, scientists may also want to return to these areas for further studies because scientific research often builds on the knowledge gained from previous expeditions. The creation of an MPA at a site preferred for further scientific investigations may restrict certain research activities.

Regulations for marine and terrestrial research in the Antarctic formulated in the Antarctic Treaty allow a broad scope of interpretation, which can lead to rather restrictive measures, as evidenced, for example, by those of the German administration for research on RV POLARSTERN in the Antarctic Ocean. Thus, the proposal by Leary (2005a,b: 178) concluding that "any future regime to be developed for the high seas could draw heavily on the experience of Antarctica", i.e., to apply the regulations developed for the Antarctic Treaty as a model for HSMPAs (see also Grant, 2005), is worrisome. Marine science, of course, does not act in a social vacuum and should not be exempted from environmental safeguarding, yet the various interests of society need balanced evaluation.

## Categorizing SPAs according to the IUCN framework of protected areas

The IUCN (1994) developed an internationally accepted Protected Areas Management Category System with six categories based on management objectives. It may seem that SPAs would be readily classified under Category Ia:

"Strict nature reserve/wilderness protection area managed mainly for science or wilderness protection, an area of land and/or sea possessing some outstanding or representative ecosystems, geological or physiological features and/or species, available primarily for scientific research and/or environmental monitoring" (IUCN/WCPA, 2004).

Although scientific research is emphasized in this category, these categories are subordinated under the general IUCN (1994) definition of protected areas, which contains no mention of scientific research:

"An area of land and/or sea especially dedicated to the protection and maintenance of biological diversity, and of natural and associated cultural resources, and managed through legal and other effective means" (IUCN, 1994).

The strict compliance with this key definition is called for again by the Fifth World Parks Congress Recommendation 19 (IUCN/WCPA, 2004):

"Therefore, PARTICIPANTS …at the Vth World Park Congress, in Durban, South Africa (8–17 September 2003) …REAFFIRM that in the application of the management categories IUCN's definition of a protected area .… must always be met as the overarching criterion."

These quotations clarify that SPAs do not fall under the overarching criterion of the internationally accepted protected area concept. SPAs deviate from this in their goals and may be defined as:

"Areas reserved for repeated and long-term oceanographic research."

In this context, it is also important to stress that in the term "SPA" the "P" does not stand for "protection" but for "priority": priority for scientific work. Other stakeholders may use the same area, if the activities are compatible with ongoing or future scientific projects. Scientists need only to claim priority of activities; exclusion of other activities is not necessary, but is to be decided on request.

Researchers do not seek a position that is in conflict with conservation and protection of marine ecosystems. SPAs would provide a certain degree of species and ecosystem protection, but would not be their main function. Rather, deep-sea science programs serve to foster human understanding of the environment. In this context, SPAs do not fit into the IUCN categories. Scientists should establish their own type of separated area, probably under the auspices of an international organization such as the Scientific Committee on Oceanographic Research (SCOR), an international non-governmental scientific organization within the International Council for Science (ICSU). SPAs could also be established and their legal basis defined by the International Seabed Authority.

## Legal provisions

Whereas UNCLOS Article 87 lists the freedoms of the high seas, including marine scientific research, it does not contain any specific provisions to safeguard special areas for long-term research. The reason for this is that until recently it seemed to be unnecessary to provide any regulation to secure research activities. The need for research is also expressed by UNCLOS Article 239 under its marine scientific research regime:

"States are obliged to promote and facilitate the development and conduct of marine scientific research."

This implies that states should also assist with securing repeated and long-term investigations.

Support was also formulated by the representatives at the 2004 meeting of UNICPOLOS (United Nations Open-ended Informal Consultative Process on Ocean Affairs and the Law of the Sea) based on a recommendation adopted at the 7th meeting of the Conference of the Parties to the Convention on Biological Diversity (UNICPOLOS, 2004):

"Encourage States, individually and in collaboration with each other or with relevant international organizations and bodies, to improve their understanding and knowledge on the deep-sea in areas beyond national jurisdiction by increasing their marine scientific research activities in accordance with the convention."

Thus, there is general support for research activities in areas beyond national jurisdiction, and repeated visits to specific areas as well as long-term observations and measurements are essential parts of ecological research. These approaches are indispensable and provisions for undisturbed research conditions must be developed. States fund research on the high seas from public resources and by this indicate the importance of and their interest in research on the high seas. Funding by single states or in cooperation with other states should receive mutual and organizational support by the international community through emphasizing and agreeing on a legal framework for securing scientific research in international waters.

To allow for the establishment of SPAs, scientists should urge their political representatives at the UNICPOLOS meetings and at the United Nations General Assembly to provide effective regulations. The process until such regulations become effective may last for many years. However, in the region of the Northeast Atlantic Ocean, the OSPAR (Oslo and Paris Convention) region, progress may be faster and should be further stimulated. WWF (ICG-MPA 05/03/1) has recently proposed to establish the HSMPA "Rainbow Hydrothermal Vent Field" under the auspices of OSPAR (ICG-MPA 05/03/1). The potential for the establishment of MPAs on the high seas by OSPAR agreements should also open a pathway for SPAs in the NE Atlantic Ocean and this may also become a model for other regions including the NW Atlantic Ocean where HAPCs (Habitat Areas of Particular Concern) are established by United States Government agencies. Bilateral or international negotiations beyond Europe (OSPAR region) should be also carried out simultaneously.

In conclusion, (1) the widely accepted IUCN definitions of MPAs cannot provide the framework for SPAs; (2) SPAs are in accord with UNCLOS directives, which should encourage states and UN authorities to promote and facilitate their scientific activities for understanding deep-sea processes; (3) States and UN authorities should establish SPAs in international cooperation on regional and/or global scales to secure scientific activities; and (4) scientists should identify themselves as stakeholders in their own and the global human community's interest to warrant continual independent research in the high seas and exclusive economic zones.

## Acknowledgments

I am most grateful to R. Y. George and the George Institute for Biodiversity and Sustainability (GIBS) for inviting me as a speaker at the symposium and for supplying travel funds. I also thank R. Lampitt and D. Leary for supplying recent information, A. Kirchner for advice on legal regulations, and D. Billett for valuable comments and for correcting the language of the manuscript draft. J. A. Koslow, M. Rex, and two anonymous reviewers kindly gave valuable advice and comments on the paper presented at the symposium.

## Literature Cited

Antarctic Treaty. 2006. Antarctic Specially Protected Areas. Available from: www.cep.aq/aspa/general/purpose and The Antarctic Protected Area System www.cep.aq/apa/introduction/information Accessed 1 February 2006.

Boetius, A. and K. Lochte. 1994. Regulation of microbial enzymatic degradation of organic matter in deep-sea sediments. Mar. Ecol. Prog. Ser. 104: 299–307.

Christiansen, B. and H. Thiel. 1992. Deep-sea epibenthic megafauna in the Northeast Atlantic: Abundance and biomass at three mid-oceanic locations estimated from photographic transects. Pages 125–138 in G. Rowe and V. Pariente, eds. Deep-sea food chains and relation to global carbon cycle. NATO ASI Series, Kluwer Academic Publ., Dordrecht.

Cripps, S. J. and S. Christiansen. 2001. A strategic approach to protecting areas on the high seas. Pages 113–121 in H. Thiel and J. A. Koslow, eds. Managing risks to biodiversity and the environment on the high seas, including tools such as marine protected areas — scientific requirements and legal aspects. Proc. Expert Workshop, International Academy for Nature

Conservation, Isle of Vilm, Germany, 27 February–4 March 2001. Bundesamt für Naturschutz, Bonn, Germany.

Devey, C. W. (InterRidge Chair). 2006a. InterRidge statement of commitment to responsible research practices at deep-sea hydrothermal vents. 4 p. Available from: www.interridge.org/node/185.

_____. (on behalf of InterRidge). 2006b. InterRidge Statement of Commitment to Responsible Research Practice at Deep-sea Hydrothermal Vents. Pages 1–5 in S. Petersen, P. Herzig, and C. L. Morgan, eds. Scientific, legal, and economic perspectives of marine mining. Book of abstracts. 36th Underwater Mining Institute, Kiel, September 24–30, 2006.

Gubbay, S. 2003. Protecting the natural resources of the high seas. Pages. 47–79 in K. M. Gjerde and C. Breide, eds. Towards a strategy for high seas marine protected areas. Proc. IUCN, WCPA and WWF Expert Workshop on High Seas Marine Protected Areas, 15–17 January 2003, Malaga. Spain.

Grant, S. M. 2005. Challenges of marine protected area development in Antarctica. Parks 15: 40–47. (High Seas Marine Protected Areas).

IUCN. 1994. Guidelines for the protected area management categories. IUCN and the World Conservation Monitoring Centre, Gland, Switzerland and Cambridge, UK. 6 p.

IUCN/WCPA. 2004. The IUCN management categories — speaking a common language about protected areas. Seventh Meeting of the Conference of the Parties to the Convention on Biological Diversity (COP 7). 9–27 February 2004, Kuala Lumpur, Malaysia. 12 p.

Juniper, S. K. 2001. Background paper on deep-sea hydrothermal vents. Pages 89–95 in H. Thiel and J. A. Koslow, eds. Managing risks to biodiversity and the environment on the high seas, including tools such as marine protected areas — scientific requirements and legal aspects. Proc. Expert Workshop, International Academy for Nature Conservation, Isle of Vilm, Germany, 27 February–4 March 2001. Bundesamt für Naturschutz, Bonn, Germany.

_____. 2003. Deep-sea hydrothermal vent and seep habitats and related governance issues. In Workshop on the Governance of High Seas Biodiversity Conservation, 16–19 June, 2003, Cairns, Australia. 7 p. Available from: www.highseasconservation.org/documents.php.

Koppelmann, R. and H. Weikert. 1993: Full-depth zooplankton profiles over the deep bathyal of the NE Atlantic. Mar. Ecol. Prog. Ser. 86: 263–272

Lampitt, R. 2004. Can deep ocean cables coexist with environmental observatories? SubCable-News 29: 33–35

Leary, D. K. 2005a. Conservation and management of vulnerable deep-water ecosystems on the high seas: Are marine protected areas all that are required? (extended abstract). International Marine Protected Area Congress, 23–28 October 2005, Geelong, Australia. 4 p. Available from: www.impacongress.org/images/IMPAC%20Proceedings%20web.pdf.

_____. 2005b. More than just bugs and bioprospecting in the abyss. Designing an international legal regime for the sustainable management of deep-sea hydrothermal vents beyond national jurisdiction. Ph.D. Thesis, Macquarie University, Australia. 501 p.

Lochte, K. 1992. Bacterial standing stock and consumption of organic carbon in the benthic boundary layer of the abyssal North Atlantic. Pages 1–10 in G. Rowe and V. Pariente, eds. Deep-sea food chains and relation to global carbon cycle. NATO ASI Ser., Kluwer Academic Publ., Dordrecht.

Mullineaux, L. S., S. K. Juniper, D. Desbruyères, and M. Cannat. 1998. Steps proposed to resolve potential research conflicts at deep-sea hydrothermal vents. EOS, Trans. Am. Geophys. Union 79: 533–538.

National Research Council. 1984. Deep seabed stable reference areas. Ocean Policy Committee, NRC/NAS, Washington, DC, 74 p.

Pfannkuche, O. 1993. Benthic response to the sedimentation of particulate organic matter at the BIOTRANS station, 47°N, 20°W. Deep-Sea Res. I (Oceanogr. Res. Pap.) 40: 135–149.

_____, A. Boetius, K. Lochte, U. Lundgreen, and H. Thiel. 1999: Responses of deep-sea benthos to sedimentation patterns in the North-East Atlantic in 1992. Deep-Sea Res. I (Oceanogr. Res. Pap.) 46: 573–596.

Soltwedel, T., E. Bauerfeind, M. Bergmann, N. Budaeva, E. Hoste, N. Jeckisch, K. von Juterzenka, J. Matthiessen, V. Mokievsky, E.-M. Noethig et al. 2005. Hausgarten: multidisciplinary investigations at a deep sea, long-term observatory in the Arctic Ocean. Oceanography 18: 46–61.

Thiel, H. 2001. Unique science and reference areas on the high seas. Pages 97–102 *in* H. Thiel and J. A. Koslow, eds. Managing risks to biodiversity and the environment on the high seas, including tools such as marine protected areas — scientific requirements and legal aspects. Proc. Expert Workshop, International Academy for Nature Conservation, Isle of Vilm, Germany, 27 February–4 March 2001. Bundesamt für Naturschutz, Bonn, Germany.

_____. 2002. Science as stakeholder — a proposal for Unique Science Priority Areas. Ocean Challenge 12: 44–47.

_____. 2003a. Anthropogenic impacts on the deep sea. Pages 427–471 *in* P. A. Tyler, ed. Ecosystems of the deep ocean. Ecosystems of the world series. Elsevier, Oxford.

_____. 2003b. Approaches to the establishment of protected areas on the high seas. Pages 169–192 *in* A. Kirchner, ed. International marine environmental law, institutions, implementations, and innovations. Kluwer Law International, The Hague.

_____ and J. A. Koslow, eds. 2001. Managing risks to biodiversity and the environment on the high seas, including tools such as marine protected areas — scientific requirements and legal aspects. Proc. Expert Workshop, International Academy for Nature Conservation, Isle of Vilm, Germany, 27 February–4 March 2001. Bundesamt für Naturschutz, Bonn, Germany, 216 p.

_____, O. Pfannkuche, G. Schriever, K. Lochte, A. J. Gooday, Ch. Hemleben, R. F. G. Mantoura, C. M. Turley, J. W. Patching, and F. Riemann. 1988/89. Phytodetritus on the deep-sea floor in a central oceanic region of the Northeast Atlantic. Biol. Oceanogr. 6: 203–239.

Tunnicliffe, V. 1990. Observations on the effects of sampling on hydrothermal vent habitat and fauna of Axial Seamount, Juan de Fuca Ridge. J. Geophys. Res. 95: 12,961–12,966.

ADDRESS: *Poppenbuettler Markt 8A, D-22399 Hamburg, Germany. E-mail: <hjalmar.thiel@ hamburg.de>, Telephone: +49 (0) 40-608-75985.*

# Ecosystem-based management as a tool for protecting deep-sea corals in the USA

LANCE E. MORGAN, CHIH-FAN TSAO, AND JOHN M. GUINOTTE

## Abstract

In the USA, deep-sea coral habitat protection occurs as a secondary issue to traditional commercial fishery management. Ecosystem-based management (EBM) would better serve the goals of fishery management and deep-sea coral protection for three reasons. First, EBM preserves all parts of the ecosystem regardless of commercial value and prevents collateral damage of fishing activities on sensitive habitats including coral areas. Second, EBM addresses ecosystem stressors, holistically providing a framework to evaluate and manage the combined impacts of fishing and non-fishing threats to deep-sea corals. Third, EBM bases adaptive management on the precautionary principle, allowing emerging science to guide management actions.

The Pew Oceans Commission (2003) and the US Commission on Ocean Policy (2004) recommended stricter protection of vulnerable coral habitats. The President's 2004 Ocean Action Plan also emphasized deep-sea coral conservation and called for further identification and protection of deep-sea coral areas. To date, the recommendations by these panels are moving slowly and to a great degree have not been initiated by the federal government, though fishery management councils have made some strides in protecting deep-sea corals under essential fish habitat designations.

While there is great diversity among deep-sea coral species, the most pressing issue they face is their vulnerability to destructive human activities that contact the seafloor or alter the deep ocean environment (Freiwald et al., 2004; Roberts and Hirschfield, 2004; Kahng and Grigg, 2005; Guinotte et al., 2006; Morgan et al., 2006; Roberts et al., 2006). These activities include both direct and indirect human impacts that are occurring over large spatial and temporal scales. These impacts interact synergistically to the detriment of deep-sea coral health.

Fishing gears that contact the seafloor inevitably disturb the seabed and pose the gravest direct threat to deep-sea corals and sponges. In recent years, several reports have documented the impacts of fishing methods to seafloor habitats including two large symposium volumes (Benaka, 1999; Barnes and Thomas, 2005) as well as the National Research Council's review of fishing impacts (National Research Council, 2002). Fishing gears that impact the seafloor include bottom trawls, bottom longlines, bottom gillnets, dredges, and pots/traps (Chuenpagdee et al., 2003; Morgan and Chuenpagdee, 2003). Among these, bottom trawling is considered by scientists, managers and fishing professionals to be the most ecologically destructive fishing method (Chuenpagdee et al., 2003; Morgan and Chuenpagdee, 2003). Bottom trawling, which targets fish living on or just above the seafloor, destroys deep-sea coral ecosystems that took centuries to millennia to form (Fosså et al., 2002; Hall-Spencer et al., 2002; Puglise et al., 2005). Large bottom trawl gear can weigh several tons (Merrett and Haedrich, 1997) and the footrope

George, R. Y. and S. D. Cairns, eds. 2007. Conservation and adaptive    39
management of seamount and deep-sea coral ecosystems. Rosenstiel
School of Marine and Atmospheric Science, University of Miami.

is further weighted to keep the net in close contact with the seafloor. The footrope can be a chain or cable and is sometimes modified with large, heavy discs that are designed to ride over obstructions and keep the net from snagging and tearing on the seafloor. The benthic impacts of bottom trawling have been compared to the clear-cutting of old-growth forests (Watling and Norse, 1998), except that ancient corals are usually not collected and/or used, and are discarded overboard as bycatch.

Bottom trawling is widespread throughout the world's oceans and there are many international examples of coral damage caused by this fishing method. In Norway, 30%–50% of pre-existing *Lophelia pertusa* (Linnaeus, 1758) reefs have been destroyed by trawling (Fosså et al., 2002) and significant trawl damage to *L. pertusa* reefs has been documented in Irish waters (Hall-Spencer et al., 2002). Bottom trawling in the Canadian Atlantic dislodges deep-sea corals, which inevitably end up in fishing nets (Mortensen et al., 2005). Recent reports state that Canadian fishermen are observing a decrease in the size and abundance of corals in their nets (Gass and Willison, 2005). Koslow and colleagues (2000) reported that trawling reduced coral cover on a Tasmanian seamount from 90% to 5%, and Anderson and Clark (2003) reported that 1 hr of trawling for or-ange roughy (*Hoplostethus atlanticus* Collett, 1889) removed 1.6 t of corals in the New Zealand fishery.

Fisheries are managed in US waters by eight regional fishery management councils. The use of trawl gear is widespread in the New England and Mid-Atlantic council re-gions. In 1996 it was estimated that an area equivalent to three to four times the size of the entire Georges Bank area was being trawled annually (Auster et al., 1996). Although deep-sea corals have been recorded in these regions since 1874, scientists suspect that the current distribution of deep-sea corals has been altered by bottom fishing, and that many of the corals in historical records have since been destroyed by fishing (Watling and Auster, 2005). Shrimp trawling has destroyed the vast majority of *Oculina* reefs in the South Atlantic region (Koenig et al., 2005). Bottom trawling affects the largest area and leaves the largest ecological footprint in the waters off the US west coast (Morgan et al., 2005). West coast trawling activity occurs in areas with significant deep-sea coral habitats and the same is true in Alaskan waters. Between 1997 and 2001, an average of 81.5 t of coral was uprooted every year by commercial fishing in the North Pacific council region; 97% of this was attributed to bottom trawls (North Pacific Fishery Man-agement Council, 2003; National Marine Fisheries Service (NMFS), 2004). At present, bottom trawling does not occur in the Caribbean and Western Pacific council regions.

**Current management authorities for deep-sea corals in the USA.**—Deep-sea cor-als occur throughout US waters (Fig. 1). The US federal agencies with deep-sea coral management authority include National Marine Fisheries Service (NMFS) (with advice from the eight regional fishery management councils), NOAA's National Marine Sanc-tuary Program, and the Minerals Management Service (MMS) in the Department of Interior. NMFS and the regional fishery management councils may protect corals by adopting regulations that restrict fishing in certain areas, but they are not mandated to do so. National Marine Sanctuaries are designed to manage multiple uses compat-ible with protection, but management of fisheries is left to fishery management councils and NMFS. Consequently, bottom trawling is allowed in most US marine "sanctuaries" (Chandler and Gillelan, 2004). MMS oversees mineral, oil/gas exploration, and extrac-tion in federal waters and is responsible for assessing the environmental impacts of these activities to natural resources, including deep-sea corals. Oil and gas development in

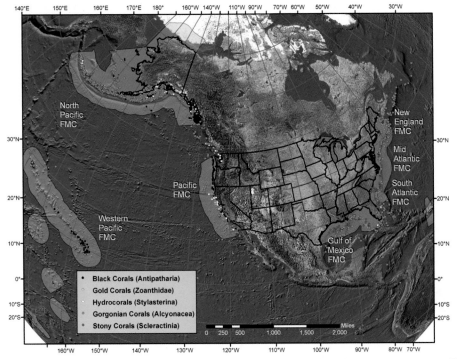

Figure 1. Overview of selected deep-sea corals in US waters with fisheries management council (FMC) regions (adapted from Morgan et al., 2006).

the Gulf of Mexico, the world's most active region for deep-water drilling (Glover and Smith, 2003; Avent, 2004), represents a threat to deep-sea corals because drilling, pipelines, and platform construction activities are occurring at depths and in areas where deep-sea corals are found (e.g., Viosca Knoll Lease Blocks 826 and 862 support diverse and abundant coral communities, S. Brooke, pers. comm.). These activities can crush corals or leak pollutants that may impact corals. Other activities which alter the seabed, including cable laying, are regulated by the Army Corps of Engineers.

**Management of deep-sea corals as essential fish habitat.**—Because fishing is the greatest current threat to deep-sea corals and most legal protections for them have been adopted in accordance with the nation's fishery management act, we focus the remaining discussion on management activities undertaken by NMFS. The authority to manage US fisheries comes from the Magnuson-Stevens Fishery Conservation and Management Act (Magnuson-Stevens Act, or MSA). Under the MSA, NMFS and the eight regional fishery management councils are required to identify and minimize impacts on essential fish habitat (EFH) of all species managed by federal fishery management plans. The MSA defines EFH as "those waters and substrate necessary to fish for spawning, breeding, feeding, or growth to maturity." The purpose of EFH is to identify areas required to support a sustainable fishery and ecosystem throughout the life cycle of the managed species. Within an EFH, the council can further designate Habitat Areas of Particular Concern (HAPCs). However, the MSA does not require protective regulations to be established for EFH and HAPC, and the HAPC designation only signifies that the habitat is a higher priority than the rest of EFH for conservation (Pautzke, 2005). National

Standard 9 of the Magnuson Stevens Act further requires the NMFS to adopt measures to minimize bycatch, including the bycatch of deep-sea corals, but this has not been used to limit fishing in coral areas.

Protection of deep-sea corals by regional fishery management councils under the EFH provisions are inconsistent and highly variable. This is a direct result of the non-binding language contained in the MSA. Some fishery management councils identify corals as a type of EFH for managed fish species, while others consider deep-sea corals themselves to be managed species. Thus some councils designate EFH specifically for deep-sea corals, while others only include deep-sea corals as EFH if a federally managed species can be shown to have a strong relationship with the coral. Often deep-sea corals are not considered in EFH designations because councils have insufficient data on the location of corals in their managed regions, or do not have sufficient information on the nature of the relationship of deep-sea corals and managed fish species.

Additionally the MSA was reauthorized in 2006, requiring NOAA to initiate a coordinated deep-sea corals research program to identify and map deep-sea coral ecosystems; it also affirms the authority of the regional fishery management councils to protect areas of the seafloor from any type of fishing gear that damages deep-sea coral habitat. The law gives the councils the unmistakable authority to move ahead with deep-sea coral protection efforts irrespective of whether coral areas have been deemed essential fish habitat. This is an important step for deep-sea coral protection, but it has not yet been acted upon by the fishery management councils.

**Limitations to EFH designations.**—The nature of single-species management or multi-species management—the current paradigm of fishery management—limits comprehensive protection for habitats and ecosystems. Many management actions, including EFH designations, are implemented through species-specific fisheries management plans, and are neither comprehensive nor primarily focused on deep-sea corals. To fully protect deep-sea corals, EFH designations must be accompanied by management actions to close areas to destructive fishing impacts. Instead, management actions are often limited by season restrictions, fishing gear, or fish species. In many cases the lack of clearly demonstrated fish—coral associations prevents EFH designation. Inconsistencies exist between fishery management councils definitions of EFH as well as their management actions. The following EFH actions highlight these inconsistencies.

Research has demonstrated the importance of deep-sea corals as fish habitat in Alaska, and managers have taken steps to protect some corals in this region. Deep-sea corals in Alaska have been described as keystone structures (Heifetz et al., 2005), and fish associations with corals are well documented (Stone, 2006). Stone (2006) reports that 85% of economically important fish and crabs, and 97% of juvenile rockfishes observed along submersible transects in the Aleutian Islands were associated with deep-sea corals or other emergent epifauna. Heifetz (2002) found Atka mackerel (*Pleurogrammus monopterygius* Pallas, 1810) and rockfish (*Sebastes* spp.) commonly associated with corals. In a study in waters of southeast Alaska, 85% of large adult rockfish were observed in and around *Primnoa* colonies (Krieger and Wing, 2002). These findings led the North Pacific Fishery Management Council to close some deep-sea coral areas to commercial fishing. In 2005 additional EFH regulations to prohibit bottom trawling in the Aleutian Islands Habitat Conservation Areas (AIHCA) (277,100 nmi$^2$) were adopted to address concerns about impacts on benthic habitat, particularly deep-sea corals, although much of the closure is deeper than trawlers operate at present. Additional protections for deep-sea corals exist for Bowers Ridge HAPC, where mobile bottom gear is prohibited, and in six small Coral Garden Marine Reserves (a total of 110 nmi[2]) currently closed to all bottom gears. These actions serve to protect some small areas and prevent the expansion of trawling, but other gears (e.g., crab pots, bottom longlines) which can crush or entangle coral are still permitted in most areas (NMFS, 2005).

Similarly, *Oculina* reefs off Florida were important breeding sites for commercially important snapper (*Lutjanus* spp.) and grouper (Serranidae) (Reed et al., 2005a). Gag grouper (*Mycteroperca microlepis* Goode and Bean, 1879) and dense spawning populations of scamp grouper (*Mycteroperca phenax* Jordan and Swain, 1884) were observed above the corals in the 70s and 80s (Koenig et al., 2005; Reed et al., 2005a). *Oculina* also provides shelter for juvenile speckled hind (*Epinephelus drummondhayi* Goode and Bean, 1878), suggesting use of the habitat as a nursery (Gilmore and Jones, 1992; Koenig et al., 2005). Reed and colleagues (2005b) examined fish on and surrounding the deep reefs of Pourtales Terrace and Miami Terrace (south Florida) and found several important commercial fishes, including blueline tilefish (*Caulolatilus microps* Goode and Bean, 1878), seabasses and groupers (serranids), red porgy (*Pagrus pagrus* Linnaeus, 1758), blackbar drum (*Pareques iwamotoi* Miller and Woods, 1988), blackbelly rosefish (*Helicolenus dactylopterus* Delaroche, 1809), scorpionfish (scorpaenids), greater amberjack (*Seriola dumerili* Risso, 1810), phycid hakes (*Urophycis* spp.), and sharks (carcharinids). In the South Atlantic region, the 300 nmi[2] *Oculina* Banks HAPC off Florida now prohibits bottom trawling, and four deep-sea coral areas have been proposed as EFH/HAPCs in the South Atlantic Fishery Management Council region.

In other regions, however, the precise nature of the association between fish and deep-sea coral habitats is less clear, even though commercial fisheries occur in deep-sea coral areas. The designation of fish habitat as essential is problematic and varies from region to region. In many cases, fish appear to prefer three-dimensional structures, but do not differentiate between corals and other living or non-living structures (Auster, 2005; Tissot et al., 2006). Other factors, including research techniques and size-, age- or species-specific fish behavior, may also influence the observed relationship. Scientists do know that habitat changes resulting from coral removal have likely influenced the behavior of fish and their population distribution (Sainsbury et al., 1997). Hence, just because we do not see fish associating with corals in the present day does not mean that they would not have associated with corals in the pre-trawling past.

In the Gulf of Maine, fish density does not seem to be influenced by the presence or absence of corals. The only fish species known to demonstrate a preference for coral areas over other structures for shelter and feeding is the oreo *Neocyttus helgae* (Holt and Byrne, 1908) (Auster, 2005; Auster et al., 2005). However, destructive fishing methods have altered coral distributions and abundance throughout the Gulf of Maine (Watling and Auster, 2005) and it is probable that the association/correlation between fish behavior and coral habitat has also been altered. EFH designation prohibits trawling for monkfish (*Lophius americanus* Valenciennes, 1837) in two submarine canyons in New England (Oceanographer and Lydonia Canyons) that are identified as HAPC for the purpose of protecting deep-sea corals. However, other bottom-tending gears not targeting monkfish still can be used to catch other fish species, and these methods negatively impact deep-sea coral ecosystems.

In the Pacific region, scientists observed large invertebrates and fishes among newly-discovered Christmas tree coral (*Antipathes dendrochristos,* Opresko, 2005) colonies off southern California, but the abundance of these invertebrates and fishes was low

(Tissot et al., 2006). Only eight of the 106 observed fish species showed a higher concentration inside the coral area as opposed to outside. A possible explanation for the lack of a demonstrable fish-coral association is that the relatively small size of most of the Christmas tree corals observed (< 50 cm) prevents them from serving as key shelter and refuge (Tissot et al., 2006). An EFH designation proposed by the fishery management council in 2005 and approved by NMFS in 2006 closed some banks to bottom trawling and all waters west of the 700 fathom depth contour to bottom trawling to protect depleted groundfish stocks. This action is intended to prevent further expansion of trawling into deeper waters, but it is unknown whether it will protect deep-sea corals since there are few known deep-sea coral records (Etnoyer and Morgan, 2005) in the region. Large closures created by "freezing the trawl footprint" may also not contain the most important deep-sea coral habitats.

Current coral management under fishery management plans has limitations; they focus on activities related specifically to certain economically valuable species, regional councils interpret fishery regulations differently, and council regulations are not law and can be rescinded. Finally, despite a dramatic increase in deep-sea coral science in recent years, the burden of proof is to demonstrate impact, rather than demonstrating no impact. This has been partially reversed with management actions that freeze the historic footprint of trawling activities, but remains a large obstacle to coral protection.

The burden of proof to determine the functional relationships between fish and corals prior to EFH designation or deep-sea coral protection is squarely at odds with the advice of the MSA to managers. MSA suggests that managers manage in a precautionary manner, but EFH designation requires that functional relationships be proven before protection is granted. This is a catch-22 situation and represents a substantial threat to deep-sea corals.

Areas with complete gear closures designated for deep-sea corals are quite small, and typically created after significant gear impacts have occurred (e.g., Aleutian Islands), and experience shows that enforcement can be a significant obstacle (e.g., illegal trawling in the *Oculina* HAPC off Florida, Reed et al., 2005a).

**The need for ecosystem-based management.**—Ecosystem-based management (EBM) will markedly improve the management of deep-sea corals and reduce current limitations on deep-sea coral conservation for three reasons. First, EBM addresses more than commercially valuable species and seeks to maintain habitats and ecosystem functions (Pikitch et al., 2004; Murawski, 2005; George et al., 2007). In EBM, deep-sea corals do not have to be proven as essential for commercial fishes before receiving protection. The habitat complexity that deep-sea corals contribute to the benthic environment is enjoyed by commercial and non-commercial fishes alike, and EBM lends itself to protecting all parts of the ecosystem regardless of their commercial value. EBM departs from the single-species approach to fisheries management, and should prevent collateral damage of fishing to the ecosystem. Deep-sea corals will gain comprehensive protection as opposed to the case-by-case/gear-by-gear management actions that have taken place to date. Access to remote locations is difficult and expensive, and using traditional coastal resource management based on individual fisheries assessments is logistically intractable. Also, with single-species management methods, a new fishery could develop and wipe out a habitat before protective measures could be enacted. Protecting an entire ecosystem allows management to be pro-active rather than reactive.

Second, EBM integrates the management of not only fishing activities but also other human impacts on the ecosystem (Rosenberg and McLeod, 2005; George et al., 2007). It provides the framework for assessing the consequences of hydrocarbon development, waste dumping, cable laying, climate change, etc., on deep-sea corals. More importantly, it allows us to evaluate and manage the synergistic impacts of all these threats combined. EBM is suitable for examining human impacts holistically and bringing together multiple government agencies with different authorities to address conservation issues.

Third, EBM bases adaptive management on the precautionary principle. EBM places the burden of proof on those whose activities affect the ecosystem (Pikitch et al., 2004; Murawski, 2005). In an EBM scenario, the fishing industry would be required to prove that fishing does not result in detrimental effects on the habitat of a particular area before fishing is allowed there. As scientific understanding progresses and ecosystem functions are better understood, management can adapt and regulate activities that are compatible (or not compatible) with specific habitats and ecosystem functions. Given that new locations/species of deep-sea coral are discovered every year and deep-sea coral research is growing at a fast pace, adaptive management would be a suitable approach for integrating new research findings into protection measures.

In recent years, management of commercial fisheries in the USA has been slowly evolving and a greater recognition of the interconnectedness of exploited species and their supporting ecosystems has taken on added importance. NMFS appears to be committed to viewing fishery management as more than the sum of multiple fishery stock assessments (Murawski, 2005). This is very good news to those concerned with biodiversity conservation, but current laws continue to provide a disservice to marine organisms that are not "proven" contributors to commercial fisheries. Deep-sea corals are not protected under the Endangered Species Act or similar legislation. EBM would also benefit sponges, echinoderms, and other cnidarians such as sea anemones, which all play important roles in benthic environments.

In summary, deep-sea corals in the USA are largely unprotected and the few seafloor habitat areas that are protected obtain a reprieve from destructive fishing methods only because they are thought to be important to a managed fishery or occur in areas where fishermen do not want to fish (e.g., rocky canyons, very deep water). Coral protection via fisheries management plans has proven to be insufficient for coral ecosystem health and a new approach is warranted. Ocean management should focus on ecosystem resilience and the maintenance of biodiversity rather than the health of commercially important fisheries alone. Fisheries and catch levels should be developed in accordance with protecting benthic habitats such as deep-sea corals and other components of seafloor ecosystems. An ecosystem-based management approach is the method by which deep-sea corals will truly receive protection from destructive fishing methods.

## Acknowledgments

We are grateful for the support of Richard and Rhoda Goldman Fund, Bullitt Foundation, National Fish and Wildlife Foundation, Curtis and Edith Munson Foundation, J. M. Kaplan Fund, and Oak Foundation for our efforts to study and conserve deep-sea corals. E. Norse, W. Chandler, H. Goldstein, and J. Palmer all provided important contributions to this paper. ESRI generously donated software. We appreciate the encouragement of R. George, and the efforts of S. Cairns to produce this volume. NOAA provided access to lots of information and data that was very helpful. We also thank the following scientists for help in accessing information and data, P. Etnoyer, P. Auster, R. Stone, L. Watling, S. Cairns, M. Stiles, J. Warrenchuk, and M. Hirshfield. Lastly, we thank the participants of the Third Symposium on Deep-Sea Corals for their infectious enthusiasm for these animals.

## Literature Cited

Anderson, O. F. and M. R. Clark. 2003. Analysis of bycatch in the fishery for orange roughy, *Hoplostethus atlanticus*, on the South Tasman Rise. Mar. Freshwat. Res. 54: 643–652.

Auster, P., R. Malatesta, R. Langton, P. Valentine, C. Donaldson, E. Langton, A. Shepard, and I. Babb. 1996. The impacts of mobile fishing gear on seafloor habitats in the Gulf of Maine (Northwest Atlantic): implications for conservation of fish populations. Rev. Fish. Sci. 4: 185–202.

Auster, P. J. 2005. Are deep-water corals important habitats for fishes? Pages 747–760 *in* A. Freiwald and J. M. Roberts, eds. Cold-water corals and ecosystems. Springer-Verlag, Berlin.

_____, J. Moore, K. B. Heinonen, and L. Watling. 2005. A habitat classification scheme for seamount landscapes: assessing the functional role of deep-water corals as fish habitat. Pages 761–769 *in* A. Freiwald and J. M. Roberts, eds. Cold-water corals and ecosystems. Springer-Verlag, Berlin.

Avent, R. 2004. Mineral Management Service environmental studies program: a history of biological investigations in the Gulf of Mexico, 1973–2000. US Department of the Interior, Minerals Management Service, New Orleans. 35 p.

Barnes, P. W. and J. P. Thomas, eds. 2005. Benthic habitats and the effects of fishing. Am. Fish. Soc. Symp. 41. American Fisheries Society, Bethesda. 890 p.

Benaka, L. R., ed. 1999. Fish habitat: essential fish habitat and rehabilitation. Am. Fish. Soc. Symp. 22. American Fisheries Society, Bethesda. 400 p.

Chandler, W. J. and H. Gillelan. 2004. The history and evolution of the National Marine Sanctuaries Act. Environmental Law Reporter News and Analysis 34: 10505–10565.

Chuenpagdee, R., L. E. Morgan, S. M. Maxwell, E. A. Norse, and D. Pauly. 2003. Shifting gears: assessing collateral impacts of fishing methods in US waters. Front. Ecol. Environ. 1: 517–524.

Etnoyer, P. and L. E. Morgan. 2005. Habitat-forming deep-sea corals in the Northeast Pacific Ocean. Pages 331–343 *in* A. Freiwald and J. M. Roberts, eds. Cold-water corals and ecosystems. Springer-Verlag, Berlin.

Fosså, J. H., P. B. Mortensen, and D. M. Furevik. 2002. The deep water coral *Lophelia pertusa* in Norwegian waters: distribution and fishery impacts. Hydrobiologia 471: 1–12.

Freiwald, A., J. H. Fosså, A. Grehan, T. Koslow, and J. M. Roberts. 2004. Cold water coral reefs: out of sight- no longer out of mind. UNEP World Conservation Monitoring Center, Cambridge. 84 p.

Gass, S. E. and J. H. M. Willison. 2005. An assessment of the distribution of deep-sea corals in Atlantic Canada by using both scientific and local forms of knowledge. Pages 223–245 *in* A. Freiwald and J. M. Roberts, eds. Cold-water corals and ecosystems. Springer-Verlag, Berlin.

George, R. Y., T. A. Okey, J. K. Reed, and R. P. Stone. 2007. Ecosystem-based fisheries management of seamounts and deep-sea coral reefs in US waters: conceptual models for proactive decisions. Pages 9–30 *in* R. Y. George and S. D. Cairns, eds. Conservation and adaptive management of seamount and deep-sea coral ecosystems. Rosenstiel School of Marine and Atmospheric Science, University of Miami. Miami. 324 p.

Gilmore, R. G. and R. Jones. 1992. Color variation and associated behavior in the epihepheline groupers, *Mycteroperca microlepis* (Goode and Bean) and *M. phenax* (Jordan and Swain). Bull. Mar. Sci. 51: 83–103.

Glover, A. G. and C. R. Smith. 2003. The deep-sea floor ecosystem: current status and prospects of anthropogenic change by the year 2025. Environ. Conserv. 30: 219–241.

Guinotte, J., J. Orr, S. Cairns, A. Freiwald, L. Morgan, and R. George. 2006. Will human-induced changes in seawater chemistry alter the distribution of deep-sea scleractinian corals? Front. Ecol. Environ. 4: 141–146.

Hall-Spencer, J., V. Allain, and J. H. Fosså. 2002. Trawling damage to Northeast Atlantic ancient coral reefs. Proc. R. Soc., Biol. Sci. 269: 507–511.

Heifetz, J. 2002. Coral in Alaska: distribution, abundance, and species associations. Hydrobiologia 471: 19–28.

_____, B. L. Wing, R. P. Stone, P. W. Malecha, and D. L. Courney. 2005. Corals of the Aleutian Islands. Fish. Ocean. 14 (suppl 1): 131–138.

Kahng, S. and R. Grigg. 2005. Impact of alien octocoral, _Carijoa riisei_, on black corals in Hawaii. Coral Reefs 24: 556–562.

Koenig, C. C., A. N. Shepard, J. Reed, F. C. Coleman, S. D. Brooke, J. Brusher, and K. M. Scanlon. 2005. Habitat and fish populations in the deep-sea _Oculina_ coral ecosystem of the Western Atlantic. Pages 795–805 _in_ P. W. Barnes and J. P. Thomas, eds. Benthic habitats and the effects of fishing. Am. Fish. Soc. Symp. 41. American Fisheries Society, Bethesda.

Koslow, J. A., G. W. Boehlert, J. D. M. Gordon, R. L. Haedrich, P. Lorance, and N. Parin. 2000. Continental slope and deep-sea fisheries: implications for a fragile ecosystem. ICES J. Mar. Sci. 57: 548–557.

Krieger, K. J. and B. L. Wing. 2002. Megafauna associations with deepwater corals (_Primnoa_ spp.) in the Gulf of Alaska. Hydrobiologia 471: 83–90.

Merrett, N. and R. Haedrich. 1997. Deep-sea demersal fish and fisheries. Chapman & Hall, London. 282 p.

Morgan, L. E. and R. Chuenpagdee. 2003. Shifting gears: addressing the collateral impacts of fishing methods in U.S. waters. Island Press, Washington. 42 p.

_____, C.-F. Tsao, and J. Guinotte. 2006. Status of deep sea corals in US waters with recommendations for their conservation and management. Marine Conservation Biology Institute, Bellevue. 64 p.

_____, P. Etnoyer, A. J. Scholz, M. Mertens, and M. Powell. 2005. Conservation and management implications of deep-sea coral and fishing effort distributions in the Northeast Pacific Ocean. Pages 1171–1187 _in_ A. Freiwald and J. M. Roberts, eds. Cold-water corals and ecosystems. Springer-Verlag, Berlin.

Mortenson, P., L. Buhl-Mortensen, D. C. Gordon, G. B. J. Fader, D. L. McKeown, and D. G. Fenton. 2005. Effects of fisheries on deepwater Gorgonian corals in the Northeast Channel, Nova Scotia. Pages 369–382 _in_ P. W. Barnes and J. P. Thomas, eds. Benthic habitats and the effects of fishing. Am. Fish. Soc. Symp. 41. American Fisheries Society, Bethesda.

Murawski, S. 2005. Strategies for incorporating ecosystem considerations in fisheries management. Pages 163–171 _in_ D. Witherell, ed. Managing our nation's fisheries II: focus on the future: proceedings of a conference on fisheries management in the United States, held in Washington, D.C., March 24–26, 2005. North Pacific Fishery Management Council, Anchorage.

National Research Council. 2002. Effects of trawling and dredging on seafloor habitat. National Academies Press, Washington DC. 126 p.

NMFS. 2004. Final programmatic supplemental groundfish environmental impact statement for Alaska groundfish fisheries. U.S. Department of Commerce, NOAA, NMFS, Alaska Region, Juneau. Unpaginated.

_____. 2005. Final environmental impact statement for essential fish habitat identification and conservation in Alaska. U.S. Department of Commerce, NOAA, NMFS, Alaska Region, Juneau. Unpaginated.

North Pacific Fishery Management Council. 2003. Stock assessment and fishery evaluation report for the groundfish resources of the Bering Sea/ Aleutian Islands region. North Pacific Fishery Management Council, Anchorage.

Opresko, D. M. 2005. A new species of antipatharian coral (Cnidaria: Anthozoa: Antipatharia) from the southern California Bight. Zootaxa 852: 1–10.

Pautzke, C. 2005. The challenge of protecting fish habitat through the Magnuson-Stevens Fishery Conservation and Management Act. Pages 19–40 _in_ P. W. Barnes and J. P. Thomas, eds. Benthic habitats and the effects of fishing: Am. Fish. Soc. Symp. 41. American Fisheries Society, Bethesda.

Pew Oceans Commission. 2003. America's living oceans: charting a course for sea change, a re-
    port to the nation. Pew Oceans Commission, Arlington. 144 p.
Pikitch, E. K., C. Santora, E. A. Babcock, A. Bakun, R. Bonfil, D. O. Conover, P. Dayton, P. Dou-
    kakis, D. Fluharty, B. Henneman, E. D. Houde, J. Link, P. A. Livingston, M. Mangel, M. K.
    McAllister, J. Pope, and K. J. Sainsbury. 2004. Ecosystem-based fishery management. Science
    305: 346–347.
Puglise, K. A., R. J. Brock, and J. J. McDonough. 2005. Identifying critical information needs and
    developing institutional partnerships to further the understanding of Atlantic deep-sea coral
    ecosystems. Pages 1129–1140 in A. Freiwald and J. M. Roberts, eds. Cold-water corals and
    ecosystems. Springer-Verlag, Berlin.
Reed, J. K., A. N. Shepard, C. C. Koenig, K. M. Scanlon, and R. G. Gilmore. 2005a. Mapping,
    habitat characterization and fish surveys of the deep-water Oculina coral reef marine protected
    area: a review of historical and current research. Pages 443–465 in A. Freiwald and J. M. Rob-
    erts, eds. Cold-water corals and ecosystems. Springer-Verlag, Berlin.
_____, S. A. Pomponi, D. Weaver, C. K. Paull, and A. E. Wright. 2005b. Deep-water sinkholes
    and bioherms of south Florida and the Pourtales Terrace- habitat and fauna. Bull. Mar. Sci. 77:
    267–296.
Roberts, J. M., A. J. Wheeler, and A. Freiwald. 2006. Reefs of the deep: the biology and geology
    of cold-water coral ecosystems. Science 312: 543–547.
Roberts, S. and M. Hirshfield. 2004. Deep-sea corals: out of sight, but no longer out of mind. Front.
    Ecol. Environ. 2: 123–130.
Rosenberg, A. A. and K. L. McLeod. 2005. Implementing ecosystem-based approaches to man-
    agement for the conservation of ecosystem services. Mar. Ecol. Prog. Ser. 241: 270–274.
Sainsbury, K. J., R. A. Campbell, R. Lindholm, and A. W. Whitelaw. 1997. Experimental manage-
    ment of an Australian multi-species fishery: examining the possibility of trawl induced habitat
    modifications. Pages 107–112 in E. K. Pikitch, D. D. Huppert, and M. P. Sissenwine, eds.
    Global trends: fisheries management. Am. Fish. Soc., Bethesda.
Stone, R. 2006. Coral habitat in the Aleutian Islands of Alaska: depth distribution, fine-scale spe-
    cies associations, and fisheries interactions. Coral Reefs 25: 229–238.
Tissot, B. N., M. M. Yoklavich, M. S. Love, K. York, and M. Amend. 2006. Structure-forming
    invertebrates as components on benthic habitat on deep banks off Southern California with
    special reference to deep sea corals. Fish. Bull. 104: 167–181.
US Commission on Ocean Policy. 2004. An ocean blueprint for the 21st century–final report of the
    U.S. Commission on Ocean Policy. US Commission on Ocean Policy, Washington. 522 p.
Watling, L. and P. J. Auster. 2005. Distribution of deep-water Alcyonacea off the northeast coast of
    the United States. Pages 279–296 in A. Freiwald and J. M. Roberts, eds. Cold-water corals and
    ecosystems. Springer-Verlag, Berlin.
_____ and E. A. Norse. 1998. Disturbance of the seabed by mobile fishing gear: a comparison
    with forest clear-cutting. Conserv. Biol. 12: 1189–1197.

ADDRESSES: (L.E.M.) *Marine Conservation Biology Institute, 14301 Arnold Dr., Suite 25, Glen
Ellen, California 95442.* (C.-F.T., J.M.G) *Marine Conservation Biology Institute, 2122 112th Ave
NE, Suite B-300, Bellevue, Washington 98004.* CORRESPONDING AUTHOR: (L.E.M.) *E-mail:
<lance@mcbi.org>.*

# A research agenda of geological, bio-physical, and geochemical aspects for deep-sea bio-buildups of the Bahamas–Florida region (Bafla)

ROBERT N. GINSBURG AND STEVEN J. LUTZ

## Abstract

Deep-sea bio-buildups (ca. 400–1300 m) are remarkably numerous on the floors of inter- and intra-bank channels and bank slopes in the Bahamas and Florida region (Bafla). Deep-sea bio-buildup is used here to encompass all local sea floor relief associated with deep-sea corals including mounds of unconsolidated deposits (bioherms, banks) and lithoherms with onion-like surficial layers that were cemented syndepositionally. Pioneering research on these bio-buildups indicates that they offer an unparalleled opportunity to address the roles of biophysical, geological, and geochemical processes in determining their locations and development. Here we focus on a research agenda for these processes. For each of a selection of key process categories, we present a brief summary of what is known and a selection of research opportunities. The categories are: regional distribution of deep-sea bio-buildups in Bafla; foundations and ages of bio-buildups; baffling of sediment by branched deep-sea corals and other invertebrates; enhanced current velocities and their interactions with bio-buildups; and syndepositional cementation of lithoherms and hardrounds.

During the past half century, pioneering research on sea floors in the Straits of Florida, around and between the shallow Bahama Banks and on the Blake Plateau (Bafla) has revealed remarkably numerous deep-sea bio-buildups (ca 400–1300 m). Here we use deep-sea bio-buildups for all those sea-floor reliefs that have accumulated in place and are associated with deep-sea branched corals and other benthos, i.e., the deep-sea coral banks, mounds, bioherms, or deep coral reefs of other authors (Fig. 1). The ranges of size and relief of bio-buildups extends from a few square meters and a few meters of relief (Neumann and Ball, 1970) to complexes of some tens of kilometers long with relief up to nearly 150 m (Paul et al., 2000; Reed et al., 2006) The two main types of bio-buildups in this region are mounds of unconsolidated sediment and branched corals also termed deep-sea banks or bioherms (Reed et al., 2006) and lithoherms that also have branched deep-sea corals and other macrobenthos, but have successive crusts of friable limestone composed of coral debris and pelagic sediment (Neumann et al., 1977; Wilber and Neumann, 1993; Paul et al., 2000; Reed et al., 2005, 2006). Paul et al. (2000) estimated that over 40,000 coral lithoherms could be present in the northern Straits of Florida and the Blake Plateau; anecdotal reports from other parts of Bafla (D. McNeill and G. Eberli, University of Miami, pers. comm.) suggest that there may be twice that many or more in the entire region.

The relief of bio-buildups in Bafla ranges from a few to > 150 m (Reed, 2002b; M. Grasmueck and G. Eberli, University of Miami, pers. comm.). They occur in areas of intermittent strong bottoms currents and have varying amounts of irregularly branched

George, R. Y. and S. D. Cairns, eds. 2007. Conservation and adaptive management of seamount and deep-sea coral ecosystems. Rosenstiel School of Marine and Atmospheric Science, University of Miami.

deep-sea corals: *Lophelia pertusa* (Linnaeus, 1758), *Enallopsammia* spp., *Madrepora* spp., and *Solenosmilia variabilis* Duncan, 1873—as well as massive sponges, gorgonians, crinoids, numerous species of fish and other macroinvertebrates (Messing et al., 1990; Reed, 2002; Reed et al., 2005, 2006). The accumulated results of research on Bafla bio-buildups, both the numerous individual papers and the thorough summaries cited below provide the foundation to address fundamental questions about the origins, processes, and history of these accumulations. We aim to stimulate research by outlining some of these fundamental questions.

A list of five research themes with a biological emphasis was identified during the Deep-sea Corals Collaboration Planning Meeting in 2002 (Puglise et al., 2005): (1) mapping the distribution of deep-sea corals; (2) ecology of organisms associated with deep-sea corals; (3) physiology of deep-sea corals—indicators and responses to change; (4) taxonomic studies; and (5) paleo-restropective analyses.

## Distribution

The generalized distribution of bio-buildups in Bafla from published information and personal communications reveals three main areas: the Straits of Florida and its northward extension, the northern slope of Little Bahama Bank and the Blake Plateau (Fig. 1). Reed (2002a,b; Reed et al., 2005, 2006) has provided the most comprehensive summary of the bio-buildups of these areas. The largest single bio-buildup is the unconslidated *Oculina* reefs at depths of 70–100 m along the eastern shelf of Florida (Fig. 1; Reed, 2002a). In depths of 490–900 m, lithoherms are locally abundant in the central and northern Straits (Neumann et al., 1977; Paul et al., 2000; Reed et al., 2006). Mounds of unconsolidated deposits with deep-sea corals are widespread, for example, on the northern slope of Little Bahama Bank. There, Mullins et al. (1981) discovered and described an extensive area of unconsolidated mounds at depths of 1000–1300 m; similar mounds occur in the Straits of Florida (Reed et al., 2006). On the Blake Plateau at depths of 640–869 m, Stetson et al. (1962) first discovered coral-capped mounds with nearly 150 m of relief. More recently, from submersible observations in the same area, Reed et al. (2006) described one of these bio-buildups, Stetson Peak as a lithoherm 153 m tall. In addition to these published reports, there are indications that many more lithoherms are likely in the southern Straits of Florida (D. McNeill, University of Miami, pers. comm.) and others at greater depths (in the 1000 m range) have been seen on seismic profiles off Key West (G. Eberli, University of Miami, pers. comm.). The presence of bio-buildups in the western end of the northeast Providence Channel, identified by examination of bottom topography (Mullins and Neumann, 1979) and from observations by Reed (Harbor Branch Oceanographic Institution, pers. comm.) as well as the occurrence of structure-forming deep-sea corals throughout the region (in the Straits of Florida and other inter-bank channels) implies that there may be many other areas of bio-buildups in Bafla. Structure-forming corals include the following species: *L. pertusa*, *Enallopsammia profunda* (Pourtalès, 1867), *Enallopsammia rostrata* (Pourtalès, 1878), *Madrepora carolina* (Pourtalès, 1871), *Madrepora oculata* (Linnaeus, 1758), and *S. variabilis*. These corals are regarded the major species associated with bio-buildups in the north Atlantic (Carins, 1979, 2000; Freiwald et al., 2004).

It is evident therefore, that this broad region extending nearly 1000 km from off Georgia to Key West and including the Blake Plateau is a major province of deep-sea bio-buildups. What is equally evident is the need for comprehensive mapping of these structures

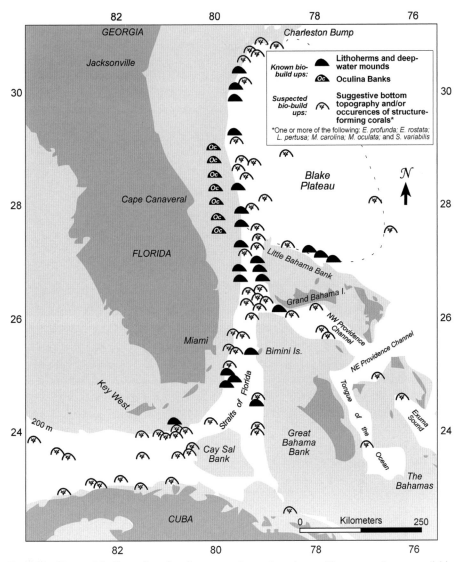

Figure 1. Chart of Bafla region showing approximate locations of known and suspected bio-buildups. Sources of information include Cairns (1979, 2000), Mullins and Neumann (1979), Mullins et al. (1981), Messing et al. (1990), Reed (2002), Reed et al. (2005, 2006), Correa et al. (2006), and Grasmueck et al. (2006).

in the known areas of occurrence as well as in other areas where they are inferred (Fig. 1). Such surveys will provide the much-needed foundation for future research and the locations for management to help ensure the conservation and protection of deep-sea coral habitat (for example, Deepwater Coral Habitat Areas of Particular Concern are currently being discussed for US water of the Bafla region by the South Atlantic Fishery. A larger view of deep-sea bio-buildups can be found in Roberts et al. (2006).

Planning the needed mapping will surely benefit from two recent developments. First is the comprehensive review of the various imaging tools (echo-sounding, side-scan sonar, multibeam, their advantages and limitations), ground-truth sampling and case histories (Fosså et al., 2005). The second more direct model for future mapping comes from

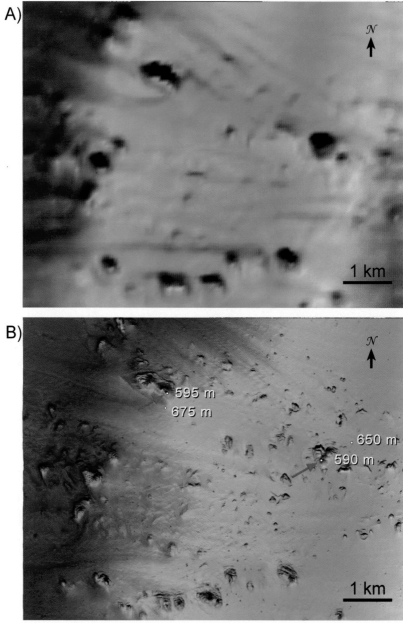

Figure 2. Shaded multibeam bathymetric maps of coral mounds in 600–700 m water depths at the toe of slope of Great Bahama Bank (both images are from the same location). (A) 12 kHz hull-mounted Kongsberg Simrad EM 120. The map with a grid resolution 50 m shows prominent mound features on the seafloor. Such maps are used for reconnaissance purposes to identify areas of interest for high resolution mapping. (B) 200 kHz Multibeam Echo Sounder map acquired with the C&C technologies C-Surveyor II ™. With a cruising altitude of 40 m above the seafloor the map grid resolution can be improved to 3 m. This map reveals the diverse morphologies of deep-sea coral mounds influenced by off-bank (E-W trending) ridges and the Florida current. Total depth range displayed is 583–719 m. Two large mounds are identified (red arrows); one is 80 m in relief, the other 60 m in relief. Depth measurements of mound tops and adjacent sea floor are included. Images and caption provided by M. Grasmueck (Univ. of Miami). See also Correa et al. (2006) and Grasmueck et al. (2006).

the results of multi-scale surveys using an AUV deployed a few tens of meters above the sea floor and equipped with multibeam in the Straits of Florida (Fig. 2).

## Foundations and ages of Bafla bio-buildups

Extensive observations and collections in the North East Atlantic have established the conditions necessary for the occurrence of *L. pertusa*, the most common and widely distributed deep-sea coral (Roberts et al., 2006), and one that is prominent in bio-buildups in Bafla (Reed et al., 2006). This coral as well as other deep-sea species require hard substrate for attachment (Wilson, 1979; Freiwald et al., 2004) and oceanic water with temperatures between 4–12 °C (Rogers, 1999). In the Bafla region, the consistent association of deep-sea corals and hard substrates has led to the view that these substrates and ocean currents that so often co-occur are necessary for both mounds and lithoherms.

The composition, origins, and ages of hard substrates are therefore of major interest in future research. Two main origins of hard substrate deserve consideration—inherited and autochthonous. On the Florida Hatteras slope, Paul et al. (2000) reported that a large lithoherm complex occurs at the seaward edge of an Eocene age outcrop. The bio-buildups on the Blake Plateau (Stetson et al., 1962) are also thought to be developed on current-swept older limestones (Pinet and Popenoek, 1985). A second major source of hard substrate for lithoherms of the Straits of Florida is probably hardgrounds, the result of syndepositional cementation of sea floor sediments (Neumann et al., 1977; Wilber and Neumann, 1993). The same interstitial precipitation of calcium carbonate that is either or both physico-chemical or mediated by microorganisms produces both the numerous lithoherms as well as their hardground substrates (Neumann et al., 1977; Wilber and Neumann, 1993; Paul et al., 2000).

For mounds that are uncemented and not associated with inherited hard substrate or hardgrounds, a quite different model is needed. One possibility is that skeletons of invertebrates or vertebrates could provide sufficient hard substrate to attract the settlement of deep-sea corals given the appropriate hydrographic conditions (Wilson, 1979; Rogers, 1999). This model is comparable to that proposed for shallow patch reefs, in which a single shell or fragment could provide substrate for coral settlement and initiate a succession of corals leading to a patch reef community (see also Enhanced Current Velocities below). Two sites to test this model are: (1) the *Oculina* reefs (Reed, 2002a), and (2) the extensive deep-sea mounds described by Mullins et al. (1981).

**Ages of bio-buildups.**—Living deep-sea corals have modern [14]C dates of 700 ± 80 yBP (Paul et al., 2000). The sediment trapped within living coral thickets has a bulk date of 3250 yBP (Paul et al., 2000) but it is probable that this sediment is a mixture of particles of different ages and that age is therefore only an average. Similarly, the bulk [14]C dates of friable crusts on the surface of some lithoherms range from 6.5 to 32.7 ky (Neumann et al., 1977; Paul et al., 2000) These bulk dates are again averages of sediment and cement that are likely of varied ages, therefore the dates may not provide accurate ages of accumulation. The ambiguity of these bulk dates can be avoided by dating coral skeletons and cements with Accelerator Mass Spectrometry. The shallower *Oculina* reefs are much younger, an estimated 980 yBP extrapolated from the [14]C date of a single coral at 22 cm below the reef surface (Reed, 2002b).

## Sediment baffling by corals and invertebrates

Most descriptions of Bafla bio-buildups with living corals and other invertebrates describe them as having capping thickets of corals with trapped sediments (Stetson et al., 1962; Neumann et al., 1977; Wilber and Neumann, 1993; Paul et al., 2000; Reed et al., 2005, 2006). Along the lower slope of Little Bahama Bank the coral thickets are preferentially developed on the southern up-current slopes of lithoherms (Wilber and Neumann, 1993). In the same area, a detailed study using a submersible described a current-related zonation of corals, zoanthids, crinoids, and alcyonarians (Messing et al., 1990).

The efficiency of sediment baffling by different benthos may have a significant influence on initiation and growth rates of bio-buildups. Questions about sediment and baffling by corals and invertebrates include: (1) Are the multi-branched deep-sea corals, especially the most common *L. pertusa,* the most effective baffle? (2) What abundance size and spacing of larger invertebrates, e.g., sponges, gorgonians and crinoids- produces an effective baffle? (3) Are different kinds of bio-baffles selective for certain hydraulically-equivalent particles?

## Enhanced current velocities

The extensive occurrence of deep-sea bio-buildups in the eastern Straits of Florida beneath the along-bank currents and their replacement with bank-derived lime muds nearby where those currents are absent (Wilber and Neumann, 1993) suggests the necessity of currents for the initiation and development of lithoherms. Two scales of submarine topography in Bafla produce convergences that accelerate current velocities. One is the Straits of Florida and channels in-between the Bahama Banks (Fig. 1) as well as the Blake Plateau beneath the Gulf Stream, all of which enhance both ocean and tidal currents. In the Straits of Florida, the current velocities reported at the same depths and area as some of the lithoherms averaged 55 cm s$^{-1}$ (Mayer et al., 1984). The currents on submarine escarpments (Paul et al., 2000) and slopes of the Little Bahama Bank are estimated to be in the same range. Mullins et al. (1981) estimated that the ambient current velocities north of Little Bahama Bank range from ~25–50 cm s$^{-1}$.

The second and smaller scale of current enhancement is the result of mounds and lithoherms themselves. Any local elevation of the sea floor enhances current velocities and produces eddies. Neumann and Ball (1970) observed "current crescents" (Potter and Pettijohn, 1963) developed on the down current side of mounds of deep-sea corals; see also Figure 2. Genin et al. (1986) documented the doubling of current velocity on the top of a seamount and similar enhancement was reported by Paul et al. (2000). Once initiated, growing bio-buildups themselves produce accelerated currents and this feedback is likely a key process in their growth from thicket to coppice (Fig. 3; Squires, 1964; Wilson, 1979). Furthermore, the interacting eddies from an area of bio-buildups may initiate new foundations by collecting coral branches that attract settlement of coral larvae.

The interactions between bio-buildups and currents are so central to the development of these structures that they deserve special attention in future research. Among the fundamental questions are: (1) What current velocities are required to resuspend pelagic sediment from inter-buildup areas and to produce a zonation of benthos as described by Messing et al. (1990)? (2) Can individual and closely-spaced buildups produce interacting eddies that can move and accumulate coral debris, which then becomes the founda-

## Bio-buildup stage                    ## Description

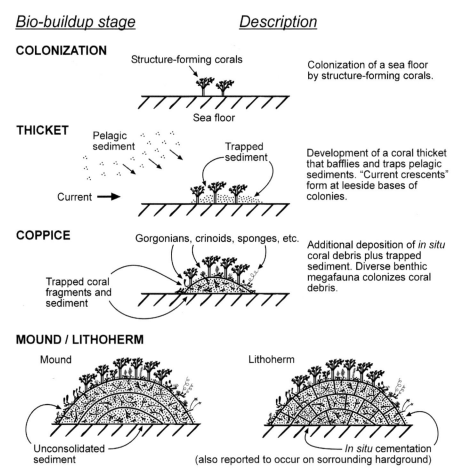

**COLONIZATION**

Structure-forming corals

Sea floor

Colonization of a sea floor by structure-forming corals.

**THICKET**

Pelagic sediment

Trapped sediment

Current →

Development of a coral thicket that baffles and traps pelagic sediments. "Current crescents" form at leeside bases of colonies.

**COPPICE**

Gorgonians, crinoids, sponges, etc.

Trapped coral fragments and sediment

Additional deposition of *in situ* coral debris plus trapped sediment. Diverse benthic megafauna colonizes coral debris.

**MOUND / LITHOHERM**

Mound

Lithoherm

Unconsolidated sediment

*In situ* cementation (also reported to occur on sorrounding hardground)

Accretion continues with a coral cap and associated megafauna. Mounds are composed of unconsolidated sediment, lithoherms by lithified structure.

Figure 3. Hypothetical development sequence for deep-sea bio-buildups adapted from Mullins et al. (1981), based on Wilson (1979), Squires (1964), Neumann and Ball (1970), and Neumann et al. (1977).

tion of new buildups? (3) Do variations in the ambient current velocities and submarine topography influence the distribution and type of bio-buildups? (4) Are there different current regimes associated with unconsolidated mounds and with lithoherms?

### Syndepositional cementation of lithoherms and hardgrounds

Neumann et al. (1977) discovered, described, and proposed the term lithoherm for mounds with crusts of friable limestone on the eastern side of the Straits of Florida. These first-studied lithoherms are ridges 30–50 m high elongated north-south with steep ca. 20–30° slopes at depths of 600–700 m. The inter-lithoherm areas are floored with hardgrounds and have small irregular lithoherms. The generalized distribution of these lithoherms (Neumann et al., 1977) extends along the eastern side of the Straits from off the Bimini Islands northward along Little Bahama Bank and with a branch eastward into the Northwest Providence Channel (Fig. 1).

The most common lithology described by Wilber and Neumann (1993) is a packstone composed of foraminifers, pteropods, and interclasts of the same composition. The interclasts testify to the alternation of cementation and erosion. In the area of lithoherms, there are rudstones and floatstones of coral fragments surrounded and infilled by packstone of foraminifers and pteropods.

The successive and friable cemented crusts, 10–30 cm thick, of both the lithoherms and the hardgrounds between them are preferentially harder at the exposed surfaces. Bulk $^{14}$C dates of nine samples range from 6.5 to 32.7 ka, but because they are averages of cement and grains that may have different ages, they are not reliable indications of the ages of accumulation. The cement is magnesium calcite with 10–14 mol % $MgCO_3$ (Wilber and Neumann, 1993). It occurs as amorphous micrite (< 4 μ), clotted and peloidal micrite, and as epitaxial overgrowths. In cavities, these cement morphologies occur in regular succession that may be the result of changes in the rate of precipitation (Wilber and Neumann, 1993).

The preferential cementation of the exposed surface of the crusts offers a special opportunity to use $^{14}$C accelerator mass spectrometry (AMS) dates of the cements in successive crusts to date cement formation. Dating the preferentially cemented tops of successive crust can provide a chronology of lithoherm accumulation. A major question about this sea floor cementation is what geochemical, microbial, or physical circumstances trigger it. Why are lithoherms and hardgrounds more abundant on the east side of the Straits of Florida, along the Florida Hatteras slope and on the Blake Plateau, all areas of topographic convergence and yet appear to be absent from the west side of the Straits?

## Initial research priorities

This paper is focused on a research agenda for the geological, physical, and geochemical processes of Bafla bio-buildups. An equally extensive agenda could be prepared for the biological aspects, which would extend from micro-organisms to fishes.
Priorities:
- Regional multibeam surveys of known and likely occurrences of Bafla bio-buildups
- Piston coring of unconsolidated mounds; lithostratigraphy and dating of recovered deposits
- Hydraulic piston-coring or core borings of lithoherms; lithostratigraphy and dating of recovered deposits
- $^{14}$C dating of the cements of crusts from lithoherms and hardgrounds and associated coral fragments
- Quantitative analysis of lithologic components of mounds and lithoherms
- Characterization of current velocity regimes of both mounds and lithoherms and the areas down-current of closely spaced bio-buildups
- Relative efficiency and selectivity of sediment trapping by different invertebrates, e.g., corals, crinoids, sponges, and alcyonarians.

## Acknowledgments

All the data and direct observations come from published sources and personal communications. We are especially indebted to J. Reed for access to his numerous publications and personal communications. Moreover, his review of the manuscript was invaluable in correcting modifying the text and Figure 1. We thank S. Cairns for advice on the occurrences of corals and A. Shepard and T. Hourigan for stimulating our interest in deep-sea bio-buildups. R. Gregory's suggestions and encouragement are much appreciated. M. Grasmueck and G. Eberli provided the revealing multibeam images of Figure 2. Partial support for the preparation of the manuscript came from the NOAA Coastal Ocean Program.

## Literature cited

Cairns, S. 1979. The deep-water Scleractinia of the Caribbean Sea and adjacent waters. Studies on the Fauna of Curacao and Other Caribbean Islands 57: 1–341.

_____. 2000. A revision of the shallow-water azooxanthellate scleractinia of the Western Atlantic. Stud. Nat. Hist. Carib. Region 75: 240 p.

Correa, T., M. Grasmueck, G. Eberli, G. Rathwell, and D. Viggiano. 2006. Morphology and distribution of deep-water coral reefs in the Straits of Florida. Project prospectus. Accessed August 2006. Available from: http://mgg.rsmas.miami.edu:8080/CSL/Medialib/correa_etal_06.pdf

Fosså, J. H., B. Lindberg, O. Christensen, T. Lundälv, I. Svellingen, P. Mortensen, and J. Alvsvåg. 2005, Mapping of *Lophelia* reefs in Norway: experiences and survey methods. Pages 359–391 *in* A. Freiwald, and J. M. Roberts, eds. Cold-water corals and ecosystems. Spring-Verlag, Berlin.

Freiwald, A., J. H. Fosså, A. Grehan, T. Koslow, and J. M. Roberts. 2004. Cold-water coral reefs. UNEP-WCMC, Cambridge. 84 p.

Genin, A., A. K. Dayton, P. F. Lonsdale, and F. N. Spiess. 1986. Corals on seamount peaks provide evidence of current acceleration over deep-sea topography. Nature 322: 59–61.

Grasmueck, M., G. P. Eberli, D. A. Viggiano, T. Correa, G. Rathwell, and J. Luo. 2006. Autonomous underwater vehicle (AUV) mapping reveals coral mound distribution, morphology, and oceanography in deep water of the Straits of Florida. Geophys. Res. Lett. 33, L23616, doi:10.1029/2006GL027734.

Mayer, D. A., K. D. Leman, and T. N. Lee. 1984. Tidal motions in the Florida Current. J. Phys. Oceanogr. 14: 1551–1559.

Messing, C. D., A. C. Neumann, and J. C. Lang. 1990. Biozonation of deep-water lithoherms as associated hardgrounds in the North-eastern Straits of Florida. Palaios 5: 15–33.

Mullins, H. T. and A. C. Neumann. 1979. Geology of the Miami Terrace and its paleoceanographic implications. Mar. Geol. 30: 205–232.

_____, C. R. Newton, K. C. Heath, and H. M. Van Buren. 1981. Modern deep-water coral mounds north of Little Bahama Bank: criteria for the recognition of deep-water coral bioherms in the rock record. J. Sediment. Res. 51: 999–1013.

Neumann, A. C. and M. M. Ball. 1970. Submersible observations in the Straits of Florida: geology and bottom currents. Geol. Soc. Am. Bull. 81: 2861–2874.

_____, G. H. Kofoed, and G. H. Keller. 1977. Lithoherms in the Straits of Florida. Geology 5: 4–10.

Paull, C. K., A. C. Neumann, B. am Ende, W. Ussler, and N. Rodriquez. 2000. Lithoherms on the Florida-Hatteras slope. Mar. Geol. 166: 83–101.

Pinet, P. R. and P. Popenoek. 1985. Shallow stratigraphy and post-Albian geologic history of the northern and central Blake Plateau. Geol. Soc. Am. Bull. 96: 627–638.

Potter, P. E. and F. Pettijohn. 1963. Paleocurrents and Basin Analysis. Academic Press, New York. 121 p.

Puglise, K. A., R. J. Brock, and J. J. Mcdonough III. 2005. Identifying critical information needs and developing partnershnips to further the understanding of Atlantic deep-sea coral ecosytems. Pages 359–391 *in* A. Freiwald and J. M. Roberts, eds. Cold-water corals and ecosystems. Spring-Verlag, Berlin Heidelberg.

Reed, J. K. 2002a. Deep-water *Oculina* coral reefs of Florida: biology, impacts and management. Hydrobiologia 471: 43–55

————. 2002b. Comparison of deep-water coral reefs and lithoherms off southeastern U.S.A. Hydrobiologia 471: 57–69.

————, D. C. Weaver, and S. A. Pomponi. 2006. Habitat and fauna of deep-water *Lophelia pertusa* coral reefs off the southeastern U.S.: Blake Plateau, Straits of Florida, and Gulf of Mexico. Bull. Mar. Sci. 78: 343–377.

————, S. Pomponi, A. Wright, D. Weaver, and C. Paull. 2005. Deep-water sinkholes and bioherms of South Florida and Pourtalès Terrace-habitat and fauna. Bull. Mar. Sci. 77: 267–296.

Roberts J. M., A. J. Wheeler, and A. Freiwald. 2006. Reefs of the deep: the biology and geology of cold-water coral ecosystems. Science 213: 543–547.

Rogers, A. D. 1999. The biology of *Lophelia pertusa* (Linnaeus, 1758) and other deep-water reef-forming corals and impacts from human activities. Int. Rev. Hydrobiol. 84: 315–406.

Stetson, T. R., D. F. Squires, and R. M. Pratt. 1962. Coral banks occurring in deep water on the Blake Plateau. Am. Mus. Novit. 2114: 1–39.

Squires, D. F. 1959. Deep sea corals collected by the Lamont Geological Observatory, Atlantic Corals. Am. Mus. Novit. 1: 1–42.

————. 1964. Fossil coral thickets in Wairarapa, New Zealand. J. Paleontol. 30: 904–915.

Wilber, J. R. and A. C. Neumann. 1993. Effects of submarine cementation on microfabrics and physical properties of carbonate slope deposits, Northern Bahamas. Pages 79–94 *in* R. Rezak and D. L. Lavoie, eds. Carbonate microfabrics. Springer, New York.

Wilson, J. B. 1979. "Patch" development of the deep-water coral *Lophelia pertusa* (L.) on Rockall Bank. J. Mar. Biol. Assoc. U.K. 59: 65–77.

ADDRESSES: (R.N.G.) *Comparative Sedimentology Laboratory, Rosenstiel School of Marine and Atmospheric Science (RSMAS), University of Miami, 4600 Rickenbacker Causeway, Miami, Florida 33139.* (S.J.L.) *Marine Conservation and Biology Institute, 600 Pennsylvania Ave. SE, Suite 210, Washington, DC 20003.* CORRESPONDING AUTHOR: (R.N.G.) *E-mail: <rginsburg@ rsmas.miami.edu>.*

# Note

# Protection of deep-water corals with the development of oil and gas resources in the U.S. Gulf of Mexico: an adaptive approach

Thomas E. Ahlfeld, Gregory S. Boland, and James J. Kendall

The occurrence of *Lophelia pertusa* (Linnaeus, 1758) in the northern Gulf of Mexico (GOM) was first documented by Louis de Pourtalès in the late 1860s. The coral specimens were found in dredge samples collected during U.S. Coast Survey cruises conducted in the Straits of Florida and between the Dry Tortugas and the Campeche Bank (Smith, 1954). An extensive deep-water reef in the GOM was discovered in the 1950s approximately 74 km east of the Mississippi River Delta (Moore and Bullis, 1960). This reef, in water depths of 420–512 m, was reported as being composed largely of *L. pertusa* with the largest portion of the reef extending to a width of 55 m and length of over 305 m (Moore and Bullis, 1960). These habitats have since been shown to be much more extensive and important to the support of diverse communities of associated fauna than previously known in the GOM. Schroeder (2002) reported observations of *L. pertusa* on the upper De Soto Slope in the northeastern GOM. Individual colonies measured from a few cm to over 1.5 m in diameter while aggregations of closely associated colonies ranged up to 1.5–2 m in height and width and 3–4 m in length (Schroeder, 2002).

Deep-sea coral reef systems generally are receiving increased attention worldwide while at the same time being threatened by a variety of activities such as commercial fishing and energy exploration. Although petroleum development activities would generally disturb only small areas of the seafloor, they could result in significant impacts to deep-sea coral communities if mitigation measures are not employed. Anchoring, structure emplacement, pipeline construction, and the remote potential of a seafloor blowout could all cause localized physical disturbances affecting deep-sea corals. Anchors from support boats and ships (or from any buoys set out to moor vessels), floating drilling units, pipelaying vessels, and pipeline repair vessels also cause severe disturbances to small areas of the seafloor with the areal extent related to the size of the mooring anchor and length of chain that would rest on the bottom. The area affected by anchoring operations will depend on the water depth, length of the chain, and size of the anchors. Discharges of drilling muds and cuttings also could cause negative impacts to deep-sea corals. Such discharges at the sea surface are dispersed widely throughout the water column and spread across broad areas of the seafloor. Detrimental effects decrease significantly with distance from the origin. A summary of the potential impacts resulting from oil and gas exploration and extraction to deep-sea coral communities is provided in Morgan et al. (2006) and the most recent MMS Environmental Impact Statement for GOM lease sales (USDOI, MMS, 2006).

George, R. Y. and S. D. Cairns, eds. 2007. Conservation and adaptive management of seamount and deep-sea coral ecosystems. Rosenstiel School of Marine and Atmospheric Science, University of Miami.

59

Although knowledge of the occurrence and distribution of deep-sea coral communities in the northern GOM has increased in the last decade, data are lacking on what regulates observed patchy distribution and how deep-sea coral banks form. Hovland et al. (1998) posed several hypotheses on coral bank formation. In addition, available data regarding coral ages and the degree to which there is an obligate deep coral fauna are equivocal. For these reasons, locating and mapping deep corals and conducting basic biological studies in these habitats are priorities established by recent reviews of deep-sea corals (McDonough and Puglise, 2003). Here we summarize the objectives and preliminary findings of recent and ongoing studies on *L. pertusa* and associated bank communities of the northern GOM continental slope. We also discuss an adaptive approach to the regulation of deep-sea oil and gas activities.

**Recent and ongoing research.**—Recent research has been undertaken to provide information needed to support the MMS management goal of environmentally sound decision-making regarding permitting and regulating offshore oil and gas activities. The need for this research is in direct response to the continued increasing development of oil and gas resources on the GOM continental slope.

The MMS now has a direct need for information that will help describe any significant ecological role that man-made structures may have in deep water of the GOM. The conversion of offshore oil and gas structures to artificial reefs has been well-accepted as a benefit to fisheries on the continental shelf of the entire GOM. A total of 244 structures were converted to artificial reefs from a total of 2207 structure removals between 1990 and June 2007. In the near future, decisions will be required for the removal of structures located in continental slope water depths. Current guidelines allow the MMS to approve alternate plans for removal of structures when the water depth is > 800 m. Options for removal at shallower depths have previously relied on the concept that the structure left behind would serve a positive fisheries enhancement or other beneficial environmental function. A recently completed study sponsored by MMS and the National Oceanic and Atmospheric Administration's Office of Ocean Exploration (NOAA OE) investigated the deep-water artificial reef effect of ships sunk in GOM waters (Church et al., 2007). Results of that project include the documentation of large (e.g., 6 m height, 1.2–1.5 m width) "thickets" of *L. pertusa* growing on a steel hulled vessel, the GULF PENN, sunk in 1942 during World War II at a depth of 538 m. This observation was important to the understanding of *L. pertusa* distribution and growth in the GOM, particularly with the known maximum growth period of 62 yrs. Smaller *L. pertusa* thickets were also observed on this ship but none were seen on other shipwrecks examined at deeper (1268–1981 m) or shallower (85–143 m) water depths.

A recently-completed research program sponsored collaboratively by MMS and the U.S. Geological Survey (USGS) is investigating six study sites with known *L. pertusa* aggregations. Submersible dives were conducted in 2004 and 2005 to characterize non-chemosynthetic megafauna and macroinfaunal communities found in GOM hard bottom areas at water depths between 310 m and 686 m. An additional objective of this project is to investigate and describe the environmental conditions that are correlated with the development and distribution of areas characterized by extensive *L. pertusa* aggregations. One of the study areas is located 53 nmi due east of the Mississippi River delta in 460–470 m of water and is the largest known *Lophelia* habitat in the Gulf of Mexico (Fig. 1).

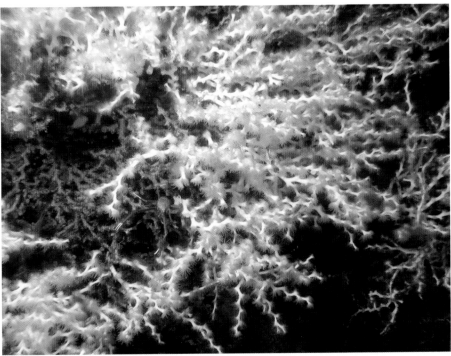

Figure 1. Dense *Lophelia pertusa* thicket at Mineral Management Service study site at a depth of 466 m (1530 ft) approximately 53 nmi east of the Mississippi River Delta. Image taken from the JOHNSON SEA LINK submersible, July 2004. White branches are living coral.

For each study site, quantitative bottom samples, video transects with integrated environmental information, and digital photomosaics were used to describe macroinfaunal communities, large-scale megafaunal distribution, substrate type, dominant fauna, topographic relief, and percentage biotic cover. Additional biological elements of this study include *L. pertusa* larval biology and behavior, temperature tolerance, feeding, microbiology, genetics, and community trophics. Near-bottom currents, sediment flux, and zooplankton availability were also studied. Results indicate a high degree of variability in coral distribution even within areas that are categorized as having a high coral density. Laboratory experiments show that *L. pertusa* collected from the GOM exhibits a borderline survival temperature of 15 °C. A mortality of 100% was observed within 24 hrs for coral fragments held at 25 °C (CSA International, Inc., 2007). Results of this study will enhance the understanding of deep-water coral distribution in the GOM and the effectiveness of management decisions intended to protect deep-water hard bottom areas with the potential of coral colonization.

A multi-year study was funded in 2006 by MMS and NOAA OE with a focus on exploration, survey, and experimental work on chemosynthetic communities and hard bottom habitats located deeper than 1000 m on the lower continental slope of the GOM. Additional study components conducted by USGS are also integrated as part of this project which began with fieldwork in March and May of 2006 including use of the research submersible ALVIN. In 2007, NOAA OE provided the remotely-operated vehicle (ROV), JASON II, for a second year of field sampling. Objectives of the study include the determination of the comparative degree of sensitivity to anthropogenic impacts through a variety of approaches such as rarity, unique taxonomy/biodiversity, or other environmental risk assessment methodologies. Although *L. pertusa* is most common world-wide between 200–500 m, it has been reported as deep as 3000 m (the deepest part of the GOM is 3840 m in Mexican waters). Other anthozoan taxa, including alcyonarians and antipatharians, are also expected on GOM lower continental slope hardbottoms.

**Current management.**—Results of deep-water studies in the GOM have been used by MMS in Environmental Impact Statements to describe the environment potentially affected by offshore oil and gas exploration and development activities and to assess the possible environmental consequences of those activities. Information produced through these studies has also been used to develop protective measures designed to prevent potential impacts or to minimize and mitigate unavoidable impacts. Such protective measures have been applied by MMS to areas with high-density assemblages of benthic organisms in areas of natural hydrocarbon seepage (Ahlfeld, 2002). One mechanism used by MMS to provide the offshore oil and gas operators with guidelines for the imple-mentation of lease stipulations and regional requirements is the Notice to Lessees (NTL). An NTL supplements regulations that govern operations on the Outer Continental Shelf (OCS) and provides clarification or interpretation of regulations and further guidance to lessees and operators in the conduct of their operations. The NTL may also serve to provide a better understanding of the scope and meaning of a regulation by explaining MMS interpretation of a requirement, for example an avoidance distance from specific types of sensitive biological communities.

In order to gain a general understanding and determine the effectiveness of the cur-rent biological review process for all potentially impacting energy development activi-ties, MMS has implemented a broad scale survey requirement throughout the GOM continental slope (NTL 2003-G03, *Remotely Operated Vehicle Surveys in Deep-water*). The GOM slope was divided into 18 segments based on three major biological depth re-gimes and an east-west division of six longitudinal sections. As activity applications are received by MMS, operators are required to perform between five and ten ROV surveys spaced throughout each particular segment or grid so that each grid is well represented by depth and spatial coverage. ROV transects are performed by operators in a pattern of six spokes approximately 100 m in length radiating in equally spaced directions from a point of origin (Fig. 2). The ROV operators are not expected to provide accurate iden-tifications but they are asked to record the types of animals present and the appearance of the bottom such as color and texture. Of particular interest is the observation of hard-bottom areas or outcrops with any attached animals. Videotapes are submitted and a general review of both transcript logs and the tapes themselves are then performed by MMS biologists. If any ROV survey information indicates that previously unknown, significant biological communities were encountered in proximity of potentially impact-ing activities, adaptive decision-making may lead to potential future alterations in the biological review process and revisions of NTL or other regulatory measures.

Following the 1984 discovery of chemosynthetic communities associated with hy-drocarbon seeps in the northern Gulf of Mexico, MMS sponsored two major multi-year studies designed to provide an understanding of their distribution and functioning. Re-sults of the first project (MacDonald et al., 1995), provided MMS with information that was used in 1998 to develop the *Deep-water Chemosynthetic Communities* NTL (NTL 98-11). Additional preliminary data from the second study (MacDonald, 2002), allowed MMS, through an adaptive approach, to develop more specific requirements for an up-dated *Deep-water Chemosynthetic Communities* NTL (NTL No. 2000-G20). All NTLs

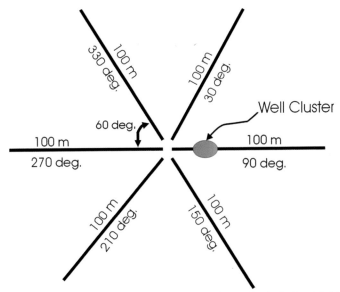

Figure 2. Suggested ROV survey pattern required to perform a limited visual habitat survey, as depicted in the Notice to Leases 2003-G03, *Remotely Operated Vehicle Surveys in Deep-water.* A minimum survey distance of 100 m is required in each of six directions radiating from the deployment location of the ROV on the operator's platform or vessel.

are accessible on the MMS website at www.gomr.mms.gov. This revised NTL requires minimum separation distances from features or areas that could support high-density chemosynthetic communities. Distances of at least 457 m for each drilling muds and cuttings discharge location and 76 m for all other proposed seafloor disturbances including anchors, anchor chains, wire ropes, seafloor template installations, and pipeline construction are required (Ahlfeld, 2002). Oil and gas companies are routinely required to survey the seafloor in the area of proposed development. Geophysical signatures that are obtained through remote sensing techniques have proven to be related to the presence of high-density communities. Required submission of this information by offshore oil and gas operators is used in a biological review process within MMS to determine the need for avoidance of the locations of "potential" communities. Due to the imperfect reliability of the correlation between geophysical remote sensing signatures and the actual presence of living communities, the conservative approach of avoidance is mandated unless operators seeking activity permits prove the absence of such communities using an in situ visual methodology (ROV, submersible, etc).

The adaptive approach used by MMS to protect chemosynthetic communities in areas of deep-water oil and gas exploration, development, and production activities serves as a model for the implementation of special measures designed to protect *L. pertusa* and associated non-chemosynthetic benthic communities. Seismic data representing the characteristics of the seabed (as opposed to deep stratigraphic sections where hydrocarbon reserves are located) are excellent for depicting hard substrate. Based on the *Lophelia* results described above, and a broader understanding of hardbottom habitats in the deep GOM, MMS is considering an augmentation of specifications of NTL 2000-G20, targeting the protection of *L. pertusa* and other rare deep-water hardbottom communities.

## Literature Cited

Ahlfeld, T. E. 2002. Cold seep research: resource management applications. Cah. Biol. Mar. 43: 231–234.

Church, R., D. Warren, R. Cullimore, L. Johnston, W. Schroeder, W. Patterson, T. Shirley, M. Kilgour, N. Morris, and J. Moore. 2007. Archaeological and biological analysis of World War II shipwrecks in the Gulf of Mexico: A pilot study of the artificial reef effect in deep-water. Final Report, Contract #73095. U.S. Dept. of the Interior, Minerals Management Service, Gulf of Mexico OCS Region, New Orleans, LA. OCS Study MMS 2007-015. 387 p.

CSA International, Inc. 2007. Characterization of northern Gulf of Mexico deep-water hard bottom communities with emphasis on Lophelia coral. U.S. Department of the Interior, Minerals Management Service, Gulf of Mexico OCS Region, New Orleans, LA. OCS Study 2007-044. 169 p + app.

Hovland, M., P. B. Mortensen, T. Brattegard, P. Strass, and K. Rokoengen. 1998. Ahermatypic coral banks off mid-Norway: evidence for a link with seepage of light hydrocarbons. Palaios 13: 189–200.

MacDonald. I. R., ed. 2002. Stability and change in Gulf of Mexico chemosynthetic communities. Prepared by Geochemical and Environmental Research Group, Texas A&M University. U.S. Dept. of the Interior, Minerals Management Service, Gulf of Mexico OCS Region, New Orleans, LA. OCS Study MMS 2002-035. 456 p.

_____, W. W. Schroeder, and J. M. Brooks. 1995. Chemosynthetic ecosystem studies final report. Prepared by Geochemical and Environmental Research Group, Texas A&M University. U.S. Dept. of the Interior, Minerals Management Service, Gulf of Mexico OCS Region, New Orleans, LA. OCS Study MMS 95-0023. 338 p.

McDonough, J. J. and K. A. Puglise. 2003. Summary: Deep-sea corals workshop. International planning and collaboration workshop for the Gulf of Mexico and the North Atlantic Ocean. Galway, Ireland, January 16–17, 2003. NOAA Tech. Memo. NMFS- SPO-60, 51 p.

Moore, D. R. and H. R. Bullis. 1960. A deep-water coral reef in the Gulf of Mexico. Bull. Mar. Sci. 10: 125–128.

Morgan, L. E., C.-F. Tsao, and J. M. Guinotte. 2006. Status of deep-sea corals in US waters, with recommendations for their conservation and management. Marine Conservation Biology Institute. Bellvue, WA. 64 p.

Schroeder, W. W. 2002. Observations of *Lophelia pertusa* and the surficial geology at a deep-water site in the northeastern Gulf of Mexico. Hydrobiologia 471: 29–33.

Smith F. G. W. 1954. Gulf of Mexico Madreporaria. Pages 291–296 *in* P. S. Galtsoff, ed. Gulf of Mexico – its origins, waters, and marine life. Fish. Bull. Fish Wildl. Serv. 55.

U.S. Dept. of the Interior. Minerals Management Service. 2006. Gulf of Mexico OCS Oil and Gas Lease Sales: 2007–2012. Western Planning Area Sales 204, 207, 210, 215, and 218, Central Planning Area Sales 205, 206, 208, 213, 216, and 222. Draft Environmental Impact Statement. U.S. Dept. of the Interior, Minerals Management Service, Gulf of Mexico OCS Region, New Orleans, LA. OCS EIS/EA MMS 2006. 852 p + app.

ADDRESSES: (T.E.A., J.J.K.) *Minerals Management Service, 381 Elden St, Herndon, Virginia 20170.* (G.S.B.) *Minerals Management Service, 1201 Elmwood Park Blvd, New Orleans, Louisiana 70123.* CORRESPONDING AUTHOR: (T.E.A.) *E-mail: <Thomas.Ahlfeld@mms.gov>.*

# Demersal fishes associated with *Lophelia pertusa* coral and hard-substrate biotopes on the continental slope, northern Gulf of Mexico

KENNETH J. SULAK, R. ALLEN BROOKS, KIRSTEN E. LUKE,
APRIL D. NOREM, MICHAEL RANDALL, ANDREW J. QUAID,
GEORGE E. YEARGIN, JANA M. MILLER, WILLIAM M. HARDEN,
JOHN H. CARUSO, AND STEVE W. ROSS

## Abstract

The demersal fish fauna of *Lophelia pertusa* (Linnaeus, 1758) coral reefs and associated hard-bottom biotopes was investigated at two depth horizons in the northern Gulf of Mexico using a manned submersible and remote sampling. The Viosca Knoll fauna consisted of at least 53 demersal fish species, 37 of which were documented by submersible video. On the 325 m horizon, dominant taxa determined from frame-by-frame video analysis included Stromateidae, Serranidae, Trachichthyidae, Congridae, Scorpaenidae, and Gadiformes. On the 500 m horizon, large mobile visual macrocarnivores of families Stromateidae and Serranidae dropped out, while a zeiform microcarnivore assumed importance on reef "Thicket" biotope, and the open-slope taxa Macrouridae and Squalidae gained in importance. The most consistent faunal groups at both depths included sit-and-wait and hover-and-wait strategists (Scorpaenidae, Congridae, Trachichthyidae), along with generalized mesocarnivores (Gadiformes). The specialized microcarnivore, *Grammicolepis brachiusculus* Poey, 1873, appears to be highly associated with *Lophelia* reefs. The coral "Thicket" biotope was extensively developed on the 500 m site, but fish abundance was low with only 95 fish per hectare. In contrast to *Lophelia* reefs from the eastern the North Atlantic, the coral "Rubble" biotope was essentially absent. This study represents the first quantitative analysis of fishes associated with *Lophelia* reefs in the Gulf of Mexico, and generally in the western North Atlantic.

The deep-water matrix-building scleractinian coral, *Lophelia pertusa* (Linnaeus, 1758) (hereafter *Lophelia* in this paper) occurs circumglobally (Rogers, 1999; Costello et al., 2005), including the Gulf of Mexico. This coral species builds large thickets and elevated banks that function as deep-water coral reefs (Rogers, 1999), providing three-dimensional habitat heterogeneity, shelter for invertebrates and fishes, feeding habitat for ambush predators and microvores, and probable spawning grounds for a few demersal fish species (Fosså et al., 2000; Reed et al., 2005). *Lophelia* habitats function as oases of macrofaunal and megafaunal biodiversity (Teichert, 1958; Jensen and Frederiksen, 1992; Mortensen et al., 1995; Fosså and Mortensen, 1998; Husebø et al., 2002; Costello et al., 2005; Reed et al., 2005) amidst the otherwise monotonous open sedimented landscape of the continental slope. *Lophelia* reefs also appear to serve as focal points that concentrate megafaunal organisms otherwise occurring in low abundance on non-coral habitats. Additionally, such reefs may also concentrate particulate food resources, as the

George, R. Y. and S. D. Cairns, eds. 2007. Conservation and adaptive management of seamount and deep-sea coral ecosystems. Rosenstiel School of Marine and Atmospheric Science, University of Miami.

elevated coral matrix intercepts bottom currents, generating eddies that entrain plankton and organic particles. While the demersal fish fauna associated with *Lophelia* reefs has been relatively well investigated in the eastern North Atlantic (Jensen and Frederiksen, 1992; Mortensen et al., 1995; Husebø et al., 2002; Costello et al., 2005), the fish fauna of those in the western North Atlantic remains essentially undocumented. The present investigation reports on the demersal fish fauna of two *Lophelia* reef study sites in the northern Gulf of Mexico.

The occurrence of living *Lophelia* in the Gulf of Mexico was first reported by Moore and Bullis (1960) from a bottom trawl sample taken on the continental slope south of Mississippi (29°05′N, 88°19′W, 420–512 m depth). In attempting to relocate the Moore and Bullis site, subsequent investigators discovered well-developed *Lophelia* colonies inhabiting topographic highs along the continental slope of the northern Gulf of Mexico (Schroeder, 2002). These topographic highs are salt diapers, partially capped by authigenic calcium carbonate (biologically precipitated in irregular layers and blocks in areas of hydrocarbon seepage). The clean, hard surface of the carbonate rock provides a settlement substrate for the larvae of diverse sessile invertebrates (anemones, sponges, bamboo corals, black corals, gorgonians, scleractinian corals), including *Lophelia*.

The present investigation was undertaken by the U.S.Geological Survey (USGS) to provide a first level community structure analysis of demersal fish species richness, abundance, and biotope affinity on *Lophelia* reefs and comparative biotopes in the Gulf of Mexico. The present fish study complements a parallel submersible study of geology, coral biology, and sessile invertebrate community ecology, targeting the same study sites undertaken by Continental Shelf Associates (CSA) and supported by the Minerals Management Service (MMS). The present study is one component of a broader suite of multi-disciplinary investigations of *Lophelia* coral reefs sponsored by USGS, and a sub-component of ongoing megafaunal community structure investigations of *Lophelia* reefs.

## Materials and Methods

The manned JOHNSON-SEA-LINK (JSL) submersibles were used to conduct two missions in July–August 2004 and September 2005. Both missions investigated the demersal fish faunas of two prominent elevated topographic features on the continental slope of the northern Gulf of Mexico (Fig. 1). Identified by reference to the MMS oil lease blocks in which they lie, these two features have been designated Viosca Knoll 826 (VK-826), and Viosca Knoll 906/907 (VK-906/907)[1] by previous researchers. Viosca Knoll 826 rises to a minimum depth of between 435–480 m; VK-906/907 to a minimum depth of between 305–340 m. The two sites represent two biologically distinct depth horizons centered on depths of 500 m and 325 m, respectively, on the continental slope in terms of resident megafaunal fishes and invertebrates. Together, they provide a distinct three-dimensional hard-substrate, live-bottom continental slope sub-biome, in contrast to the dominant, essentially two-dimensional, soft-substrate, open slope biome.

The 2004 submersible mission was largely devoted to site exploration, specimen collection for taxonomic identification, and video documentation to characterize and differentiate biotopes utilized by demersal fishes (all fishes regularly associated with the benthic boundary layer, whether benthic or benthopelagic). The term *biotope* as used herein specifies a distinct physical (substrate, topography) and biological (sessile invertebrate assemblage) environment inhabited by the resident demersal fish fauna. The initial 2004 mission was also used to establish parameters (lighting, camera settings, submersible logistics) to enable consistent video transect methodology. Only five quantitative transects useful for analysis were accomplished in 2004. However, our portion of the

---

[1]The study area has subsequently been resolved as located more precisely within blocks VK-906/862.

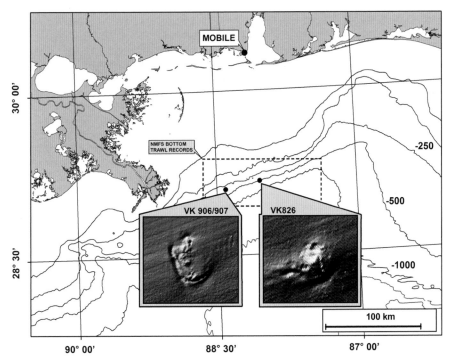

Figure 1. Location of two Viosca Knoll-826 submersible *Lophelia* reef study sites in the northern Gulf of Mexico, and location of comparative NOAA bottom trawl records (open rectangle). Depth contours are in meters.

2005 submersible mission was devoted largely to definitive video transects to enable a fundamental analysis of fish species composition, diversity, abundance, and habitat associations. All dives were conducted during daylight hours, although complete darkness prevails at the depths of our submersible operations. During both missions a small number of fishes were collected in situ using the JSL suction sampler and manipulator, and small baited traps deployed by the submersible. Additional fishes were collected by bottom trawl on a supplementary remote sampling cruise in 2005.

Sites for submersible dives and for deployment of traps and trawls were determined using a pre-existing three-dimensional topographic map of the VK-906/907 site accomplished by the U.S. Navy submersible NR-1, and a composite map prepared from an oil industry single-beam echosounder transect survey of the VK-826 site (Schroeder, 2002). Additionally, we conducted single-beam acoustic transect surveys between submersible dives, and at night during both USGS submersible missions. Acoustic profiles were obtained using a SIMRAD EQ50 color echosounder at a frequency of 38 kHz, tuned to detect the characteristic acoustic reflection of the hard coral matrix, and a Knudsen 320 B/R oceanographic monochromatic echosounder at a frequency of 3.5 kHz. Pulse interval and gain were adjusted to maximize erratic acoustic reflection from coral structures, contrasting with the continuous strong reflection defining the hard substrate seafloor. Ship's position was determined via differential GPS, accurate to within 5 m. Submersible position on the bottom was estimated via Trackpoint II "Integrated Positioning System" (ORE Offshore) using dual acoustic beacons interpreted topside by HBOI submersible operations personnel. Only returns with signal strength above a pre-determined threshold were accepted in plotting the most probable bottom positions of the submersible.

**Quantitative submersible transects.**—Bottom transects were accomplished largely opportunistically during bottom exploration or during transits between target coral collection sites. The fundamental method was a "belt transect", conducted with the submersible cruising as slowly as possible (typically 0.3 kt) into the direction of bottom current to maintain constant speed and consistent course direction. Altitude was held as close above the substrate as possible (i.e., bottom of JSL skimming over the bottom). Color video was obtained using a Sony DX2 3000A 3-chip CCD camera, with 6–48 mm zoom lens, mounted on an extensible arm on the port side of the submersible sphere, 1.37 m above the bottom of the vehicle. The "belt", or area of substrate being transected, was typically illuminated by two high intensity 400 W, 5600 °K HMI lights affixed to the submersible's forward upper work bar, and by four additional individually-selectable HMI lights surrounding the video camera. Additionally, a 1000 W, 6000 °K xenon arc light mounted on the starboard side of the JSL upper work bar was used for forward reconnaissance to illuminate the intended transect path ahead of JSL (usually not illuminating the very near field used for fish analysis). Video and audio information was recorded to a mini-DV format tape recorder and an S-digital recorder. The S-digital recorder was used to obtain very high-quality video (with no data overlay) to enable preparation of high-quality frame grabs to facilitate species identification. During video transecting, the extensible support arm supporting the video camera was kept fully retracted and focal length was maintained at 6 mm (i.e., full wide angle). The camera was panned inward (toward the transect centerline) 15°, and tilted downward 45°. Pan, tilt, and zoom were held in this pre-determined configuration throughout designated transects. With the submersible transecting parallel to the bottom, minimum distance between the camera lens and a fish situated on flat substrate directly ahead of camera was 1.94 m. Targeted transect duration was 5.0 min (sometimes truncated by limiting topography, video tape change-out, or JSL operational exigencies). During transecting, data including time (hr:min:sec) and depth (ft) data were continuously overlaid onto the video record.

The areal field of view available for analysis during standardized moving transects was determined by deployment on the substrate of a 1.78 by 1.22 m wire panel ("hogwire") painted white, with its outer frame painted orange. This panel was subdivided into smaller rectangles of known dimensions, with two rectangular reference grids of 1.0 m by 0.5 m dimensions delineated in black. Additionally, a 0.5 m diameter Secchi disk type signpost (one half of disk painted flat white, the other half neutral gray) was deployed to estimate the distance at which fishes could be recognized from the background, and at which fishes could be viewed well enough to be positively identified. The submersible was backed away from the grid panel until the panel lay within the illuminated field available for fish identification and enumeration. The submersible was similarly backed away from the signpost until the gray, and then the white halves merged into the background from the perspective of the scientist within the sphere. At each of these two points, the submersible's ranging sonar was used to resolve respective distances to the signpost. The submersible video camera mounting was also equipped with two lasers that projected parallel beams 25 cm apart, used as a reference scale to determine size of objects and fishes.

**Non-transect video segments.**—For all 2004 and 2005 dives, video obtained when the submersible was slowly traversing bottom, but the video camera not standardly configured for transecting (as above), was utilized for a second type of analysis of rank order by species occurrence. Only segments with a wide angle perspective were utilized for this second type of analysis.

**Supplementary video.**—In addition to analysis of the video records documenting the two USGS submersible missions, video records from the parallel CSA 2004–2005 submersible missions (S. Viada, CSA, pers. comm.) from the Viosca Knoll region (12 of 16 dives on VK-826, VK-862, and VK-906/907, representing approximately 30 hrs total) were also examined. Submersible video records from four additional Viosca Knoll dives were obtained from a NOAA Exploration mission in 2005 (W. Schroeder, Dauphin Island Marine Laboratory, pers. comm.). Examination of supplementary video was undertaken to qualitatively scan for potential additional fish taxa contributing to the demersal fauna, but not recorded during USGS dives.

**Video analysis.**—In the laboratory, all original mini-DV tapes were copied onto DVD as VOB (Video Object) files (MPEG2 compatible format) at full quality for subsequent preparation of frame grabs and to create a backup video data archive. The entire video record of each dive (ca 2 hrs total bottom time) was then converted from DVD to sequential still frames (0.9 mb each), separated by an interval of approximately one second (0.996 s) using VideoCharge™ 3.0 frame-grab software (which requires DVD input). Analysis of the resulting full-quality images (in uncompressed .tiff format) proceeded as follows:

*Analysis I.*—Quantitative species abundance and rank order from standardized transects: Step 1) The original mini-DV was initially viewed using a Sony® GV-D9000 DVR linked to a Sony Trinitron commercial-quality high resolution monitor to establish identities of all demersal fish species recorded (identification by senior author), denote time segments for capture as still images to document species identifications, empirically define biotopes encountered, and record the starting and ending times of each quantitative transect. Separate high-quality frame grabs documenting individual fish species were assembled into a taxonomic identification reference library used by project personnel. Step 2) Using the sequential frame grab record, each designated transect was viewed (on monitors with either 1660 × 1200 pixel or 1280 × 1024 pixel resolution, 0.26 mm pixel pitch) advancing frame-by-frame using the Microsoft Windows™ software "Picture and Fax Viewer" utility. Each transect was analyzed by each sequential 1-s still frame, building an Excel spreadsheet file recording dive number, transect number, frame grab file number, date, time, depth, fish occurrence by species, major biotope, and sub-biotope designations. The number of frame grabs was totaled for each transect to yield the total time analyzed (i.e., total number of grabs, multiplied by 0.996 s$^{-1}$ per grab).

Deployment on the substrate of the wire mesh panel of known dimensions resolved the typical video camera illuminated field of view useful for analysis per frame grab during moving transects as 15.0 m$^2$ (range 12.0–16.0 m$^2$). However, at a speed over ground of 0.3 kt (0.15 m s$^{-1}$), the actual area for scoring demersal fish counts per 1-second frame was approximately 1.0 m$^2$. A fish of typical total length (0.25–0.75 m) was in the illuminated field of view (lower two-thirds of the video screen) for a maximum of 15 s, and crossed the video frame margin in < 3 s. However, a fish was only scored when it left the field of view and intersected the video frame margin (bottom, left, or right). Since each fish scored in a 0.996 s$^{-1}$ frame grab occurred within a scoring area of 1.0 m$^2$, the total area analyzed per transect could then be determined. There was a very low probability of counting the same fish in the same scoring area again in sequential frames. The record of frame grab fish scores revealed only one instance where sequential frames with fish of the same species were within 3 s of one another.

To minimize recounts of individual fish swimming along with the submersible, appearing in more than one frame grab, and/or re-entering the field of view, each fish was counted only once, as it left the field of view. Leaving the field of view was defined as exiting the frame by crossing the bottom, left, or right video margin (i.e., fish leaving the video field of view as the submersible advanced forward). Species abundance scores were totaled per transect to determine rank order. Scores for all species were totaled to estimate population density per unit area.

*Analysis II.*—Species occurrence and relative rank order from non-transect video segments: Step (1) Individual occurrences of each species (regardless of number of individuals of that species simultaneously in the field of view) were recorded per each 1-s frame over the entire frame grab record for that dive, excluding transects, but including time intervals when the submersible was stationary, and when the camera was panning or zoomed in upon the substrate and/or on sessile invertebrate assemblages. The occurrence of a species was positively scored if that species was present within the analyzed field of view (lower two-thirds of the total field of view). Step (2) Scores were summed by species to determine rank order by frequency of occurrence for the total pool of analyzed frames. For both abundance and occurrence analyses, each data entry included scoring of major biotope category, and the presence or absence of *Lophelia* coral. Taxon abundance and occurrence data were analyzed for all dives per each of the two Viosca Knoll depth horizons (325 m and 500 m).

**Remote sampling.**—In addition to taxonomic voucher fish specimens collected with the JSL, others were obtained during the 2004 submersible mission using small bottom trawls (3-m and 4-m footrope otter trawls) and 1.0 m mouth-opening benthic sled deployed remotely from the submersible mothership. Remote sampling was also conducted using the R/V TOMMY MUNRO (Gulf Coast Research Laboratory) in June 2005, sampling soft-substrate open-slope areas immediately adjacent to the two USGS Viosca Knoll study sites. Specimens were obtained both to help validate taxonomic identifications of fishes obtained by the JSL, and to document the comparative fauna of the open slope away from *Lophelia* coral biotopes. During the same mission, baited commercial fisheries Caribbean "Z" traps (also known as Antillean "Z" traps and "Chevron" traps) (FAO Corporate Document Repository) were also deployed over structured substrate to capture reef-associated fishes. Traps were 1.5 m long with two funnels, a time-release escape panel, and covered with plastic-coated 4 cm wire mesh).

**Taxonomic Validation.**—Opportunistically during both submersible dives (but not during quantitative transects), the submersible was stopped and high-quality close-up images of individual fish specimens were obtained using the JSL digital still camera and videocamera (employing the zoom function). The digital still camera was mounted atop the forward collection basket, with illumination provided by one fixed HMI light and/or an accessory strobe light. Close-up images were used to assist in validating species identifications. Additional voucher specimens for taxonomic reference were obtained from both JSL in situ collections, and from surface-deployed traps, bottom trawls, and a benthic sled. High quality voucher specimens were prepared and photographed at sea to accompany underwater images and physical specimens documenting the fauna. Specimens were examined in the laboratory to yield definitive species identifications (J.H.C. and K.J.S.). The senior author is responsible for all taxonomic identifications from videotapes and DVDs, except for elasmobranch identifications provided by J. Castro (Mote Marine Laboratory), and Cynoglossidae identifications provided by T. Munroe (NOAA Fisheries Systematics Laboratory). Voucher specimens documenting this investigation are currently maintained at Florida Integrated Science Center, Gainesville, Florida. These specimens will ultimately be curated in the cataloged fish collection of the Florida Museum of Natural History, University of Florida, Gainesville, Florida, and in the fish collection of the U.S. National Museum of Natural History, Smithsonian Institution.

Species recorded during the USGS 2004 and 2005 Viosca Knoll submersible missions were contrasted with demersal fish species reported from historical National Marine Fisheries Service (NOAA) trawl surveys (Springer and Bullis, 1956; Bullis and Thompson, 1965) and the NOAA SEAMAP bottom trawl database (NOAA Fisheries Mississippi Laboratories, Pascagoula, Mississippi, data received 2004). To confine the comparison to the immediate area of the Viosca Knoll study sites, we included only species recorded between 300–550 m depth, and within a rectangle bounded by lat. 28°55′–29°20′N, longitude 87°29′–88°40′W (Fig. 1) (n = 265 NOAA trawl stations).

Biotope affinities of the overall Viosca Knoll demersal fish fauna were documented by taxon for both depth horizons combined, expressed as frequency of occurrence of each taxon among the various biotope categories. The null hypothesis that key numerically dominant fish species were randomly distributed, regardless of biotope, was tested by a $\chi^2$ goodness of fit test of observed vs expected frequencies of occurrence from video data analysis.

## Results

Twenty submersible dives were accomplished on target *Lophelia* sites; 10 in 2004, and 10 in 2005, with 12 on the VK-826 site, and eight on the VK-906/907 site (Fig. 1). Dive tracks on each site, largely targeting known or suspected *Lophelia* coral areas, intersect-

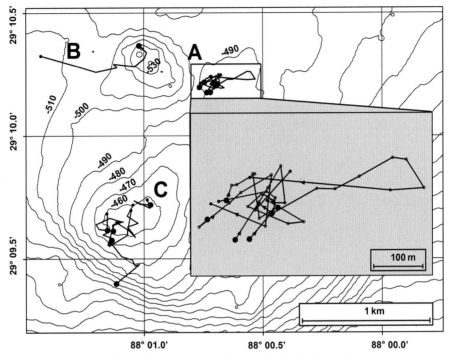

Figure 2. Bathymetric chart (10-m isobaths) of Viosca Knoll-826 *Lophelia* reef study site, showing tracks of 12 USGS submersible dives undertaken in 2004–2005: A = "Big Blue Reef" on northeastern sector of overall feature; B = 100 m deep depression; C = main knoll on southwestern sector of feature (with *Lophelia*). Inset shows detail of eight dives conducted on "Big Blue Reef". Key: Large dots = beginning of bottom time; small dots = Trackpoint II navigation fixes during the course of a dive, including final fix at end of bottom time; arrowheads indicate direction of submersible movement.

ed multiple times (Figs. 2, 3), the intersection matrix essentially pin-pointing the areas populated by *Lophelia* and other colonial particulate-feeding invertebrates (anemones, sponges, bamboo corals, black corals, gorgonians). Only relatively limited portions of the flanks and crests of selected ridges were found to be colonized by extensive *Lophelia* reef (Fig. 4). Remote deployments to sample fishes included four fish traps, four bottom trawl stations and two benthic sled stations (Table 1). Additional fish specimens were selectively captured on 12 occasions using the submersible manipulator/suction sampler. Total submersible bottom time was 44 hrs, 45.4 min, all of which was used to document demersal fish species identifications. However, due to division of bottom time activities among multi-disciplinary tasks, only a limited portion of that time was available for dedicated moving video transects. Thirty-two transects from seven dives were accomplished to support Analysis I; total transect time was 141 min (Table 1). Additional non-transect segments used for Analysis II totaled an additional 115 min of video.

Initial video analysis enabled an empirical differentiation of the overall demersal environment of the Viosca Knoll study area into two depth horizons by fish species occurrence, one centered on 325 m, another on 500 m depth. Video analysis yielded four major empirically-defined biotope categories based on terrain, relief, and development of *Lophelia* coral (Table 2), "Open"—open sedimented soft substrate (Figs. 5A,B); "Plate"—flat, low-relief hard substrate biotope (Figs. 5C,D); "Rock"—sculpted, fragmented, and/or eroded high-relief biotope (Fig. 6A,B); and "Thicket"—soft or hard

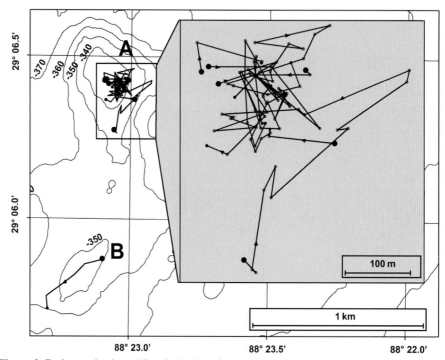

Figure 3. Bathymetric chart (10-m isobaths) of Viosca Knoll-906/907 *Lophelia* reef study site, showing tracks of eight USGS submersible dives undertaken in 2004–2005: A = area of live-bottom development, including *Lophelia* coral; B = area visited on one exploratory dive. Key: Large dots = beginning of bottom time; small dots = Trackpoint II navigation fixes during the course of a dive, including final fix at end of bottom time; arrowheads indicate direction of submersible movement.

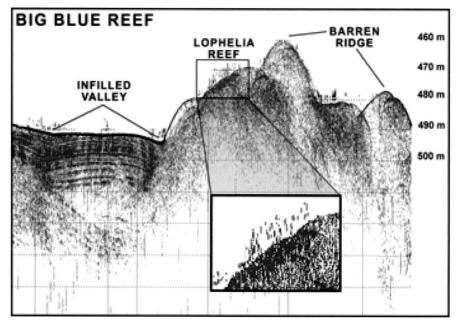

Figure 4. Knudsen echosounder single beam acoustic (3.5 kHz) profile of *Lophelia pertusa* coral reef, Big Blue Reef, on flank of a ridge, northeastern sector of VK-826 study site.

Table 1. Submersible dives and surface vessel bottom sampling stations conducted by USGS during three Viosca Knoll cruises, 2004–2005. Key: BS = benthic sled, BT = bottom trawl, FC = submersible fish collection, FT = baited fish trap, V = submersible video documentation.

| USGS Cruise number | Station number | Study site | Depth (m) | Sample type | Video bottom time (hh:mm:ss) | Number of video transects | Transect time (hh:mm:ss) |
|---|---|---|---|---|---|---|---|
| 2004-03 | JSL-4744 | VK-906/907 | 315 | V and FC | 2:44:46 | 0 | 0:00:00 |
| 2004-03 | JSL-4745 | VK-906/907 | 336 | V and FC | 0:58:01 | 0 | 0:00:00 |
| 2004-03 | JSL-4746 | VK-906/907 | 345 | V and FC | 2:01:58 | 0 | 0:00:00 |
| 2004-03 | JSL-4747 | VK-906/907 | 316 | V and FC | 2:58:00 | 0 | 0:00:00 |
| 2004-03 | JSL-4748 | VK-826 | 446 | V and FC | 2:24:17 | 0 | 0:00:00 |
| 2004-03 | JSL-4749 | VK-826 | 511 | V | 2:29:23 | 0 | 0:00:00 |
| 2004-03 | JSL-4750 | VK-826 | 528 | V and FC | 2:32:01 | 4 | 0:19:09 |
| 2004-03 | JSL-4751 | VK-826 | 462 | V and FC | 2:46:07 | 0 | 0:00:00 |
| 2004-03 | JSL-4752 | VK-826 | 469 | V | 2:40:44 | 0 | 0:00:00 |
| 2004-03 | JSL-4753 | VK-826 | 475 | V | 2:37:41 | 1 | 0:05:08 |
| 2004-03 | USGS-9004 | VK-906/907 | 327 | BT | NA | NA | NA |
| 2004-03 | USGS-9007 | VK-826 | 536 | BT | NA | NA | NA |
| 2004-03 | USGS-9013 | VK-826 | 457 | BS | NA | NA | NA |
| 2004-03 | USGS-9014 | VK-826 | 382 | BS | NA | NA | NA |
| 2004-03 | USGS-9017 | VK-826 | 308 | BT | NA | NA | NA |
| 2004-03 | USGS-9018 | VK-826 | 325 | BT | NA | NA | NA |
| 2005-03 | USGS-0017/0073 | VK-906/907 | 360 | FT | NA | NA | NA |
| 2005-03 | USGS-0018/0074 | VK-906/907 | 360 | FT | NA | NA | NA |
| 2005-03 | USGS-0025/0075 | VK-826 | 486 | FT | NA | NA | NA |
| 2005-03 | USGS-0027/0076 | VK-826 | 486 | FT | NA | NA | NA |
| 2005-04 | JSL-4873 | VK-906/907 | 315 | V | 1:49:18 | 0 | 0:00:00 |
| 2005-04 | JSL-4874 | VK-906/907 | 315 | V | 1:43:31 | 6 | 0:25:18 |
| 2005-04 | JSL-4875 | VK-906/907 | 337 | V | 2:19:49 | 5 | 0:22:23 |
| 2005-04 | JSL-4876 | VK-906/907 | 312 | V | 2:47:16 | 6 | 0:27:00 |
| 2005-04 | JSL-4877 | VK-826 | 479 | V | 2:28:35 | 0 | 0:00:00 |
| 2005-04 | JSL-4878 | VK-826 | 465 | V | 1:02:06 | 0 | 0:00:00 |
| 2005-04 | JSL-4879 | VK-826 | 454 | V and FC | 2:29:28 | 4 | 0:12:00 |
| 2005-04 | JSL-4880 | VK-826 | 455 | V | 2:25:50 | 6 | 0:29:56 |
| 2005-04 | JSL-4881 | VK-826 | 451 | V | 2:31:18 | 0 | 0:00:00 |
| 2005-04 | JSL-4882 | VK-826 | 478 | V | 0:55:17 | 0 | 0:00:00 |
| Totals | | | 315–536 | | 44:45:26 | 32 | 2:20:54 |

substrate extensively covered with *Lophelia* coral (Fig. 6C). A fifth biotope category, "Rubble" (Fig. 6D), occurred only rarely on Viosca Knoll, but was included to provide comparability with major *Lophelia* biotopes identified in investigations from other geographic regions (e.g., Mortensen et al., 1995; Freiwald et al., 2002). "Rubble" was defined as: live and/or dead coral branches and fragments lying on the substrate. To score each frame grab according to a specific category, > 50% of the analyzed field of view (lower two-thirds of the total field of view) had to correspond to one of the five categories.

Video analysis yielded 37 distinct demersal fish taxa (Table 3A), plus one taxon positively identified visually (*Scyliorhinus retifer*), but not captured on video, and two additional tentatively identified taxa. A few taxa that could be identified only to the family or genus level included more than one similar species not readily distinguishable on

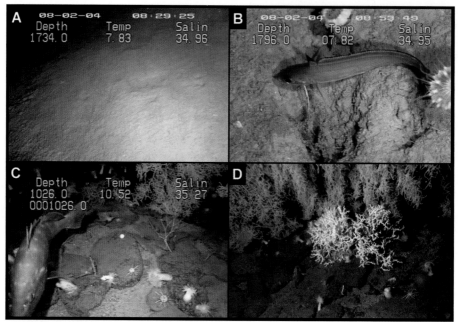

Figure 5. Examples of Viosca Knoll biotopes: (A) "Open" biotope, 528 m; (B) "Open" biotope, 547 m, with the hake *Laemonema goodebeanorum* and tube-dwelling cerianthid anemones; (C) "Plate" biotope, 316 m, with *Lophelia pertusa* hard coral and *Leiopathes* black coral bushes; (D) "Plate" biotope, 312 m, with *Epinephelus niveatus* snowy grouper.

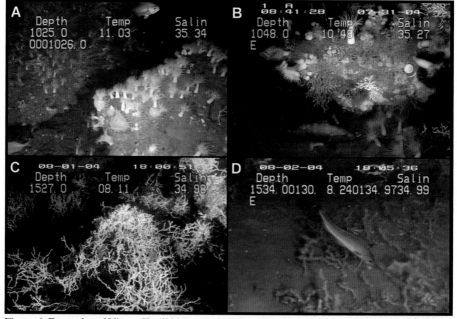

Figure 6. Examples of Viosca Knoll biotopes: (A) "Rock" biotope, 312 m, with *Hyperoglyphe perciformis* barrelfish; (B) "Rock" biotope, 320 m, supporting a diverse assemblage of sessile invertebrates, *Epinephelus niveatus* beneath; (C) "Thicket" biotope, 465 m, a monoculture of *Lophelia pertusa*, (D) "Rubble" biotope, rare on Viosca Knoll, 467 m, with *Laemonema goodebeanorum*.

Table 2. Biotope categories and descriptions, as applied to analysis of Viosca Knoll study sites.

| Biotope Category | Criteria (Biotope Category covering > 50% of analyzed field of view; lower two-thirds of video screen = 15.0 m²). |
|---|---|
| Open (non-coral) | Terrain flat or undulating, comprised of deep soft sediment, often hummocky with obvious biogenic burrows and mounds. Key indicator taxa: black cerianthiid anemones (burrowers). |
| Plate (non-coral) | Terrain flat or terraced hard-pan, or hard-pan with a thin veneer of sediment. Maximum relief < 10 cm. Substrate is typically populated by attached sessile invertebrates. Key indicator taxa: white anemones, glass sponges, gorgonians, bamboo corals, black corals. |
| Rock (non-coral) | Terrain uneven and either highly eroded, sculpted, or fragmented, with outcropping edge, and large crevices or pockets. Maximum relief > 10 cm. Substrate barren, or sparsely to densely populated by sessile invertebrates. Key indicator taxa: white anemones, glass sponges, gorgonians, bamboo corals, black corals. |
| Rubble (coral debris) | Terrain either hard or soft, but with live and/or dead *Lophelia pertusa* coral branches and fragments covering > 50% of field of view. |
| Thicket (live coral) | Terrain either hard or soft, predominantly live (white) coral developed into expanses of tall, extensively-branched bushes covering > 50% of field of view. |

video. Twelve video species identifications were validated using submersible-caught specimens. Twenty-seven species were documented from remote traps, trawls, and benthic sleds (Table 3A), including 15 species not documented in submersible videos or by submersible collections. Comparison of the species list from USGS cruises with the list from NOAA Fisheries bottom trawl records within a comparative depth and geographic range (Fig. 1) yielded 30 species common to both databases (Table 3A). The NOAA database contained an additional 23 demersal fish species not documented during the present study (Table 3B), although four of these species may be identical in both databases (due to use of different taxonomic names for potentially synonomous taxa). Most of these additional NOAA species are fishes typically associated with soft substrate on the open slope (e.g., Macrouridae, Gadidae, Merluccidae, Rajidae, Alepocephalidae), away from hard bottom and reef biotopes. Examination of CSA video for 12 additional dives on Viosca Knoll sites added no further species to the overall USGS faunal list.

Deployment of the Secchi-disk signpost target resolved maximum horizontal visibility to a scientist in the submersible sphere as 12.2 m for the neutral gray half (representing most fishes), and 19.8 m for the flat white half (representing white or silver fishes). However, by comparative video vs sonar reference to both the metal frame and the signpost at various distances from the submersible, it was determined that fishes viewed by the videocamera could be reliably identified to species only within a distance of 5 m ahead of the sphere (i.e., approximately 3 m ahead of the camera).

**Quantitative video analyis.**—Video records for 32 moving transects from seven USGS submersible dives were converted into 8486 1-s frame grabs. Analysis of these frames documented at least 37 total demersal species identified from video (Table 3A). Additional taxa could be resolved only to a higher taxonomic level. Of the total frame grab record, 4498 frames represented transects accomplished on the 325 m depth ho-

Table 3A. Demersal fish taxa documented by USGS submersible (JSL) video and collections on Viosca Knoll study sites, versus those from comparative USGS trawl, sled, and trap collections, and NOAA bottom trawls collections. NOAA records are from 300–550 m between lat. 28°55′–29°20′N, long. 87°29′–88°40′W. Key: DFG = digital frame grab from JSL video, LP = Layout digital image, UDP = JSL digital still image, VS = visual JSL record (not tallied in totals), X = positive record, XX = taxon recorded both in present study and NOAA bottom trawl database, XXO = taxon also recorded in NOAA trawl database, but under an earlier species name, XXX = taxon recorded in present study both by JSL and in trawl, trap, or sled collections, ?? = tentative record (not tallied in totals).

| Demersal fish taxa: | USGS Video record | USGS JSL coll. n = 12 | USGS Trawl coll. n = 4 | USGS Sled coll. n = 2 | USGS Trap coll. n = 4 | NOAA Trawl coll. n = 265 | USGS Voucher image |
|---|---|---|---|---|---|---|---|
| *Anthias nicholsi* Firth, 1933 | ?? | | | | | | |
| *Argentina striata* Goode and Bean, 1896 | X | | X | | | | |
| *Bassogigas* sp. | X | X | | | | | LP |
| *Bathygadus melanobranchus* Vaillant, 1888 | | | XX | | | XXO | |
| *Bathypterois* cf. *bigelowi* Mead, 1958 | XX | | | | | XX | |
| *Bembrops anatirostris* Ginsburg, 1955 | | | XX | | | XX | LP |
| *Caristius sp.* | X | | | | | | DFG |
| *Chlorophthalmus agassizi* Bonaparte, 1840 | | | XX | | | XX | LP |
| *Coelorinchus caribbaeus* (Goode and Bean, 1885) | XX | | XX | X | | XX | LP, DFG |
| *Conger oceanicus* (Mitchill, 1818) | XX | | XX | | XX | XX | LP, DFG |
| *Cyttopsis rosea* (Lowe, 1843) | X | | X | | | | DFG |
| *Dibranchus atlanticus* Peters, 1876 | XX | | XX | | | XX | LP, DFG |
| *Epigonus pandionus* Goode and Bean, 1881 | XX | | | | | XX | DFG |
| *Epinephelus niveatus* (Valenciennes, 1828) | X | | | | | | DFG |
| *Facciolella* sp. | X | X | | | | | LP |
| *Gephyroberyx darwini* (Johnson, 1866) | X | | | | | | DFG,UDP |
| *Glossanodon* sp. | XX | | | | | XX | DFG |
| *Gnathagnus egregius* (Jordan and Thompson, 1905) | | | XX | | | XX | LP |
| *Grammicolepis brachiusculus* Poey, 1873 | XX | XX | | | | XX | LP, DFG |
| *Helicolenus dactylopterus* (Delaroche, 1809) | XX | XX | XX | | XX | XX | LP, DFG |
| *Hemanthias aureorubens* (Longley, 1935) | X | | X | | | | LP, DFG |
| *Hoplostethus mediterraneus* Cuvier, 1829 | ?? | | ?? | | | XX | DFG |
| *Hoplostethus occidentalis* Woods, 1973 | XX | XX | | | | XX | LP, DFG |
| *Hymenocephalus* sp. | | | XX | | | XX | LP |
| *Hyperoglyphe perciformis* (Mitchill, 1818) | X | | | | | | DFG |
| *Laemonema goodebeanorum* Meléndez and Markle, 1997 | XXX | XXX | XXX | | | XXO | LP, DFG |
| *Lophius gastrophysus* Miranda Ribeiro, 1915 | XX | XX | | | | XX | LP, DFG |
| *Lopholatilus chamaeleonticeps* Goode and Bean, 1879 | X | | | | | | DFG |
| *Malacocephalus occidentalis* Goode and Bean, 1885 | | | XX | | | XXO | LP |
| *Monomitopus* sp. | X | X | | | | | LP |
| *Neobythites marginatus* Goode and Bean, 1896 | X | X | | | | | LP |
| *Nezumia aequalis* (Günther, 1878) | XX | | | | | XX | DFG |
| *Odontaspis ferox* Risso, 1810 | X | | | | | | DFG |
| *Paralichthys albigutta* Jordan and Gilbert, 1882 | | | | X | | | |
| *Parasudis truculentus* (Goode and Bean, 1896) | | | XX | | | XX | LP |
| *Physiculus karrerae* Paulin, 1989 | XXX | XXX | | | X | XXO | LP, DFG |
| *Poecilopsetta beani* Goode, 1881 | | | X | | | | LP |
| *Polyprion americanus* (Bloch and Schneider, 1801) | X | | | | | | DFG |
| *Pontinus longispinis* Goode and Bean, 1896 | | | | X | | | LP |

Table 3A. Continued.

| Demersal fish taxa: | USGS Video record | USGS JSL coll. n = 12 | USGS Trawl coll. n = 4 | USGS Sled coll. n = 2 | USGS Trap coll. n = 4 | NOAA Trawl coll. n = 265 | USGS Voucher image |
|---|---|---|---|---|---|---|---|
| *Pontinus rathbuni* Goode and Bean, 1896 | XXX | XXX | XXX | | | | LP |
| *Pseudomyrophis nimius* Böhlke, 1960 | | | X | | | | LP |
| Scorpaenidae spp. | | | X | | | | UDP* |
| *Scyliorhinus retifer* (Garman, 1881) | VS | | | | | XX | VS |
| *Setarches guentheri* Johnson, 1862 | | | XX | | | XX | LP |
| *Squalus asper* Merrett, 1973 | X | | | | X | | DFG |
| *Squalus cubensis* Howell Rivero, 1936 | XX | | | | | XX | DFG |
| *Steindachneria argentea* Goode and Bean, 1896 | | | XX | | | XX | LP |
| *Symphurus marginatus* (Goode and Bean, 1886) | | | XX | | | XXO | LP |
| *Synagrops bellus* (Goode and Bean, 1896) | XX | | XX | | | XX | DFG |
| *Synaphobranchus* sp. | XX | | | | | XX | DFG |
| *Trachyscorpia cristulata* Poey, 1873 | X | | | | | | DFG |
| *Urophycis cirrata* (Goode and Bean, 1896) | XX | | XX | | XX | XX | DFG |
| *Urophycis floridana* (Bean and Dresel, 1884) | XX | XX | | | XX | XX | LP, DFG |
| Translucent Neobythitinae | X | | | | | | DFG |
| Unknown Pomacentridae-like fish | X | | | | | | DFG |
| Total taxa: 53 | | | | | | | |
| Totals by data source: | 37 | 12 | 26 | 1 | 6 | 30 | |

*This taxon has been determined to include *Idiastion kyphosus* Eschmeyer, 1965 (VK-826, Station JSL-2004-03-4748), the first record of this species from the Gulf of Mexico.

Table 3B. Demersal fish taxa recorded in NOAA bottom trawl database, but not recorded by USGS, 2004–2005, in either submersible video or suction samples, or in trawl and sled samples. Limits of records: 28°55′–29°20′N, 87°22′–88°40′W, 300–550 m (Fig. 1).

| Demersal fish taxa | Comment of species identification |
|---|---|
| *Bathygadus macrops* Goode and Bean, 1885 | may = *B. favosus* Goode and Bean, 1886 |
| *Bembrops gobioides* (Goode, 1880) | probably = *B. anatirostris* |
| *Beryx splendens* Lowe, 1834 | |
| *Breviraja spinosa* Bigelow and Schroeder, 1950 | |
| *Chaunax pictus* Lowe, 1846 | probably = *C. suttkusi* Caruso, 1989 |
| *Dipterus oregoni* (Bigelow and Schroeder, 1958) | reported as *Raja oregoni* |
| *Etmopterus virens* Bigelow, Schroeder and Springer, 1953 | |
| *Fenestraja sinusmexicanus* (Bigelow and Schroeder, 1950) | reported as *Breviraja sinusmexicanus* |
| *Gadella imberbis* (Vaillant, 1888) | reported as *Brosmiculus imberbis* |
| *Gadomus arcuatus* (Goode and Bean, 1886) | |
| *Galeus area* (Nichols, 1927) | |
| *Hydrolagus alberti* Bigelow and Schroeder, 1951 | |
| *Laemonema barbatulum* Goode and Bean, 1883 | probably = *L. goodebeanorum* |
| *Malacocephalus laevis* (Lowe, 1843) | reported as *Ventrifossa occidentalis* |
| *Merluccius albidus* (Mitchill, 1818) | |
| *Nezumia bairdii* (Goode and Bean, 1877) | |
| *Nezumia* sp. | |
| *Peristedion gracile* Goode and Bean, 1896 | |
| *Peristedion* sp. | |
| *Physiculus fulvus* Bean, 1885 | probably = *P. karrerae* |
| *Synagrops spinosus* Schultz, 1940 | |
| *Talismania* sp. | |
| *Xenodermichthys* sp. | |
| Total = 23 species | |

Table 4. Abundance scores per taxon for Viosca Knoll study sites demersal fish taxa from transect frame-by-frame analysis for USGS 2004–2005 submersible video records. Taxon list is coordinated with that in Table 5 for comparison of abundance and occurrence scores.

| Depth horizon | 325 m | 325 m | 500 m | 500 m |
|---|---|---|---|---|
| Abundance | n | Rank | n | Rank |
| Taxa | | | | |
| *Hyperoglyphe perciformis* | 579 | 1 | 0 | |
| *Gephyroberyx darwini* | 55 | 2 | 0 | |
| *Epinephelus niveatus* | 0 | | 0 | |
| Unidentified Scorpaenidae | 0 | | 0 | |
| *Urophycis* + *Laemonema* | 1 | 4.5 | 10 | 1 |
| *Conger oceanicus* | 1 | 4.5 | 6 | 2.5 |
| *Helicolenus dactylopterus* | 5 | 3 | 1 | 6.5 |
| *Cyttopsis rosea* | 0 | | 0 | |
| *Polyprion americanus* | 0 | | 0 | |
| *Physiculus karrerae* | 0 | | 0 | |
| Unidentified Gadiformes | 0 | | 1 | 6.5 |
| *Lopholatilus chamaeleonticeps* | 0 | | 0 | |
| Macrouridae | 0 | | 1 | 6.5 |
| *Hoplostethus occidentalis* | 0 | | 2 | 4 |
| *Grammicolepis brachiusculus* | 0 | | 6 | 2.5 |
| *Trachyscorpia cristulata* | 0 | | 1 | 6.5 |
| *Squalus* spp. (2 species) | 0 | | 0 | |
| *Lophius gastrophysus* | 0 | | 0 | |
| All other identified taxa | 2 | | 1 | |
| Images unidentifiable to taxon | 4 | | 8 | |
| Totals all taxa | 648 | | 38 | |
| Database | | | | |
| Total 1-s frame grabs | 4,498 | | 3,988 | |
| Total area sampled (m$^2$) | 4,516 | | 4,004 | |
| Population density (fish ha$^{-1}$) | 1,435 | | 95 | |

rizon (VK-906/907), yielding 648 individual fishes scored among seven species (Table 4). Frames representing the five biotope categories were scored as follows: "Open"–383 frames; "Plate"–2800; "Rock"–1315; Rubble–0, "Thicket"–0. Only 51 "Rock" biotope frames contained substantial (but < 50%) *Lophelia* coral cover. Thus "Plate" and "Rock" hard-substrate biotopes without *Lophelia* (but almost always populated by anemones, sponges, bamboo corals, and black corals) appeared to dominate the VK-906/907 landscape, accounting for all but three of 648 fish scores. The 3988 frames representing the 500 m depth horizon (VK-826) yielded 38 individual fishes scored among 10 species (Table 4). Frames representing biotope categories were scored as follows: "Open"–1671 frames; "Plate"–1052; "Rock"–335; "Rubble"–0; "Thicket"–925; plus six frames over open space as the submersible crested a ridge top. *Lophelia* coral was much more prevalent on the 500 m depth horizon, and was typically developed as dense coral monoculture thickets. This three-dimensional coral biotope contrasted dramatically with the dominant low-relief, sparsely populated (sessile invertebrates other than *Lophelia*) "Open" and "Plate" biotopes. Dominant fish biotopes were "Open" (non-coral) (18 fishes scored) and "Thicket" (15 fishes scored). For both depth horizons, "Rubble" biotope was essentially absent.

In the general absence of food habits data for deep-living fishes that occur on *Lophelia* reefs, we have hypothesized probable assignments to trophic guilds by analogy with the known food habits of better-known shallow-water reef and shelf fishes. Large, mobile, schooling fishes that are probable macrocarnivores or mesocarnivores (sensu guilds defined by Ebeling and Hixon, 1991) dominated the 325 m depth horizon (submersible visual observations), including *Hyperoglyphe perciformis* (barrelfish) and *Epinephelus niveatus* (snowy grouper), but the former was under-represented and the latter un-represented in the transect analysis abundance summary (Table 4). Both species typically remained just outside the analyzed field of view during moving transects, as did *Polyprion americanus*, also un-represented in Table 4 relative to submersible visual observations. Other numerically dominant species included the probable sit-and-wait ambush mesocarnivores, *Helicolenus dactylopterus*, and *Conger oceanicus* (this cryptic species also under-represented during moving transects, although frequently observed by the diving scientists), and the apparent hover-and-wait mesocarnivore, *Gephyroberyx darwini*. All of these species were highly associated with hard substrate biotopes, "Plate" and "Rock", particularly in areas with extensive sessile invertebrate live cover. Overall population density of demersal fishes on the VK-906/907 study site estimated from summary of 1-s frame grabs (each representing approximately 1.0 m$^2$) was 1435 fish ha$^{-1}$ (Table 4).

On the 500 m depth horizon, large cruising predators were essentially absent. Moreover, the suite of dominant species scored (Table 4) was more diverse in probable feeding modes. The top-ranking taxon was comprised of three species of benthic euryphagous "hakes" (*Laemonema goodebeanorum, Urophycis cirrata, and Urophycis floridana*), which probably feed opportunistically as both mesocarnivores and microcarnivores. Also important was the apparent ambush predator, *C. oceanicus*, the hover-and-wait strategist, *Hoplostethus occidentalis,* and a morphologically very specialized epifaunal picker, *Grammicolepis brachiusculus,* all three species closely associated with *Lophelia* "Thicket" biotope on VK-826. Four species of Scorpaenidae and Gadiformes completed the dominant species list. Overall population density of demersal fishes on the VK-826 study site estimated from summary of 1-s frame grabs (each representing approximately 1.0 m$^2$) was 95 fish ha$^{-1}$ (Table 4). Thus, despite extensive three-dimensional habitat in the form of *Lophelia* thickets, population density on the deeper study site was lower by a factor of 15.

**Non-transect frames analysis.**—A total of 6879 frame grabs were analyzed for demersal fish occurrence (presence of a given taxon in the analyzed field of view) from the 20 dives, documenting at least 3 distinct species (Table 5). The 325 m depth horizon analysis included 2368 frame grabs, yielding 598 fish occurrences (622 individual fish) among 16 species. The 500 m depth horizon analysis included 4512 frame grabs yielding 230 fish occurrences (233 individual fish) among 23 species. All individuals of all identified species recorded during video analyses appeared to be adults or subadults; no obvious juveniles were observed. Nor were obvious juveniles observed in close-up imaging using the digital still camera. Dominant species, determined via frequency of occurrence in non-transect segments of dives, are given in Table 5. On the 325 m depth horizon, dominant taxa (orders and families) included the Perciformes (Stromateidae, Serranidae, Polyprionidae, Scorpaenidae), Beryciformes (Trachichthyidae), Zeiformes (Grammicolepidae, Zeidae), Gadiformes (Gadidae, Moridae), and Anguilliformes (Congridae). On the 500 m depth horizon, faunal composition by major taxa was similar, except that the large mobile foraging Stromateidae, Serranidae, and Polyprionidae dropped out, and

Table 5. Dominance rank by total occurrences per taxon for Viosca Knoll study sites demersal fish taxa from frame-by-frame analysis of non-transect time segments of all USGS 2004–2005 submersible dive video records.

| Taxa | 325 m horizon Occurrences | 325 m horizon Rank | 500 m horizon Occurrences | 500 m horizon Rank |
|---|---|---|---|---|
| *Hyperoglyphe perciformis* | 213 | 1 | | |
| *Gephyroberyx darwini* | 74 | 2 | 5 | 10 |
| *Epinephelus niveatus* | 45 | 3 | | |
| Unidentified Scorpaenidae (2 spp.) | 14 | 4 | 17 | 5 |
| *Urophycis* + *Laemonema* (3 spp.) | 9 | 5 | 26 | 2.5 |
| *Conger oceanicus* | 7 | 6 | 56 | 1 |
| *Helicolenus dactylopterus* | 3 | 7 | 16 | 6 |
| *Cyttopsis rosea* | 2 | 8 | | |
| *Polyprion americanus* | 1 | 11 | | |
| *Physiculus karrerae* | 1 | 11 | | |
| Unidentified Gadiformes | 1 | 11 | | |
| *Lopholatilus chamaeleonticeps* | 1 | 11 | | |
| Macrouridae (2 spp.) | 1 | 11 | 9 | 9 |
| *Hoplostethus occidentalis* | | | 26 | 2.5 |
| *Grammicolepis brachiusculus* | | | 19 | 4 |
| *Trachyscorpia cristulata* | | | 14 | 7 |
| *Squalus* spp. (2 species) | | | 13 | 8 |
| *Lophius gastrophysus* | | | 3 | 11 |
| All other identified taxa | | | 8 | |
| Images unidentifiable to taxon Database | 17 | | 25 | |
| Total 1-s frame grabs | 2,368 | | 4,512 | |
| Total fish occurrences (N grabs) | 368 | | 230 | |
| Total identified fish taxa | 16 | | 23 | |
| Occurrences on coral biotopes | 14 | | 153 | |
| Occurrences on non-coral biotopes | 354 | | 77 | |

the open-slope Macrouridae (rattails) and Squalidae (dog sharks) entered as important contributors to the fauna (Table 5). Despite the similarity in composition among major taxa at the two depth horizons, there was substantial faunal transition between these horizons at the species level (Table 5). Species dominance rank by occurrence during non-transect video frames roughly paralleled dominance rank by abundance during moving transect frames. However, many more species were documented during the non-transect video segments, particularly including shy, cryptic, and smaller fish species. Such species were more readily documented when the submersible was stationary and the video camera used to zoom in on the substrate or the *Lophelia* thicket.

*Lophelia* colonies were sparse and poorly developed at the 325 m depth horizon. *Lophelia* coral largely occurred as small isolated bushes within an assemblage of mixed sessile invertebrates (sponges, anemones, black corals). No "Thicket" biotope was scored among all frame grabs analyzed. Fish taxa were primarily found on non-coral biotopes. Species occurrences were overwhelmingly scored from non-coral biotopes (Table 5: 354 of 368 frames). In contrast, the reverse situation was observed at the 500 m depth horizon where abundant coral "Thicket" biotope was scored during video frame analysis (Table 5).

**Biotope associations among dominant fish species.**—As advanced by Elliott (1977), a hypothesis of random distribution is appropriate for low density populations, a model which obtains for fish taxa inhabiting Viosca Knoll biotopes. The $\chi^2$ goodness of fit test of observed vs expected counts from occurrence data for 12 key taxa (occurrences ≥ 10) from non-transect segments analyzed revealed that no taxon was randomly distributed among the comparative biotopes (Table 6).

**Remote sampling results.**—Twenty-seven species were captured in bottom trawl and benthic sled collections (Table 3A). Traps returned six species, adding three different species, making the total of 30 remotely collected species. Remote sampling added 15 new species to our overall Viosca Knoll taxonomic list (Table 3A), yielding a total of at least 53 species documented by the USGS study.

## Discussion

**Methodological limitations.**—Moving quantitative belt transects conducted as per pre-defined criteria from the JSL submersible have inherent limitations that affect video estimates of demersal fish diversity and population density. During a moving transect, the JSL cannot deviate from its course for the purpose of identifying or photographing an individual fish. Nor can the submersible stop to collect a fish specimen to validate species identification. The video camera cannot be turned or zoomed in, when the objective is to maintain a consistent field of view to score species abundances. Thus, certain individual fishes cannot be identified to species, genus, or family. Furthermore, small species, juveniles, and cryptic fishes may be under-represented in species scores, or go undetected. The lights, sounds, and motion of the JSL are unusual disturbances in the typically dark, quiet, and still environment of the deep slope. Qualitatively, some species appeared to be repelled by the submersible, at least initially (e.g., *E. niveatus*, *P. americanus*), or more continuously (e.g., *H. perciformis*, *Hemanthias aureorubens*). Some may slowly retreat into cover as the submersible advances (e.g., *H. occidentalis*). A few species appear to be attracted to the submersible (e.g., *E. niveatus* and *C. oceanicus*) following a period of accommodation, such that individuals following the JSL during moving transects may be under-counted initially, then over-counted later in the same dive. Schooling species with large numbers of fish constantly moving (e.g., *H. perciformis*) are difficult to score, and individuals may re-enter and leave the video field multiple times. Despite such limitations, fish species occurrence and abundance on *Lophelia* reefs have previously been successfully quantified using underwater video (Mortensen et al., 1995; Fössa et al., 2002; Costello et al., 2005). The present study, however, is the first to conduct quantitative fish faunal structure on *Lophelia* and associated biotopes based upon tightly-standardized submersible moving transects rendered into sequential equal-time, equal-area high-quality digital still frames for objective scoring. It is also the first quantitative analysis of fish community structure for *Lophelia*-associated biotopes in the Gulf of Mexico and western Atlantic.

The number of dives per each of the two study sites was limited. Under perfect conditions, two dives per day were possible. However, due to competition with other objectives for bottom time, the number of quantitative transects that could be undertaken per dive was limited. Moreover, effort was very unevenly apportioned per biotope category since the hard-bottom and coral areas were the central focus of multidisciplinary objectives in the overall USGS program of investigations. Thus, although hard-substrate biotopes

Table 6. Chi-square test of observed vs expected count data by biotope for 12 key Viosca Knoll demersal fish taxa (n ≥ 10). Observed count data are from frame-by-frame analysis of all non-transect time segments. Expected counts for a hypothesized random distribution of a given taxon across biotopes were determined as the proportion of total frame grabs times the total observed count for that taxon. Critical value = 11.14 (4 df), P > 0.05.

| Biotope | Counts n | Open | Plate | Rock | Rubble | Thicket | Total | $\chi^2$ |
|---|---|---|---|---|---|---|---|---|
| Frame grabs (n) | | 775 | 2,671 | 1,628 | 1 | 1,797 | 6,872 | |
| Proportion of Total Count | | 0.1128 | 0.3887 | 0.2369 | 0.0001 | 0.2615 | 1.0000 | |
| *Hyperoglyphe perciformis* | obs | 0 | 126 | 86 | 0 | 1 | 213 | |
| | exp | 24 | 83 | 50 | 0 | 56 | | 6,699.26 |
| *Gephyroberyx darwini* | obs | 0 | 13 | 62 | 0 | 4 | 79 | |
| | exp | 9 | 31 | 19 | 0 | 21 | | 2,543.92 |
| *Conger oceanicus* | obs | 0 | 5 | 7 | 0 | 51 | 63 | |
| | exp | 7 | 24 | 15 | 0 | 16 | | 1,685.05 |
| *Epinephelus niveatus* | obs | 0 | 15 | 30 | 0 | 0 | 45 | |
| | exp | 5 | 17 | 11 | 0 | 12 | | 544.44 |
| *Urophycis + Laemonema* | obs | 16 | 9 | 1 | 0 | 9 | 35 | |
| | exp | 4 | 14 | 8 | 0 | 9 | | 219.66 |
| Unidentified Scorpaenidae | obs | 9 | 10 | 3 | 2 | 7 | 31 | |
| | exp | 3 | 12 | 7 | 0 | 8 | | 58.57 |
| *Hoplostethus occidentalis* | obs | 4 | 2 | 2 | 0 | 18 | 26 | |
| | exp | 3 | 10 | 6 | 0 | 7 | | 209.61 |
| *Helicolenus dactylopterus* | obs | 4 | 9 | 0 | 0 | 6 | 19 | |
| | exp | 2 | 7 | 5 | 0 | 5 | | 27.38 |
| *Grammicolepis brachiusculus* | obs | 1 | 1 | 4 | 0 | 13 | 19 | |
| | exp | 2 | 7 | 5 | 0 | 5 | | 106.83 |
| *Trachyscorpia cristulata* | obs | 1 | 9 | 1 | 0 | 3 | 14 | |
| | exp | 2 | 5 | 3 | 0 | 4 | | 18.80 |
| Macrouridae | obs | 8 | 2 | 0 | 0 | 0 | 10 | |
| | exp | 1 | 4 | 2 | 0 | 3 | | 63.24 |
| Squaloid sharks (4 species) | obs | 4 | 3 | 0 | 0 | 3 | 10 | |
| | exp | 1 | 4 | 2 | 0 | 3 | | 14.80 |

("Plate" and "Rock") without *Lophelia* coral (but almost always populated by anemones, sponges, bamboo corals, and black corals) appeared to comprise the dominant landscape of VK-906/907, video footage was skewed toward such biotopes. The same was true on the 500 m depth horizon, where *Lophelia* "Thicket" biotope appeared as a prevalent biotope. Comparative areas of "Open" soft-substrate biotope away from reef influence were less frequently traversed during all dives on the Viosca Knoll study sites, particularly when coral collection was the primary objective. However, open soft substrate does appear to be relatively rare on the elevated, carbonate-capped topographic features of Viosca Knoll. Only one dive into the 100 m deep depression on the northwestern corner of VK-826 (Fig. 2, reference "B") encountered extensive soft substrate throughout the dive. The relative rarity of many otherwise dominant open-substrate, middle-slope taxa (e.g., Macrouridae, Halosauridae, Synaphobranchidae) tends to confirm that soft-substrate is disproportionately unavailable in the study area. Thus, comparison of fish assemblages associated with hard-bottom and *Lophelia* biotopes, vs those associated with open soft substrate biotopes is basically limited to the comparative lists of USGS submersible documented taxa vs USGS trawl/sled and NOAA trawl taxa (Tables 3A,B).

**Demersal fish faunal and trophic structure.**—The deep slope biotopes investigated, including *Lophelia* reefs, are sparsely populated with demersal fishes. Only 686 total fishes were scored over 141 min during 32 moving transects, averaging < 5 fish min$^{-1}$. Species biotope affinities were better revealed during opportunistic non-transect intervals, which allowed for closer observations of fishes and their habitats, including observations with the submersible stationary and the video camera free to pan, tilt, and zoom.

Biotopes populated by sessile invertebrates differed substantially between the two depth horizons. On the 325 m depth horizon, a broad suite of sessile invertebrates (anemones, glass sponges, black corals, gorgonians, and *Lophelia*) contributed substantially to forming mixed live cover. Large expanses of "Rock" and "Plate" biotope were densely populated with this type of cover. *Lophelia* occurred primarily as individual small bushes, within the mix of sessile invertebrates. When it occurred in isolation, *Lophelia* was found mostly on bare hard substrate, varying in size from small sprigs with < 10 polyps to bushes up to 1 m high and 1 m in diameter. Typically, such small bushes were composed entirely of live white coral. No "Thicket" biotope was observed on the shallower VK-906/907 site. In contrast, *Lophelia* was the dominant sessile invertebrate on the 500 m depth horizon at the VK-826 site. In places, it formed extensive monospecific thickets covering ridge flanks and crests, sometimes in sequential parallel windrow formations. Thickets were alternatively developed atop thick soft sediment, or on carbonate pavement coated with a thin veneer of sediment. Typically, thickets ended abruptly, giving way to barren sediment or pavement without transitional habitat. Coral rubble was scarce, but sometimes found immediately at the base of thickets. Wherever found, it had the appearance of having been rapidly degraded. Elsewhere, *Lophelia* existed as isolated colonies on otherwise barren carbonate rock.

In terms of taxonomic and probable trophic diversity, the demersal fish fauna of the Viosca Knoll sites, including *Lophelia* reefs, appears rather rarified compared to shallower reef systems. The total fauna documented in this study included 53 species, 37 of which were documented from hard-substrate or coral biotopes. However, only a few were common or abundant, and fewer still highly associated with *Lophelia* "Thicket" biotope. On the 325 m depth horizon site, the fauna was dominated by Serranidae (1 species),

Stromateidae (Centrolophidae) (1), Beryciformes (1), Congridae (1), and Gadiformes (1). Faunal composition of the deeper site was similar, except that the Serranidae and Stromateidae (plus Polyprionidae) dropped out. Thus, large, highly-mobile, benthopelagic visual predators were prominent only on the shallower site where ambient light must still be sufficient to sustain a strategy of visual predation. Nonetheless, in terms of foraging guilds recognized among coral reef fishes (Ebeling and Hixon, 1991) large macrocarnivores and mesocarnivores dominated both *Lophelia* depth horizons in the northern Gulf of Mexico. Among predatory fishes of shallow coral reefs, Hobson (1975, 1979) distinguished five categories [summary based on Hixon (1991), adapted here for Gulf of Mexico *Lophelia* reefs with examples from the present study]: (1) open-water pursuers (*H. perciformis, Squalus* spp., *Odontaspis ferox*), (2) cryptic ambush predators (*H. dactylopterus, Trachyscorpia cristulata*), (3) tactical predators (*E. niveatus, P. americanus*), (4) slow stalkers (*H. occidentalis, G. darwini, G. brachiusculus*), (5) crevice predators (*C. oceanicus*). All five were present on the shallow *Lophelia* depth horizon; category 1 was greatly depleted in abundance and occurrence on the deeper horizon; category 3 was absent; categories 2 and 5 were important on both depth horizons.

Aside from macrocarnivores and mesocarnivores that appear to depend on vision, several trophic categories were absent from our *Lophelia* study reefs compared to shallower reef systems. The absence of herbivores below the depth of photosynthesis is unsurprising. However, the fundamental absence of microplanktivores (aside from rare individuals of *H. aureorubens* and *Anthias nicholsi*) is remarkable since microplanktivores represent a characteristic component of the world reef fish fauna (Hobson, 1991). Planktivores dominated numerically on the deep reef at Enewetak Atoll, Marshall Islands, down to 300 m (Thresher and Colin, 1986), and also dominated on northern Gulf of Mexico shelf-edge reefs, at least to 180 m (Weaver et al., 2002). Undoubtedly, structurally complex *Lophelia* reefs function in the same way as shallow reefs in concentrating particulate matter and plankton (Wolanski and Hamner, 1998), accelerating the delivery of such food items, and providing shelter from predation (Hobson, 1991). The abundance of planktonic prey on the Viosca Knoll sites is evidenced by the diversity and density of sessile particulate-feeding invertebrates populating these sites. But, sessile invertebrates are stationary contact feeders or filter feeders. In contrast, planktivorous fishes must actively select individual prey animals from the water column, and feed via discrete visual strikes (Zaret, 1972; Confer and Blades, 1975; Durbin, 1979). At the depth of Gulf of Mexico *Lophelia* reefs, ambient light is apparently insufficient to support this feeding strategy. Thus despite abundant shelter available in the form of anemone, sponge, black coral forests, and *Lophelia* thickets, planktivorous fishes are absent from the Viosca Knoll sites. The notable absence of juvenile fishes from our *Lophelia* biotopes, and from comparative eastern North Atlantic reefs (Husebø et al., 2002; Costello et al., 2005) is perhaps similarly explained, since early juveniles of most marine fishes typically depend on a plankton diet (Durbin, 1979). In contrast to our findings for the Viosca Knoll fauna, microplanktivory has been reported by Costello et al. (2005) among fishes inhabiting well-developed *Lophelia* reefs on the Sula Ridge off Norway. Shoals of *Sebastes* spp. are reported to hover over the reef tops, facing into the current at 230–320 m depth. Apparently, sufficient ambient light is available on Norwegian *Lophelia* reefs to enable visual microplanktivory. A diet consisting entirely of zooplankton has been confirmed for *Sebastes* spp. in a separate *Lophelia* reef study off southwestern Norway (Husebø et al., 2002).

A further notable attribute of the fauna of the Viosca Knoll study sites is the rarity of epifaunal croppers and benthivores. Much of the fish diversity of shallow coral reefs consists of species that either crop sessile invertebrates (Harmelin-Vivien, 2002) or exploit small benthic invertebrates (Choat and Bellwood, 1991). Sessile megafaunal invertebrates, including *Lophelia*, are abundant on northern Gulf of Mexico hard-bottom slope biotopes. The diversity and abundance of benthic and epibenthic invertebrates (e.g., crustaceans, mollusks) has yet to be assessed for *Lophelia* reefs in the Gulf of Mexico. However, *Lophelia* reefs in the eastern North Atlantic are reported to sustain a high diversity of benthic/epibenthic macrofaunal invertebrates (Jensen and Frederiksen, 1992; Mortensen et al., 1995; Fosså and Mortensen, 1998; Husebø et al., 2002; Costello et al., 2005), with population densities up to three times higher than on adjacent soft substrate (Mortensen et al., 1995). Additionally, Reed (2002) reported that *Lophelia* reefs along the Florida-Hatteras slope support large, but unstudied, populations of sponges, gorgonians, and small macroinvertebrates.

The apparent absence of demersal microcarnivores (epifaunal croppers) among the fish fauna of Viosca Knoll may be a consequence of the limiting energetic cost of processing low-quality prey in a cold-water regime (Harmelin-Vivien, 2002). The negative correlation between increasing latitude and diversity in the world's shallow-water fish fauna (Mead, 1970; Briggs, 1974; Ehrlich, 1975; Springer, 1982; Ebeling and Hixon, 1991; Hobson, 1994; Harmelin-Vivien, 2002) has previously been explained by the progressive loss at higher latitudes of trophic specialists (Ebeling and Hixon, 1991; Harmelin-Vivien, 2002). Sessile invertebrate croppers are diverse and important on tropical reefs (Randall, 1967; Hobson, 1974; Harmelin-Vivien, 1979), but apparently absent from temperate reefs (Harmelin-Vivien, 2002). Low-quality invertebrate prey is energetically expensive to process for low-caloric return (Brey et al., 1988), and may contain high concentrations of anti-predator metabolites (Hay, 1996). Utilizing low-quality prey such as sessile invertebrates (Cummins and Wuycheck, 1971; Brey et al., 1988) may have evolved only on tropical reefs where intense competition for high-quality resources has favored trophic radiation, and only in shallow tropical waters where sustained high temperatures facilitate metabolism of refractory food resources (Harmelin-Vivien, 2002). Tropical reefs are dominated by perciform and tetraodontiform fishes (Randall et al., 1990; Ebeling and Hixon, 1991). These taxa include the most recently evolved and most highly derived forms, including almost all fishes adapted to feed as herbivores or sessile invertebrate croppers (Harmelin-Vivien, 2002). The trophic rarity of benthic microcarnivores on deep cold-water reefs reflects the phylogenetic rarity of percomorph taxa on these reefs. Among more ancient groups occurring on deep reefs, only the Zeiformes seem to contain species adapted for specialized microcarnivory.

Specialized reef microcarnivores that pick small mobile crustaceans off the substrate or off sessile invertebrates appear to be largely absent from the Viosca Knoll demersal fish fauna. There is one notable exception, the zeiform species *G. brachiusculus*, highly adapted morphologically (deep, strongly compressed body), behaviorally (slow, deliberate maneuvering using dorsal and anal fin undulation), and trophically (small tubular mouth with fixed funnel-like opening) to prey upon small reef-dwelling prey, probably small epibenthic and hyperbenthic crustaceans sheltering within *Lophelia* reefs. A second zeiform fish, *Neocyttus helgae* (Holt and Byrne, 1908), from eastern North Atlantic deep reefs (Costello et al., 2005) may represent a trophic analog to *G. brachiusculus*.

**Habitat and apparent trophic associations.**—The microcarnivore *G. brachiusculus* is non-randomly distributed, and appears to be associated primarily with the "Thicket" biotope. Despite its large body size and weak swimming abilities (one specimen was plucked from open water using the JSL manipulator claw), *G. brachiusculus* adults have rarely been collected in bottom trawls (eight total records over six decades of NOAA Fisheries bottom trawling in the Gulf and Caribbean. This is probably due to a high association with reef and rock biotopes that are difficult to trawl. All three Gulf of Mexico bottom trawl records came from within the Viosca Knoll region rectangle, suggesting an association with continental slope reefs in the northern Gulf of Mexico. Such an association is consistent with USGS submersible data, in which 68% of *G. brachiusculus* video records were from the "Thicket" biotope, and 21% from the high-relief "Rock" biotope. Aside from this sole specialist, medium-sized generalized macro- and meso-carnivores (Ebeling and Hixon, 1991), to which category we would tentatively assign beryciform species (*H. occidentalis, G. darwini*) and hake-like gadiform species (*L. goodebeanorum, Physiculus karrerae*, and species of *Urophycis*), appear to be the predominant predators of small benthic and epibenthic organisms on Viosca Knoll hard-substrate biotopes. However, the hake-like gadiform species appear equally at home on soft-substrates, with roughly equivalent occurrences on "Open" soft-substrate biotope vs hard-substrate and structured biotopes.

A first-order statistical test of habitat affinities of 12 numerically-dominant Viosca Knoll demersal fish species via $\chi^2$ goodness of fit revealed that none of these key species were randomly distributed across the four biotopes. However, three taxa, *H. dactylopterus, Trachyscorpia cristulata*, and squaloid sharks (4 spp.) closely approached the critical value for a random distribution. Bias in the frequency of occurrence data by taxon (observed vs expected) in Table 6 suggests the habitat affinities of individual taxa. Departure from randomness was greatest for *H. perciformis* and *G. darwini*, both of which occurred predominantly on "Plate" and "Rock" biotopes (i.e., non-reef hard substrates), never on the "Open" biotope. *Epinephelus niveatus* displayed a similar pattern, never occurring on the "Open" or *Lophelia* "Thicket" biotopes. *Conger oceanicus* occurred disproportionately on the "Thicket" biotype, corresponding to its observed behavior of burrowing into the base of *Lophelia* bushes. *Grammicolepis brachiusculus* also occurred disproportionately on the "Thicket" biotope. Macrouridae displayed an affinity for unstructured low-relief biotope ("Open" and "Plate"). Extensive coral rubble was recorded only once among 6879 Viosca Knoll video frames analyzed. Thus, no association with this rare biotope was documented in the data.

**Fish faunas from comparative investigations.**—The fish fauna of *Lophelia* reefs in the western North Atlantic, including the Gulf of Mexico, has previously been reported only incidentally. In an appendix to their report on deep-water lithoherms (some topped by *Lophelia*) of the northeastern Straits of Florida, Messing et al. (1990) noted these species observed from submersible: a small macrourid, *P. americanus, Chaunax* cf. *pictus* Lowe, 1846, *Polymixia* sp., *Beryx decadactylus, Odontaspis noronhai*, and an ophidiid/bythitid. Of these species, only *P. americanus* has also been documented from Viosca Knoll in the present study. Reed et al. (2005) reported on the invertebrate and fish faunas inhabiting deep-water sinkholes and bioherms off South Florida, none of which were populated by *Lophelia*. However, these authors noted the following species common to Pourtalès Terrace bioherms populated by stylasterid corals, and to Blake Plateau *Lophelia* reefs (based on unpublished data): *H. dactylopterus, Hoplosthethus* sp,

*Laemonema melanurum* Goode and Bean, 1896, *Chlorophthalmus agassizi*, *Nezumia* spp., and *Xiphias gladius* Linnaeus, 1758. Only the first two species listed also occurred frequently on Viosca Knoll coral biotope. While *L. melanurum* was not observed during USGS missions on Viosca Knoll, three potential ecological analogs (species of *Urophycis* and *Laemonema*) were recorded on coral biotope, but more frequently on unstructured "Open" and "Plate" biotopes. *Chlorophthalmus agassizi* and *Nezumia* spp. are characteristic open-slope species. These species were not recorded by us from "Thicket", "Rock" or "Plate" biotopes, and are probably not highly associated with such biotopes. Other demersal fish species reported by Reed et al. (2005) from three-dimensional deep-water habitats off South Florida, and shared with the Viosca Knoll fauna, include *G. darwini*, *E. niveatus*, and congrid eels (probably *C. oceanicus*).

In contrast to the poorly-known fish fauna of western Atlantic *Lophelia* reefs, that of eastern North Atlantic reefs has been relatively well studied (Jensen and Frederiksen, 1992; Mortensen et al., 1995; Husebø et al., 2002; Costello et al., 2005). Based on multiple imaging data, Costello et al. (2005) found 25 fish species in 17 families inhabiting *Lophelia*-associated habitats (coral reef, transition zone, coral debris zone) at eight sites over a depth range of 39–1015 m off Ireland, the Faroe Islands, and Scandinavia. The transition zone of patchy coral was earlier defined by Mortensen et al. (1995) and Freiwald et al. (2002). Both zones are essentially lacking from Viosca Knoll *Lophelia* reefs. Costello et al. (2005, table 4) reported considerable overlap in habitat affinities among species recorded from four natural seafloor habitats: *Lophelia* reef (16 total species), transitional habitat (21), coral debris habitat (18), and open seabed (21). (Note that the original totals by respective habitat in Costello et al.'s table 4 are each erroneously summed). Only one species (*N. helgae*) was exclusively associated with reef. Only two species each were exclusively found on transitional or coral debris habitats, and only three exclusively on open seabed. Eleven species were found in common among all four habitats. However, no single species reported in Costello et al. (2005) that was found on all three coral-associated habitats was not also found on open seabed habitat. Gadoid fishes predominated, along with the Scorpaenidae (Sebastidae). Species associated exclusively or more consistently with open seabed habitats were typified by families Macrouridae, Rajidae, Lophiidae, and Pleuronectidae.

No species reported from eastern North Atlantic *Lophelia* reefs were shared with Gulf of Mexico reefs. However, prominently contributing to the faunas of both eastern North Atlantic *Lophelia* reefs and Viosca Knoll study sites were the Gadiformes (Gadidae, Moridae), Beryciformes (Trachichthyidae), and Scorpaeniformes (Scorpaenidae). All may be trophic generalists. The gadiform and scorpaeniform taxa exploit both open and structured biotopes on the continental slope, while beryciform taxa are more consistently associated with structured biotopes. All appear to be facultatively associated with *Lophelia* reefs. However, Husebø et al. (2002) reported that long-line catches yielded seven times more scorpaenid species (*Sebastes marinus* Linnaeus, 1758) from coral vs non-coral habitats, and nearly twice as many of two gadid species (*Brosme brosme* Ascanius, 1772; *Molva molva* Linnaeus, 1758). No single demersal fish species in the eastern North Atlantic has been reported to be an obligate *Lophelia* associate (Husebø et al., 2002), matching similar findings among reef-associated invertebrates (Burdon-Jones and Tambs-Lyche, 1960; Jensen and Frederiksen, 1992). In the Gulf of Mexico, however, at least the highly-specialized zeiform fish, *G. brachiusculus*, may be an obligate *Lophelia* inhabitant.

Although bottom trawl sampling added 15 species to the overall USGS missions demersal fish species list for Viosca Knoll, most additions were fishes not typically associated with three-dimensional biotopes. Accordingly, most of these same species pertain to families characteristic of the two-dimensional open slope biome, and otherwise broadly and ubiquitously distributed. These include the Macrouridae (*Hymenocephalus* sp., *Malacocephalus occidentalis*), Steindachneriidae (*Steindachneria argentea*), Chlorophthalmidae (*C. agassizi, Parasudis truculentus*), Paralichthyidae (*Paralichthys albigutta*), Poecilopsettidae (*Poecilopsetta beani*), Cynoglossidae (*Symphurus marginatus*), Percophidae (*Bembrops anatirostris*), Scorpaenidae (*Pontinus longispinis, Setarches guentheri*), Uranoscopidae (*Gnathagnus egregius*), and Ophichthyidae (*Pseudomyrophis nimius*). Many of the fish species identified from eastern North Atlantic *Lophelia* reefs (Jensen and Frederiksen, 1992; Mortensen et al., 1995; Husebø et al., 2002; Costello et al., 2005) similarly pertain to taxa more generally characteristic of the open slope biome than to *Lophelia* reefs. When found on deep coral biotopes, such taxa may be considered as either facultative or incidental, i.e., not distinctly associated with coral habitat. Such typical open-slope taxa (e.g., Macrouridae, Synaphobranchidae, Ophidiidae, Ipnopidae, Halosauridae) were barely represented on Viosca Knoll where soft substrate is uncommon. Thus, direct ecological interaction between coral-associated fishes and typical open-slope deep-sea fishes may be limited. Midwater fishes were also very rarely observed during USGS Viosca Knoll dives, again suggesting limited interaction between the hard-bottom fauna and the mesopelagic deep-sea fauna.

Trawl samples from the Viosca Knoll vicinity also returned juveniles of at least one species, *C. oceanicus,* that inhabits *Lophelia* "Thicket" biotope as adults. Thus, at least for this species, the absence of juveniles from coral biotope can be explained. Furthermore, an ontogenetic linkage has been documented between the soft-substrate and coral biomes of the continental slope. Populated by macrocarnivores to a large extent, *Lophelia* reefs in all regions may represent a high predation risk habitat for juvenile fishes.

**Regional biotope contrasts.**—A striking difference between Viosca Knoll *Lophelia* reefs and eastern North Atlantic *Lophelia* reefs is the virtual absence of the coral rubble and patch reef transition zones (Mortensen et al., 1995; Freiwald et al., 2002) on the northern Gulf of Mexico reefs, and the apparently very high proportion of living white coral in the Gulf of Mexico (Schroeder, 2002). Both *Lophelia* rubble and dead coral have been reported to be important high-density, high-diversity invertebrate habitats in the eastern North Atlantic (Wilson, 1979; Jensen and Frederiksen, 1992; Mortensen et al., 1995; Costello et al., 2005). In the western North Atlantic, Messing et al. (1990) reported that the upcurrent ends of *Lophelia*-topped lithoherms in the Florida Straits were covered with *Lophelia* rubble. Rubble was reported to extend beyond the foot of the lithoherms forming a talus apron, much like the rubble zones described for Lophelia reefs in the eastern North Atlantic. Among Norwegian bioherms studied, dead coral has been reported to cover an average basal area nearly eight-fold larger than that occupied by living coral (Mortensen et al., 1995). *Lophelia* rubble is also utilized as habitat by demersal fish species (Costello et al., 2005), and may form a distinct biotope for species such as *Lophiodes beroe* Caruso, 1981 and *Chaunax stigmaeus* Fowler, 1946, both found preferentially on *Lophelia* rubble on the Blake Plateau (Caruso et al., 2007).

In contrast, in the northern Gulf of Mexico, there is typically a dramatic and abrupt discontinuity between live *Lophelia* bushes or *Lophelia* reef thicket and adjacent barren substrate. Among 8486 frame grabs analyzed from 32 moving video transects on the

Viosca Knoll study sites, not one frame was scored as representing the "Rubble" biotope. Among 6879 additional still frames analyzed from non-transect video segments, only one frame was scored as containing > 50% rubble substrate in the field of view.

The remarkable rarity of *Lophelia* rubble from northern Gulf of Mexico reefs begs explanation. Among hypotheses that could be advanced, we offer the following alternatives: (1) The reefs are very young, as suggested by the preponderance of living white coral, such that time has been insufficient for extensive accumulation of rubble; (2) In the hydrocarbon seep environment of Gulf of Mexico salt diapers, rubble is rapidly degraded chemically, biologically, or both; (3) Active bottom currents continuously or episodically sweep rubble from the underlying hardpan substrate, transporting it down-ridge to be buried in sediment-filled valleys. None of these hypotheses has yet been tested.

Our finding that the shallower VK-906/907 depth horizon had 15-fold greater abundance of demersal fishes than the deeper depth horizon corresponds with a similar bathymetric trend in fish abundance for the faunas of eastern North Atlantic *Lophelia* reef habitats (Costello et al., 2005). As depth increases, fish trophic diversity and abundance both decline, paralleling findings for the invertebrate macrofauna (R.A.B., unpubl. data). Two trophic guilds of demersal fishes predominate on the deeper Viosca Knoll study site, large macrocarnivores and medium-sized opportunistic mesocarnivores.

**New regional faunal records.**—Documentation of *P. americanus* in the present study represents the first record of this species from the Gulf of Mexico, although *P. americanus* is known from deep habitats off the adjacent southeastern U.S. (Messing et al., 1990; Sedberry et al., 1999; Sedberry, 2002, Reed et al., 2007). This species and may utilize *Lophelia* biotope for spawning (Reed et al., 2007). The Viosca Knoll video record of the shark *O. ferox* (smalltooth sand tiger) is the third from the western Atlantic, second from the Gulf of Mexico (Bonfil, 1995), and second from within the U.S. EEZ (Sheehan, 1998). The video record of *Caristius* sp. (Table 3A) appears to represent the second record of this taxon from the Gulf (Trolley et al., 1990).

## Acknowledgments

We thank our USGS *Lophelia* research team colleagues for conducting transects and obtaining species images during their JSL dives. Special thanks go to J. Berg, B. Albert, S. C. Keitzer, J. Rochello, J. C. Carr, and M. M. Cheung for assistance in the field and laboratory. Additional JSL dive DVDs for comparative study were contributed by S. Viada, CSA. W. Schroeder, Dauphin Island Marine Laboratory, provided an advance multibeam map of VK-906/907, facilitating the 2004 submersible mission; he also contributed comparative NOAA Ocean Exploration dive videos. Taxonomic identifications for shark images were contributed by J. Castro, Mote Marine Laboratory, for barrelfish images by G. Sedberry, South Carolina Dept. Natural Resources, and of Cynoglossidae images by T. Munroe, NOAA Fisheries Systematics Laboratory, Smithsonian Institution. Historical trawl records of demersal fishes were provided by NOAA Fisheries, Mississippi Laboratories, Pascagoula, Mississippi. We thank the captain, ship's crew, and submersible operations team of the R/V SEWARD JOHNSON and R/V SEWARD JOHNSON II, Harbor Branch Oceanographic Institution, for their effective support of our research effort. We also thank the captain and crew of the R/V TOMMY MUNRO, Gulf Coast Research Laboratory. This investigation was supported by the USGS Outer Continental Shelf Ecosystem Program, and was sponsored and facilitated by the Minerals Management Service.

## Literature Cited

Bonfil, R. 1995. Is the ragged-tooth shark cosmopolitan? First record from the western North Atlantic. J. Fish Biol. 47: 341–344.

Brey, T., H. Rumohr, and S. Ankar. 1988. Energy content of macrobenthic invertebrates: general conversion factors from weight to energy. J. Exp. Mar. Biol. Ecol. 117: 271–278.

Briggs, J. C. 1974. Marine biogeography, McGraw-Hill, New York.

Bullis, H. R., Jr. and J. R. Thompson. 1965. Collections by the exploratory fishing vessels Oregon, Silver Bay, Combat, and Pelican made during 1956 to 1960 in the southwestern North Atlantic. U.S. Fish Wildl. Serv. Spec. Sci. Rep. Fish. 510: 1–130.

Burdon-Jones, C. and H. Tambs-Lyche. 1960. Observations on the fauna of the North Brattholmen stone-coral reef near Bergen. Årbo Univ. Bergen, Mat. Naturv. Ser. 4: 1–24.

Caruso J. H., S. W. Ross, K. J. Sulak, and G. R. Sedberry. 2007. Deep-water chaunacid and lophiid anglerfishes (Pisces: Lophiiformes) off the Southeastern United States. J. Fish Biol. 70: 1015–1026.

Choat, J. H. and D. R. Bellwood. 1991. Reef-fishes: their history and evolution. Pages 39–66 *in* P. F. Sale, eds. The ecology of fishes on coral reefs. Academic Press, San Diego.

Confer, J. L. and P. I. Blades. 1975. Omnivorous zooplankters and planktivorous fish. Limnol. Oceanogr. 20: 571–579.

Costello, M. J., M. McCrea, A. Freiwald, T. Lundälv, L. Jonsson, B. J. Bett, T. C. E. van Weering, H. de Haas, J. M. Roberts, and D. Allen. 2005. Role of cold-water *Lophelia pertusa* coral reefs as fish habitat in the NE Atlantic. Pages 771–805 *in* A. Freiwald and J. M. Murray, eds. Coldwater corals and ecosystems. Springer-Verlag, Berlin.

Cummins, K. W. and J. C. Wuycheck. 1971. Caloric equivalents for investigations in ecological processes. Int. Assoc. Theor. Appl. Limnol. Commun. 18: 1–158.

Durbin, A. G. 1979. Food selection by plankton-feeding fishes. Pages 203–218 *in* H. E. Clepper, ed. Predator-prey systems in fisheries management. Sport Fish. Inst., Washington, D.C.

Ebeling, A. W. and M. A. Hixon. 1991. Tropical and temperate reef fishes: comparison of community structures. Pages 509–563 *in* P. F. Sale, ed. The ecology of fishes on coral reefs. Academic Press, San Diego.

Ehrlich, P. R. 1975. The population biology of coral reef fishes. Annu. Rev. Ecol. Syst. 6: 211–247.

Elliott, J. M. 1977. Some methods for the statistical analysis of samples of benthic invertebrates. Freshwater Biol. Assoc., Spec. Publ. No. 125.

FAO Corporate Document Repository [Internet]. How to make various types of traps and pots. Chapter 6 IN: Fishing with traps and pots. Available from: http://www.fao.org/docrep/004/x2590e/x2590e07.htm). Accessed 4 October 2006.

Fosså, J. H. and P. B. Mortensen. 1998. Artsmangfoldet på *Lophelia*-korallrev langs norskekysten. Forekomst og tilstand. Fisken og Havet 17: 1–95 (in Norwegian).

_____, _____, and D. M. Furevik. 2000. *Lophelia* korallrev langs norskekysten. Forekomst og tilstand. Fisken og Havet 2: 1–94 (in Norwegian).

_____, _____, and _____. 2002. The deep-water coral *Lophelia pertusa* in Norwegian waters: distribution and fishery impacts. Hydrobiologia 471: 1–12.

Freiwald, A., V. Hühnerback, B. Lindberg, J. B. Wilson, and J. Campbell. 2002. The Sula Reef complex, Norwegian Shelf. Facies 47: 179–200.

Harmelin-Vivien, M. L. 1979. Ictyofaune des récifs coiralines de Tuléar (Madagascar): ecologie et relations trophiques. Ph.D. Thesis, Université Aix-Marsielle. 165 p.

_____. 2002. Energetics and fish diversity on coral reefs. Pages 265–274 *in* P. F. Sale, ed. Coral reef fishes, dynamics and diversity in a complex ecosystem. Academic Press, Amsterdam.

Hay, M. E. 1996. Marine chemical ecology: what's known and what's next? J. Exp. Mar. Biol. Ecol. 9: 414–416.

Hixon, M. A. 1991. Predation as a process structuring coral reef fish communities. Pages 475–508 *in* P. F. Sale, ed. The ecology of fishes on coral reefs. Academic Press, San Diego.

Hobson, E. S. 1975. Feeding patterns among tropical reef fishes. Am. Sci. 63: 382–392.

_____. 1979. Interactions among piscivorous fishes and their prey. Pages 231–242 *in* H. E. Clepper, ed. Predator-prey systems in fisheries management. Sport Fish. Inst., Washington, D.C.

_____. 1991. Trophic relationships of fishes specialized to feed on zooplankters above coral reefs. Pages 69–95 *in* P. F. Sale, ed. The ecology of fishes on coral reefs. Academic Press, San Diego.

_____. 1994. Ecological relations in the evolution of acanthopterygian fishes in warm-temperate communities of the northern Pacific. Environ. Biol. Fishes 40: 49–90.

Husebø, Å., L. Nøttestad, J. H. Fosså, D. M. Furevik, and S. B. Jørgensen. 2002. Distribution and abundance of fish in deep-sea coral habitats. Hydrobiologia 471: 91–99.

Jensen, A. and R. Frederiksen. 1992. The fauna associated with the bank-forming deepwater coral *Lophelia pertusa* (Scleractinaria) on the Faroe shelf. Sarsia 77: 53–63.

Mead, G. W. 1970. A history of South Pacific fishes. Pages 236–251 *in* W. S. Wooster, ed. Scientific exploration of the South Pacific. Natl. Acad. Sci., Washington, D.C.

Messing, C. D., A. C. Neumann, and J. C. Lang. 1990. Biozonation of deep-water lithoherms and associated hardgrounds in the northeastern Straits of Florida. Palaios 5: 15–33.

Mortensen, P. B., M. Hovland, T. Brattegard, and R. Farestveit. 1995. Deep water bioherms of the scleractinian coral *Lophelia pertusa* (L.) at 64° N on the Norwegian shelf: structure and associated megafauna. Sarsia 80: 145–158.

Moore, D. R., and H. R. Bullis, Jr. 1960. A deep-water coral reef in the Gulf of Mexico. Bull. Mar. Sci. Gulf Carib. 10: 125–128.

Randall, J. E. 1967. Food habits of reef fishes of the West Indies. Stud. Trop. Oceanogr. 5: 665–847.

_____, G. R. Allan, and R. C. Steene. 1990. Fishes of the Great Barrier Reef and Coral Sea. Crawford House, Bathurst, Australia.

Reed, J. C. 2002. Comparison of deep-water coral reefs and lithoherms off southeastern USA. Hydrobiologia 471: 57–69.

Reed, J. K., C. C. Koenig, and A. N. Shepard. 2007. Impacts of bottom trawling on a deep-water *Oculina* coral ecosystem off Florida. Bull. Mar. Sci. 81: 481–496.

_____, S. A. Pomponi, D. Weaver, C. K. Paull, and A. E. Wright. 2005. Deep-water sinkholes and bioherms of South Florida and the Pourtalès Terrace – Habitat and fauna. Bull. Mar. Sci. 77: 267–296.

Rogers, A. D. 1999. The biology of *Lophelia pertusa* (Linnaeus ,1758) and other deep-water reef-forming corals and impacts from human activities. Int. Rev. Hydrobiol. 84: 315–406.

Schroeder, W. W. 2002. Observations of *Lophelia pertusa* and the surficial geology at a deep-water site in the northeastern Gulf of Mexico. Hydrobiologia 471: 29–33.

Sedberry, G. R. 2002. Polyprionidae, wreckfishes (giant sea basses). Pages 1297–1298 *in* K. Carpenter, ed. The living marine resources of the western Central Atlantic, FAO Species identification guide for fishery purposes, vol. 2, FAO, Rome.

_____, C. A. P. Andrade, J. L. Carlin, R. W. Chaman, B. E. Luckhurst, C. S. Manooch II, G. Menezes, B. Thomsen, and G. F. Ulrich. 1999. Wreckfish (*Polyprion americanus*) in the North Atlantic: fisheries, biology, and management of a widely distributed and long-lived fish. Pages 27–50 *in* J. A. Musick, ed. Life in the slow lane: ecology and conservation of long-lived marine animals, Am. Fish. Soc. Symp. 23.

Sheehan, T. F. 1998. First record of the ragged-tooth shark, *Odontaspis ferox*, off the U.S. Atlantic Coast. Mar. Fish. Rev. 60: 33–34.

Springer, S. and H. R. Bullis, Jr. 1956. Collections made by the Oregon in the Gulf of Mexico. U.S. Fish Wildl. Serv. Fish. Bull. 65: 581–624.

Springer, V. G. 1982. Pacific plate biogeography, with special reference to shore-fishes. Smithson. Contrib. Zool. 367: 1–182.

Teichert, C. 1958. Cold- and deep-water coral banks. Bull. Am. Assoc. Petrol. Geol. 42: 1064–1084.

Thresher, R. E. and P. L. Colin. 1986. Trophic structure, diversity and abundance of fishes of the deep reef (30–300 m) at Enewetak, Marshall Islands. Bull. Mar. Sci. 38: 253–272.

Trolley, S. G., M. M. Leiby, and J. V. Gartner, Jr. 1990. First record of the family Caristiidae (Osteichthyes) from the Gulf of mexico. NE Gulf Sci. 11: 159–162.

Weaver D. C., G. D. Dennis III, and. K. J. Sulak. 2002. Community structure and trophic ecology of demersal fishes on the Pinnacles Reef tract. U.S. Department of the Interior, U.S. Geological Survey Biological Sciences Report USGS BSR 2001-0008; Minerals Management Service, OCS Study MMS-2002-034. 94 p + 4 app.

Wilson, J. B. 1979. The distribution of the coral *Lophelia pertusa* (L.) [*L. prolifera* (Pallas)] in the north-east Atlantic. J. Mar. Biol. Assoc. U.K. 59: 149–164.

Wolankski, E. and W. M. Hamner 1998. Topographically controlled forces in the ocean and their biological influence. Science 241: 177–181.

Zaret, T. M. 1972. Predators, invisible prey, and the nature of polymorphism in the Caldocera (class Crustacea). Limnol. Oceanogr. 17: 171–184.

ADDRESSES: (K.J.S., K.E.L., A.D.N., M.R., G.E.Y., J.M.M, W.M.H.) *U.S. Geological Survey, Florida Integrated Science Center, 7920 NW 71st St., Gainesville, Florida 32653.* (R.A.B.) *ENSR International, 9700 16th St. N., St. Petersburg, Florida 33716.* (J.H.C.) *Department of Biological Sciences, University of New Orleans, 2000 Lake Shore Drive, New Orleans, Louisiana 70148.* (A.J.Q.) *Naval Research Laboratory, Code 7332, Stennis Space Center, Mississippi 39529.* (S.W.R.) *U.S. Geological Survey, Florida Integrated Science Center, 600 4th St, St. Petersburg, Florida 33701.* CORRESPONDING AUTHOR: (K.J.S.) *E-mail: <ksulak@usgs.gov>.*

# Linking deep-water corals and fish populations

PETER J. AUSTER

## Abstract

The role of emergent fauna as physical habitat used by fish populations has been examined for a number of fish species in deep-water environments. Deep-water corals have been a central focus of such work during the past decade due to their sensitivity to human disturbance, slow recovery rates, and limited distribution. Some authors have suggested corals are important for mediating population processes of fishes while others have demonstrated minimal associations of fishes with corals. Further, the co-occurrence of fishes with corals does not necessarily mean there is a functional link to population processes. Expanded observational studies that include corals as well as non-coral features as shelter, sources of benthic prey, and sites with accelerated flows to enhance zooplankton prey delivery, are required to better understand the role that deep-water corals play in mediating the distribution and abundance of fishes. Studies are best designed to test a series of alternatives (or predictions) rather than simply testing for cases of no response (i.e., to a null hypothesis). The use of spatial replicates across regions and sampling over daily and seasonal time frames will produce information at space and time scales that can be applied to management of both major taxonomic groups.

Understanding the functional relationships between seafloor fauna and associated fishes (as habitat, competitors, and prey) is central to managing exploited demersal populations in a sustainable manner and implementing ecosystem-based approaches to management. In particular, the role of emergent fauna as physical habitat used by fish populations has been examined for a number of continental shelf fish species. The association of fishes with emergent biotic structure occurs along a gradient of habitat complexity from animal-formed depressions and tubes (e.g., Auster et al., 1995, 1997) and rocky reefs (Ebeling and Hixon, 1991; Stein et al., 1992; Auster et al., 2003) to tropical coral reef systems (Hixon, 1993). Such associations enhance survivorship by providing cover from predators (Tupper and Boutilier, 1995; Carr et al., 2002) and sites for enhanced capture of prey (Genin, 2004). Syntheses of the results of studies, conducted across wide geographic ranges have allowed managers to generalize the importance of conserving such habitats and integrating habitat conservation objectives into the decision-making process. However, our understanding of the ecological and demographic linkages between deep-water (outer shelf, slope, and deep-sea) structure forming taxa (e.g., corals, sponges) and fishes is still in an exploratory phase.

An expanding and recent literature documents patterns in the use of deep-water coral habitats by fishes (i.e., those deeper than 200 m on continental slopes, submarine canyons, seamounts, and in the abyss) although the linkages between coral-associated fishes and their more widely distributed populations remain undefined. For example, species richness and abundance of fishes was greatest on *Lophelia pertusa* (Linnaeus, 1758) reefs when compared to other non-living-coral habitats (i.e., reef transitional zone, coral debris, and

George, R. Y. and S. D. Cairns, eds. 2007. Conservation and adaptive management of seamount and deep-sea coral ecosystems. Rosenstiel School of Marine and Atmospheric Science, University of Miami.

off-reef habitats) along the continental margin in the Northeast Atlantic (Costello et al., 2005). However, *L. pertusa* reef did not support a unique fish community structure and depth (above and below 400–600 m) was the most significant parameter correlated with species composition. In the deep basins of the Gulf of Maine (Northwest Atlantic), the density of fishes associated with dense octocoral thickets and other dense epifauna habitats was not statistically different, but densities in both emergent fauna habitat types were higher than in other less spatially complex habitats (Auster, 2005). In contrast to fish communities associated with *L. pertusa* reefs, species richness in these octocoral habitats was not the highest of all habitats surveyed. In all cases the density of coral associated fishes, when viewed in light of frequency dependent habitat selection theory, suggest deep-water corals are demographically important habitats based on high density aggregations of fishes (MacCall, 1990). However, if the spatial extent of such habitats is relatively small, as well as the proportion of fish populations that occur there, the functional role of coral habitats may be minimal in sustaining populations (Auster, 2005).

There appears to be differential use of complex coral and other biogenic habitats by fishes on areas of the continental slope and on seamounts. In the Aleutian Islands region and at the head of Pribilof Canyon in the Bering Sea, there were positive associations between multiple taxa of fishes (many in the genus *Sebastes*) and octocorals having both fan and sea whip morphologies (Brodeur, 2001; Stone, 2006). Habitats with tall morphotypes of deep-water corals (e.g., *Gerardia* sp.) generally supported higher fish densities than non-coral areas in the Hawaiian Islands (Parrish, 2006). However, after accounting for the effects of bottom relief in the analysis, any statistically significant association between fish and corals was removed, suggesting that both taxa may aggregate in areas of enhanced flows for increased rates of prey delivery with little actual interaction (Parrish, 2006). Tissot et al. (2006) also found that fishes and structure-forming invertebrates (including corals) off California co-occur in the same types of habitats with accelerated flows but found no demonstrable functional relationship between individual invertebrates and fishes. While orange roughy (*Hoplostethus atlanticus* Collett, 1889) off New Zealand generally occur in large spawning and feeding aggregations over seamount summits that support dense coral assemblages (Koslow et al., 2000), there is little direct association between fishes and these invertebrates. Observations of orange roughy off the Bay of Biscay in the northeast Atlantic revealed that fishes occurred in dense aggregations over the seafloor in a submarine canyon, in the absence of structure-forming invertebrates, to likely exploit prey delivered by enhanced flows (Lorance et al., 2002). No aggregations were observed elsewhere in areas of lower flow regime suggesting that habitats that produce enhanced flows and increased encounter rates with prey, even in the absence of increased biogenic structure, were utilized.

A common classification scheme for seafloor habitats would aid in comparing studies and combining data sets for understanding the role of structure-forming invertebrates in mediating the distribution and abundance of fishes. For example, Auster et al. (2005) developed a hierarchical fish habitat classification scheme from observations on the New England Seamount chain in the North Atlantic. The scheme integrated geological and biological attributes of the structural components of seafloor habitats as well as flow regime at multiple spatial scales. Initial observations from depths of 2500–1100 m, showed the only coral-associated fish, the false boarfish *Neocyttus helgae* (Holt and Byrne, 1908), used fan shaped corals (*Paragorgia* sp.) and depressions in basalt pavement habitats as shelter or flow refuge, suggesting that both biological and geological habitats may be functionally equivalent.

Overall, we must acknowledge that the studies cited above are but snapshots in time. One explanation for the lack of fish associations with corals and other structure-forming invertebrates in some studies is that observation programs are missing the time period when particular fishes (e.g., juveniles, spawning fishes) use such habitats. This can reflect variations in habitat use over both daily and seasonal time periods. (It is interesting to note that none of the studies focused on diel patterns of habitat use.) Another explanation is that all highly structured habitats are not utilized in a similar manner and use by fishes is more spatially constrained, or more stochastic in nature. The only way to resolve this issue is to conduct studies over wide areas (i.e., use of spatial replicates) and over both daily and seasonal time frames. Studies are best designed to test a series of alternatives (or predictions) rather than simply testing for the case of no response (i.e., null hypothesis). There are a number of alternative predictions that can be tested to better explain the functional role of coral habitats for fishes: (1) fishes use corals as cover from predators (i.e., to enhance survivorship), (2) fishes co-occur with corals for feeding opportunities on coral associated taxa, (3) fishes use corals (and the area of vortices around them) for sites to enhance prey capture (e.g., to minimize physiological requirements for station-keeping while maintaining access to enhanced flows delivering prey, as sites that enhance flow and prey delivery superior to non-coral features, as sites for ambush predation), (4) both fishes and corals co-occur in areas of high flows for enhanced prey delivery but have no direct association, and (5) fishes use functionally equivalent biogenic features that include corals for particular ecological functions.

These are not mutually exclusive predictions. In order to differentiate the functional role of corals, sampling for gut contents of fishes, small and large scale flow measurements (around coral colonies, other habitat forming taxa, and around geologic features), prey field samples (of coral associates, epibenthos, planktonic prey), imagery to determine comparative patterns of habitat use (based on use of spatial replicates), and time series imagery around coral shelter sites (to determine diel and seasonal patterns of use) will be required. Sequential sets of observations or experiments may be needed to determine if one or more predictions operate at the same time.

There also is a separate set of alternative predictions that can be tested to explain the role that corals might play in mediating the demographic patterns of fishes and their potential linkages to coral habitats: (1) deterministic processes mediate habitat use (corals and other habitat features) at specific or all life history stages, (2) deterministic processes mediate connectivity of fish recruitment into coral habitats, (3) stochastic processes mediate patchiness in habitat use, and (4) stochastic processes mediate patchiness in habitat use and corals are but one of a range of habitats occupied.

These alternatives are obviously quite general and should be refined when considering particular species or taxonomic groups. Further, any sampling program should be nested within some spatial context based on our current understanding of the life history of the species and the oceanographic regime in which it exits. Figures 1 and 2 illustrate a range of population configurations and habitat linkages that could be exhibited by particular fish taxa. These address population ranges that transcend or are constrained by the shelf-slope boundary, have obligate or a gradient of facultative use of coral habitats, and have different processes mediating connectivity between coral habitats. At best, testing the predictions outlined above could involve a coordinated macro-ecological approach that includes a number of species with variable distributions (i.e., such as the range illustrated in Figure 1A) as well as contrasting species or populations with distributions that occupy variable percentages of coral habitat (i.e., contrasting distributions illustrated in Figure

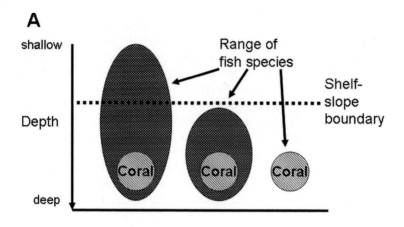

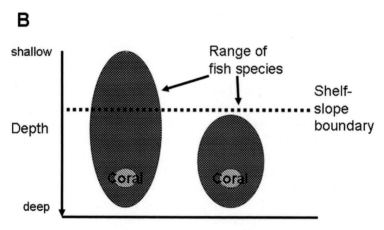

Figure 1. Conceptual illustration of alternative ranges of fish species and relative area of range that includes deep-water corals. (A) Fish species that transcend shelf and slope regions (left), occur only on the slope (center), and occur only in coral habitats (right). (B) Fish species where corals are only a minor type of habitat within the overall range.

1A and B). A range of scenarios regarding the connectivity of particular life history stages of fishes to patches of coral habitat can also be proposed. For example, depth-related ontogenetic shifts in the distribution of a number of species in the genus *Sebastes* in the Gulf of Alaska are common (Love et al., 2002) and could serve as a foundation for linking demographic patterns to use of biogenic habitats including corals (Fig. 2A). Questions of connectivity (Fig. 2B) between patches of coral habitat could focus on up-stream-downstream relationships related to recruitment of propagules as well as move-ments of juveniles and adults. For example, do juveniles and adults exhibit directional movement related to density-dependent processes related to shelter or prey resources?

While many issues discussed here can be partially resolved by various types of sam-pling within and between coral habitats, experimental approaches can validate results from pattern recognition studies. For example, Koenig et al. (2005) found significant increases in density of a number of deep coral-associated species along a gradient of habitat complexity composed of artificial reef units deployed at three different densities (i.e., 5, 10, and 20 units m$^{-2}$). The results of this study also suggest that there are density-

**A**

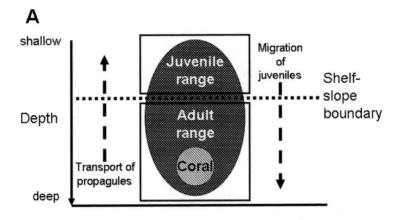

**B**

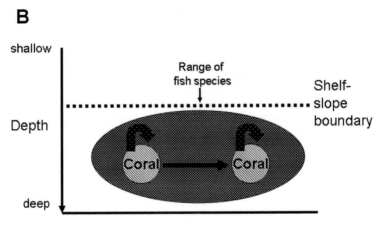

Figure 2. Conceptual illustration of two alternative patterns of fish movement that link populations to deep-water coral habitats. (A) Ontogentic linkages across the shelf and slope. (B) Upstream-downstream linkages as propagules, juveniles or adults.

dependent factors that limit fish use of particular habitats as fish density increased between 5 and 10 unit sites but not between 10 and 20 unit sites. Adaptive management strategies can also contribute to developing a mechanistic understanding of the ecological role of coral habitats when responses must be measured at the scale of fish populations or communities (e.g., Sainsbury et al., 1997).

Knowing if we can generalize our understanding of habitat linkages to fish populations and communities on the continental shelf to those in the deep-sea will depend on efforts to gain a better spatial and temporal coverage to elucidate patterns of habitat use and studies designed specifically to test the role of particular mechanisms mediating distribution and abundance.

### Acknowledgments

This paper was modified from an initial contribution by the author to the ICES Working Group on Deep-water Ecology and resulted from studies supported by NOAA's National Undersea Research Program and Office of Ocean Exploration as well as the Census of Marine Life Gulf of

Maine Program. The contents of this paper benefited from earlier discussions on this topic with many colleagues but especially those with L. Watling, L. Kaufman, and F. Parrish as well as comments by two anonymous reviewers. The opinions expressed herein are those of the author and do not necessarily reflect the opinions of NOAA, the Census of Marine Life, or its sub-agencies.

## Literature Cited

Auster, P. J. 2005. Are deep-water corals important habitats for fishes? Pages 747–760 *in* A. Freiwald and J. M. Roberts, eds. Cold-water corals and ecosystems. Springer-Verlag, Berlin Heidelberg.

_____, J. Lindholm, and P. C. Valentine. 2003. Variation in habitat use by juvenile Acadian redfish, *Sebastes fasciatus*. Environ. Biol. Fish. 68: 381–389.

_____, R. J. Malatesta, and S. C. LaRosa. 1995. Patterns of microhabitat utilization by mobile megafauna on the southern New England (USA) continental shelf and slope. Mar. Ecol. Prog. Ser. 127: 77–85.

_____, _____, and C. L. S. Donaldson. 1997. Distributional responses to small-scale habitat variability by early juvenile silver hake, *Merluccius bilinearis*. Environ. Biol. Fish. 50: 195–200.

_____, J. Moore, K. Heinonen, and L. Watling. 2005. A habitat classification scheme for seamount landscapes: assessing the functional role of deepwater corals as fish habitat. Pages 761–769 *in* A. Freiwald and J. M. Roberts, eds. Cold-water corals and ecosystems. Springer-Verlag, Berlin Heidelberg.

Brodeur, R. D. 2001. Habitat-specific distribution of Pacific ocean perch (*Sebastes alutus*) in Pribilof Canyon, Bering Sea. Cont. Shelf Res. 21: 207–224.

Carr M. H, T. W. Anderson, and M. A. Hixon. 2002. Biodiversity, population regulation, and the stability of coral-reef fish communities. Proc. Natl. Acad. Sci. USA 99: 11241–11245.

Costello, M. J., M. McCrea, A. Freiwald, T. Lundalv, L. Jonsson, B. J. Bett, T. van Weering, H. de Haas, J. M. Roberts, and D. Allen. 2005. Role of cold-water *Lophelia pertusa* coral reefs as fish habitat in the NE Atlantic. Pages 771–805 *in* A. Freiwald and J. M. Roberts, eds. Cold-water corals and ecosystems. Springer-Verlag, Berlin Heidelberg.

Ebeling, A. W. and M. A. Hixon. 1991. Tropical and temperate reef fishes: comparison of community structures. Pages 509–563 *in* P. F. Sale, ed. The ecology of fishes on coral reefs. Academic Press, Inc., San Diego.

Genin, A. 2004. Bio-physical coupling in the formation of zooplankton and fish aggregation over abrupt topographies. J. Mar. Syst. 50: 3–20.

Hixon, M. A. 1993. Predation, prey refuges, and the structure of coral reef fish assemblages. Ecol. Monogr. 63: 77–101.

Koenig, C. C., A. N. Shepard, J. K. Reed, F. C. Coleman, S. D. Brooke, J. Brusher, and K. M. Scanlon. 2005. Habitat and fish populations in the deep-sea *Oculina* coral ecosystem of the western Atlantic. Am. Fish. Soc. Symp. 41: 795–805.

Koslow, J. A., G. Boehlert, J. D. M. Gordon, R. L. Haedrich, P. Lorance, and N. Parin. 2000. Continental slope and deep-sea fisheries: implications for a fragile ecosystem. ICES J. Mar. Sci. 57: 548–557.

Lorance, P., F. Uiblein, and D. Latrouite. 2002. Habitat, behaviour and colour patterns of orange roughy *Hoplostethus atlanticus* (Pisces: Trachichthyidae) in the Bay of Biscay. J. Mar. Biol. Assoc. U.K. 82: 321–331.

Love, M. S., M. Yoklavich, and L. Thorsteinson. 2002. The Rockfishes of the Northeast Pacific. University of California Press, Berkeley. 404 p.

MacCall, A. D. 1990. Dynamic Geography of Marine Fish Populations. University of Washington Press, Seattle. 153 p.

Parrish, F. A. 2006. Precious corals and subphotic fish assemblages. Atoll Res. Bull. 543: 425–438.

Sainsbury, K. J., R. A. Campbell, R. Lindholm, and W. Whitelaw. 1997. Experimental management of an Australian multispecies fishery: examining the possibility of trawl-induced habitat modification. Am. Fish. Soc. Symp. 20: 107–112.

Stein, D. L., B. N. Tissot, M. A. Hixon, and W. Barss. 1992. Fish-habitat associations on a deep reef at the edge of the Oregon continental shelf. Fish. Bull. 90: 540–551.

Stone, R. P. 2006. Coral habitat in the Aleutian Islands: depth distribution, fine scale species associations, and fisheries interactions. Coral Reefs 25: 229–238.

Tissot, B. N., M. M. Yoklavich, M. S. Love, K. York, and M. Amend. 2006. Structure-forming invertebrates as components of benthic habitat on deep banks off southern California with special reference to deep sea corals. Fish. Bull. 104: 167–181.

Tupper, M. and R. G. Boutilier. 1995. Effects of habitat on settlement, growth, and post-settlement survival of Atlantic cod (*Gadus morhua*). Can. J. Fish. Aquat. Sci. 52: 1834–1841.

ADDRESS: *National Undersea Research Center and Department of Marine Sciences, University of Connecticut at Avery Point, Groton, Connecticut 06340. E-mail: <peter.auster@uconn.edu>.*

# Patterns of groundfish diversity and abundance in relation to deep-sea coral distributions in Newfoundland and Labrador waters

Evan N. Edinger, Vonda E. Wareham, and Richard L. Haedrich

**Abstract**

The degree of association between groundfish and corals in Newfoundland and Labrador waters was analyzed on spatial scales of hundreds of kilometers. Groundfish diversity and abundance of ten groundfish species and two invertebrate species were compared with deep-sea coral distributions using standardized stock assessment trawl surveys conducted between September 2003 and October 2005. Standardized trawl survey data were stratified by depth and by five coral classes defined by large gorgonians, small gorgonians, seapens and/or cup corals, soft corals, and absence of all corals. Groundfish species richness was highest in sets containing small gorgonians. Various fish species were most abundant in coral-defined classes in specific depth ranges, but no coral class had significantly higher fish abundances than other coral classes at all depths. For several species, numerical abundance was greatest in one coral class at shallow depth, but wet weight per tow was greatest in a different coral class or depth, or both. Coral-structured environments may be important to fish and crustaceans at different life history stages. Although relationships between corals and groundfish or invertebrates are not obligate and may result from coincidence, conservation areas established for corals may effectively protect populations of groundfish, including some commercial species.

Possible relationships between deep-water corals and a variety of groundfish, especially commercial and endangered species, have received considerable attention in the Eastern Atlantic (Husebø et al., 2002; Costello et al., 2005), the Florida and Gulf of Mexico region (Reed et al., 2005; Sulak et al., 2007) and in the Pacific (Kreiger and Wing, 2002), but relatively little in the Northwest Atlantic (Auster et al., 2005; Mortensen et al., 2005). Fish utilization of coral habitat can include juvenile fish habitat, feeding areas, resting areas, or refuges from predation (Auster, 2005; Costello et al., 2005). Alternatively, certain fish species may be more abundant in sites with abundant corals than in otherwise similar areas only by coincidence, i.e., habitat conditions that favor corals also favor abundance of certain fish species, but with no direct functional connection between the corals and the fish (cf. Auster et al., 2005).

In this contribution, we map coral abundance hotspots in Newfoundland and Labrador waters, and compare groundfish diversity and abundance among coral classes within an initial dataset of standardized trawl surveys. Our results are a preliminary analysis based on the first 2 yrs of data collected in the ongoing deep-sea coral research program in Newfoundland and Labrador.

George, R. Y. and S. D. Cairns, eds. 2007. Conservation and adaptive management of seamount and deep-sea coral ecosystems. Rosenstiel School of Marine and Atmospheric Science, University of Miami.

## Methods

**Coral distributions and definition of coral classes.**—Coral distributions were mapped based on two data sources (Wareham and Edinger, 2007): (1) 2003–2005 Department of Fisheries and Oceans (DFO) trawl survey by-catch, covering the area south of latitude 58°N (NAFO zone 2H and south); (2) 2005 shrimp industry scientific survey of northern Labrador and the Davis Strait, covering the area north of latitude 58°N (NAFO zones 2G and 0B), totalling 1614 sets. DFO trawl surveys and the Northern Shrimp Survey both followed a stratified random sampling design. Bycatch was examined for corals by trained technicians. All coral samples were identified by the authors. The DFO and shrimp industry trawl survey by-catch data indicate presence or absence of corals in samples (Wareham and Edinger, 2007). Mean depth and bottom temperature were recorded for each tow using a CTD mounted on the net.

Coral species were divided into four functional groups, and each set was assigned a coral class on the basis of the presence of the functional groups (Table 1). Group 4 included the large long-lived gorgonians: *Primnoa resedaeformis* (Gunnerus, 1763), *Paragorgia arborea* (Linnaeus, 1758), *Keratoisis ornata* (Verrill, 1878), *Acanthogorgia armata* (Verrill, 1878), *Paramuricea grandis* (Verrill, 1884) and/or *Paramuricea placomus* (Linnaeus, 1758), and antipatharians: *Bathypathes arctica* (Lütken, 1871; two growth forms, which may represent separate species). Group 3 included the small gorgonians *Acanella arbuscula* (Johnson, 1862) and *Radicipes gracilis* (Verrill, 1884). Group 2 included the sea pens *Distichophyllum gracile* (Verrill, 1882), *Umbellula lindahli* (Kölliker, 1875), *Funiculina quadrangularis* (Pallas, 1766), *Pennatula grandis* (Ehrenberg, 1834), *Pennatula* sp., *Halipteris finmarchia* (Sars, 1851), and five other species of sea pens as yet unidentified, and/or the cup corals: *Flabellum alabstrum* (Moseley, 1873), *Vaughanella margaritata* (Jourdan, 1895), and *Desmophyllum dianthus* (Esper, 1794) (Wareham and Edinger, 2007). Group 1 included the soft corals *Gersemia rubiformis* (Ehrenberg, 1834), *Duva (Capnella) florida* (Verrill, 1869), and *Anthomastus grandiflorus* (Verrill, 1878). Most large and small gorgonians occurred in multiple species assemblages, as did many of the sea pens and cup corals, while the nephtheid soft corals (*G. rubiforms*, and/or *D. florida*) often occurred as single records, particularly in shallow water. Coral class scores were assigned on a hierarchical scale of 0–4, in which 4 indicates presence of large gorgonians or antipatharians, usually accompanied by corals of

Table 1. Coral class definitions. Coral classes are a hierarchical nested series defined by occurrence of species indicated but often including species from classes of lower number.

| Coral class | Definition |
| --- | --- |
| 4 | Sets containing 1 or more large gorgonian or antipatharian corals *Primnoa resedaeformis, Paragorgia arborea, Keratoisis ornata, Acanthgorgia armata, Paramuricea placomus* or *Paramuricea grandis, Bathypathes arctica*, usually with some corals from classes 1 to 3. |
| 3 | Sets containing 1 or more small gorgonian corals, *Acanella arbuscula, Radicipes gracilis*, usually with soft corals, sea pens, or cup corals. |
| 2 | Sets containing 1 or more sea pens (*Distichophyllum gracile, Funiculina quadrangularis, Halipteris finmarchia, Pennatula grandis, Pennatula* sp., *Umbellula lindahli*, and five other species of sea pens as yet unidentified), sometimes with cup corals (*Flabellum alabastrum, Vaughanella margaritata*, or *Desmophyllum dianthus*, often with soft corals. |
| 1 | Sets with soft corals (*Gersemia rubiformis, Gersemia* sp., *Duva (Capnella) florida, Anthomastus grandiflorus*, but no other coral groups. |
| 0 | No corals |

other functional groups, 3 indicates presence of small gorgonians, often accompanied by sea pens, cup corals or soft corals, but not large gorgonians, 2 indicates sea pens or cup corals, and possibly soft corals, but no gorgonians, 1 indicates presence of soft corals, but no other functional groups, 0 indicates absence of corals (Table 1). Presence or absence of each coral species in each set, coral species richness, total wet weight of corals, wet weight of each class of coral, and coral class scores were determined for each set.

**Groundfish abundance data analysis.**—Abundance of groundfish was compared among depth-stratified sets defined by coral classes in the 2003–2005 DFO groundfish scientific surveys in Newfoundland and southern to central Labrador and the 2005 Northern Shrimp Survey in northern Labrador and the Davis Strait. All sets (n = 1614) were 15 min tows, covering a distance of 1.4 km, and conducted with a Campellen 1800 shrimp trawl equipped with rockhopper footgear. Groundfish were identified and quantified by DFO fisheries technicians using standard protocols, and trained in fish identification by DFO fisheries scientists. Abundance of groundfish was compared among the four coral classes: sets containing large gorgonian or antipatharian corals (n = 57), sets containing small gorgonians but not large gorgonians or antipatharians (n = 50), sets containing sea pens, and/or cup corals but not large or small gorgonian corals (n = 42), sets containing soft corals but no other functional groups (n = 270), and sets without corals (n = 1195).

Groundfish species chosen for analysis are characteristic of and often dominant on the slope (Koslow et al, 2000), and include commercial species, potential (Devine et al., 2006) or recognized (COSEWIC, 2005) species-at-risk, and sedentary species (O'Dea and Haedrich, 2002) known to be associated with benthic structures (Table 2). To account for variation in fish abundance with depth, sets were divided into five depth classes: 0–200 m, 200–400 m, 400–600 m, 600–1000 m, and 1000+ m (Table 3). The 400 m isobath is of particular relevance in Newfoundland as this depth defines the shelf break at the edge of the deep Northeast Newfoundland Shelf (Fig. 1).

Because fish catch data were not normally distributed, numbers and weights of each fish species were square-root transformed prior to analysis. Fish numerical abundance per tow and wet weight per tow were compared among depth classes and coral classes using two-way ANOVA of square-root transformed numerical abundances and wet weights, followed by Tamhane's post-hoc test assuming unequal variances. Because

Table 2. Groundfish and commercial invertebrate species analyzed in 2003–2005 DFO and Northern Shrimp Survey data.

| Common name | Scientific name |
| --- | --- |
| Atlantic Cod | *Gadus morhua* |
| Roughhead grenadier | *Macrourus berglax* |
| Rock grenadier | *Coryphaenoides rupestris* |
| Deepwater redfish | *Sebastes mentella* |
| Golden redfish | *Sebastes marinus* |
| Greenland halibut | *Reinhardtius hippoglossoides* |
| Wolffish | *Anarhichas* spp. |
| Lumpfish | *Cyclopterus lumpus* |
| Eelpouts | *Lycodes* spp. |
| Witch flounder | *Glyptocephalus cynoglossus* |
| Northern shrimp | *Pandalus borealis*, *Pandalus montagui* |
| Snow crab/Queen crab | *Chionecetes opilio* |

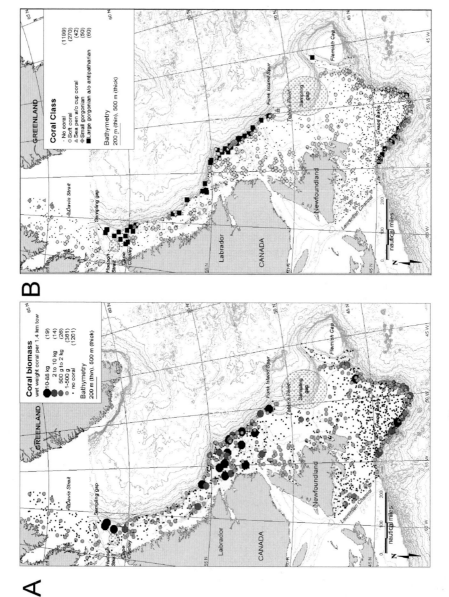

Figure 1. A. Maps showing distribution of coral biomass per tow (A) and distribution of coral classes (B) in 2003–2005 scientific survey data. Refer to Table 1 for definition of coral classes.

Table 3. Depth ranges and numbers of sets in coral classes for each depth interval in concurrent DFO-NSS data. $N_0$ = 0–200 m, $N_{200}$ = 200–400 m, $N_{400}$ = 400–600 m, $N_{600}$ = 600–1,000 m, $N_{1,000}$ = > 1,000 m. Data source: 2003–2005 DFO and Northern Shrimp Survey data.

| Coral class | Symbol | Definition | Depth range | $N_{tot}$ | $N_0$ | $N_{200}$ | $N_{400}$ | $N_{600}$ | $N_{1000}$ |
|---|---|---|---|---|---|---|---|---|---|
| 4 | LC | Large gorgonians or antipatharians, plus other spp. | 168–1,433 | 57 | 4 | 8 | 13 | 16 | 16 |
| 3 | AR | Small gorgonians, plus other species | 210–1,433 | 50 | 0 | 4 | 9 | 20 | 17 |
| 2 | SP | Sea pens and/or cup corals | 150–1,400 | 42 | 4 | 10 | 14 | 5 | 9 |
| 1 | SC | Soft corals | 48–1,433 | 270 | 119 | 93 | 36 | 11 | 11 |
| 0 | NC | No corals | 46–1,433 | 1,195 | 616 | 327 | 146 | 56 | 50 |

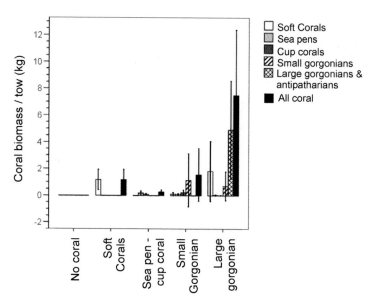

Figure 2. Average wet weight per tow among coral classes. Wet weights include both skeleton and tissue. Error bars: 95% confidence intervals.

per tow and wet weight per tow were compared among depth classes and coral classes using two-way ANOVA of square-root transformed numerical abundances and wet weights, followed by Tamhane's post-hoc test assuming unequal variances. Because square-root transformation did not normalize all data, and variance was not homogeneous, comparisons of fish abundance between coral classes were repeated using the non-parametric Kruskall-Wallis test on untransformed numerical abundance and weight data. Correlations between the abundance of each fish species and coral species richness, coral biomass, coral class, depth, and bottom temperature were estimated using Spearman rank correlation.

### Results

**Coral biomass hotspots.**—Peaks of coral wet weight per tow (Fig. 1A) were found along the southwestern edge of the Grand Banks, along the northeast edge of the Northeast Newfoundland Shelf, especially the tip of Funk Island Bank (51°N/50°30′W), along the continental shelf and slope of central and southern Labrador, near Cape Chidley, Labrador (60°N/61°W), and in the Davis Strait (62°N/61°W). Coral species richness and coral wet weight per tow were significantly correlated (Spearman's rho = 0.422, P < 0.001, n = 434). Wet weight of all corals was significantly higher in sets containing large gorgonians and antipatharians than in sets containing small gorgonians, which in turn had higher weights than sets containing only soft corals, sea pens, or cup corals (one-way ANOVA: $F_{(1,3)}$ = 10.95, P < 0.0001; Fig. 2).

There were two sampling gaps in the scientific surveys (Fig. 1). The larger sampling gap along the edge of the Northeast Newfoundland Shelf, roughly between the Funk Island Spur (51°N/50°30′W) and the Sackville Spur (48°N/47°W) was unintentional, resulting from mechanical failure of the research vessel. DFO sampling extended across the Flemish Pass and the westernmost portion of the Flemish Cap, but most of the Flemish Cap was sampled only by the Fisheries Observer Program (Wareham and Edinger,

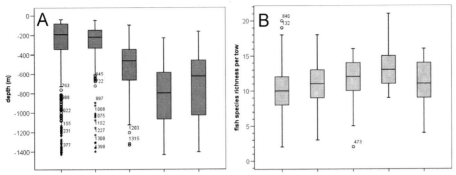

Figure 3. (A) Depth distributions of sets in four coral classes. Sets containing gorgonian corals (class 4, class 3) and sea pens (class 2) were significantly deeper than sets with soft corals only or sets without corals. Boxes: interquartile range; error bars: 95% confidence intervals. (B) Fish species richness and coral species richness per tow in sets with no coral, single soft coral, multiple soft coral, or habitat-forming corals. Coral classes: 0: No coral, 1: soft corals, 2: sea pens and cup corals, 3: small gorgonians, 4: large gorgonians and antipatharians. Boxes: interquartile range; error bars: 95% confidence intervals. Fish species richness was highest in sets defined by small gorgonians ($F_{(1,4)}$ = 16.70, P < 0.001).

**Depth distributions, coral diversity, and fish diversity.**—Sets characterized by sea pens and/or cup corals (class 2), small gorgonians (class 3), or large gorgonian or antipatharian corals (class 4) were deeper than sets containing only soft corals (class 1) or sets without corals (Fig. 3A). Coral species richness per tow was significantly higher in sets containing small gorgonians (mean 3.34) and sets containing large gorgonians (mean 3.22) than in sets containing soft corals (mean 1.09), or sea pens and cup corals (mean 1.38; one-way ANOVA: $F_{(1,3)}$ = 104.9, P < 0.0001). The site with the highest coral species richness (10 coral species per set) occurred on the southwestern Grand Banks and was dominated by large gorgonians, but had several species of pennatulaceans (Wareham and Edinger, 2007). Fish species richness was significantly higher in sets with corals than in sets without corals, and was highest in sets defined by small gorgonians (one-way ANOVA: $F_{(1,4)}$ = 16.70, P < 0.0001; Fig. 3B). Fish species richness and coral species richness were weakly, but significantly, correlated (Spearman rank correlation: rho = 0.189, P < 0.001, n = 434).

**Groundfish abundance variation with coral class and depth.**—Atlantic cod, *Gadus morhua* (Linnaeus, 1758), abundance did not vary significantly among coral classes, except at shallow (0–200 m) depths, where *G. morhua* were most abundant in non-coral sets, and in 200–400 m depth sets, where *G. morhua* numbers were greatest in soft coral sets (Fig. 4). *G. morhua* weights did not vary significantly among coral classes (Table 4). Numerical abundance and weight of rock grenadier, *Coryphaenoides rupestris* (Gunnerus, 1765) were greatest in waters deeper than 1000 m (Fig. 5, Table 4), where *C. rupestris* appeared to be most abundant in sets defined by sea pens and/or cup corals (Fig. 5), but Kruskall-Wallis analysis found that *C. rupestris* abundances were not significantly different among coral classes in tows deeper than 1000 m (Table 5). In intermediate depth waters (400–600 m), *C. rupestris* was most abundant in sets containing gorgonian corals (Fig. 5, Tables 4, 5), but in 600–1000 m, *C. rupestris* was less abundant in soft coral sets than in all other sets (Fig. 5, Table 5). Roughhead grenadier, *Macrourus berglax* (Lacépède, 1802) was most abundant in sets defined by large gorgonians or an-

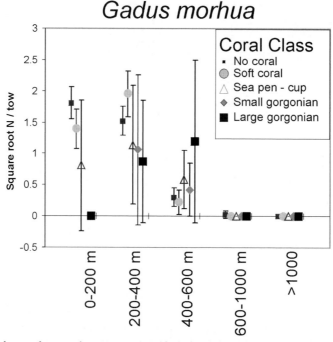

Figure 4. *Gadus morhua* number per tow, stratified by depth class and coral class in 2003–2005 scientific survey data. *Gadus morhua* in the 200–400 m depth range were significantly more abundant in soft coral sets than in all other coral classes, but in shallow water, cod were most abundant in non-coral sets. *Gadus morhua* numbers in the 400–600 m depth class and deeper did not vary significantly among coral classes. Error bars: 95% confidence intervals.

tipatharians at 200–400 m depths, but at 400–600 m and 600–1000 m depths they were most abundant in soft coral sets (Fig. 6, Tables 4, 5).

Deep-water redfish, *Sebastes mentella* (Travin, 1951), weight per tow was greatest in the small gorgonian coral class sets in the 200–400 m depth range, followed by sea pen sets in the same depth range (Fig. 7, Table 5). Differences in *S. mentella* weight among coral classes in the 400–600 m depth range followed a similar pattern (Fig. 7, Table 5). There were no significant patterns in the abundance of golden redfish, *Sebastes marinus* (Linnaeus, 1758) in relation to depth, coral class, or the interactions term between depth and coral class (Tables 4, 5).

In shallow waters (< 200 m), Greenland halibut, *Reinhardtius hippoglossoides* (Walbaum, 1792), was more abundant in sets with corals of various classes than in sets without corals. In the 200–400 m depth range, *R. hippoglossoides* was most abundant in large gorgonian sets or soft coral sets, followed by small gorgonian or non-coral sets, and were least abundant in sea pen sets (Fig. 8; Table 5). By contrast, in the 400–600 m depth range, *R. hippoglossoides* had greatest numbers in sets without corals, and abundance decreased with higher coral classes (Fig. 8, Table 5).

Wolffish, *Anarhichas* spp. , numbers and weights were highest in sets with small gorgonians in 200–400 m depth (Fig. 9, Table 5). A similar but weaker pattern was evident in the 400–600 m depth range (Table 5), but there was a weak trend toward higher numbers in soft coral and sea pen sets in the 600–1000 m depth range (Table 5). Lumpfish, *Cyclopterus lumpus* (Linnaeus, 1758), abundance displayed no significant patterns with

Table 4. Results of two-way ANOVA analysis comparing effects of depth and coral class on fish numerical abundance per tow (N) and wet weight per tow (Wt) in 2003–2005 scientific survey data. All data were square-root transformed, but typically displayed heterogeneity of variance. Relative differences determined by Tamhane's test assuming unequal variances. Significance values: *: $P < 0.05$, **: $P < 0.01$, ***: $P < 0.001$. Refer to Table 1 for definition of coral classes, and Table 3 for definition and numbers of depth classes.

| Fish species | | Depth class n = 1,614 | Coral class n = 1,614 | Interactions n = 1,614 |
|---|---|---|---|---|
| *Gadus morhua* | N | * | NS | NS |
| | Wt | NS | NS | NS |
| *Coryphaenoides rupestris* | N | *** | LG = AR > NC = SC = SP*** | *** |
| | Wt | *** | LG = AR > NC = SC = SP*** | *** |
| *Macrourus berglax* | N | *** | LG = AR > NC = SC = SP*** | *** |
| | Wt | *** | LG = AR > SC = SP > NC*** | *** |
| *Sebastes marinus* | N | NS | NS | NS |
| | Wt | NS | NS | NS |
| *Sebastes mentella* | N | *** | NS | NS |
| | Wt | *** | NS | *** |
| *Reinhardtius hippoglossoides* | N | *** | NS | ** |
| | Wt | *** | NS | *** |
| *Cyclopterus lumpus* | N | NS | NS | NS |
| | Wt | NS | NS | NS |
| *Anarhichas* spp. | N | *** | SC > SP ≥ LG = AR = NC** | *** |
| | Wt | 0.064 | NS | 0.051 |
| *Lycodes* spp. | N | 0.069 | SC > NC > SP = AR = LG* | NS |
| | Wt | NS | SC > NC > SP = AR = LG*** | NS |
| *Glyptocephalus cynoglossus* | N | *** | SP > AR = NC > SC = LG*** | NS |
| | Wt | *** | SP > AR = NC = SC = LG*** | * |
| *Pandalus borealis* and *Pandalus montagui* | N | NS | NS | ** |
| | Wt | *** | NS | * |
| *Chionecetes opilio* | N | NS | NS | * |
| | Wt | NS | SP ≥ NC = SC = LG ≥ AR* | ** |

respect to depth or coral class, except in the 200–400 m depth range where it was most abundant in soft coral sets (Table 5).

Witch flounder, *Glyptocephalus cynoglossus* (Linnaeus, 1758), which was most abundant along the margins of the Laurentian Channel, was most numerous in sets with sea pens at all depths (Fig. 10A, Table 5), but generally had greater wet weight per tow in shallow to intermediate-depth water (Fig. 10B). *Glyptocephalus cynoglossus* was generally least abundant in sets with large gorgonians (Fig. 10). Eelpouts, *Lycodes* spp. (Sabine, 1824), were most numerous and had highest weights per tow in soft coral sets at all depths except > 1000 m, although abundance was also high in sets with large skeletal corals at 200–400 m depth (Fig. 11, Table 5).

Shrimp, *Pandalus borealis* (Kroyer, 1838) and *Pandalus montagui* (Leach, 1814) numbers were highest in soft coral sets in water shallower than 200 m, with roughly equal numbers in the few shallow water sets containing large corals, followed by sets without corals (Fig. 12A, Tables 4, 5). By contrast, *P. borealis* and *P. montagui* weights were greatest in sets with soft corals at 200–400 m, and lowest in sea pen and small gorgonian sets in that depth range (Fig. 12B; Table 4). Snow crab, *Chionecetes opilio* (Fabricius, 1788), numbers followed a similar pattern (Fig. 13A), but *C. opilio* weights

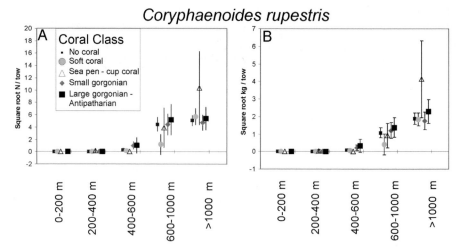

Figure 5. *Coryphaenoides rupestris* average number (A) and wet weight (B) per tow, square-root-transformed, stratified by depth class and coral class, in 2003–2005 scientific survey data. Error bars: 95% confidence intervals.

gorgonian sets in that depth range (Fig. 12B; Table 4). Snow crab, *Chionecetes opilio* (Fabricius, 1788), numbers followed a similar pattern (Fig. 13A), but *C. opilio* weights were greater in sets with sea pens in the 200–400 m depth range and highest in sets with large gorgonian corals in waters deeper than 400 m (Fig. 13B, Tables 4, 5).

Numbers and weights of *G. morhua*, *C. rupestris*, *M. berglax*, and *R. hippoglossoides* were significantly correlated with coral species diversity and total coral wet weight per tow (Table 6), as were weights, but not numbers, of *Anarhichas* spp., and *P. borealis* and *P. montagui*. The fish species that showed positive correlations with corals had stronger correlations with depth and/or bottom temperature than with variables relating directly to corals.

## Discussion

**Fish species richness and abundance in sets with and without corals.**—Nearly all fish species considered were more abundant in coral classes than in non-coral classes in at least one depth range, but only the witch flounder showed greatest abundances in the same coral class at all depth ranges. Many fish species were more abundant in soft coral sets, particularly in waters shallower than 400 m. Very few fish species were most abundant in large gorgonian/antipatharian sets. Greatest fish species richness was observed in small gorgonian sets. Taken together, these observations suggest that the fish and invertebrate species considered do not have strong or obligate relationships with large skeletal corals, but also suggest that the importance of soft corals, sea pens, and small gorgonians as potential fish and invertebrate habitat, especially for juveniles, should not be overlooked.

Groundfish abundance patterns among coral and depth classes reflect the depth distributions of the species considered. Grenadiers are slope-dwelling species, while cod, lumpfish, wolffish, and eelpout are shelf-dwelling species that occur in much smaller numbers at slope depths > 400 m (Scott and Scott, 1988). The apparent association between witch flounder and sea pens is intriguing, but may be purely coincidental: witch flounder were most abundant along the margins of the Laurentian Channel, a soft-sedi-

Table 5. Kruskall-Wallis test results comparing groundfish numerical abundance per tow (N) and wet weight per tow (Wt) among coral classes, stratified by depth range, in 2003–2005 scientific survey data. NS: Not Significant, *: P < 0.05, **: P < 0.01, ***P < 0.001. Coral classes: NC: No coral, SC: Soft coral, SP: sea pen, AR: small gorgonian (*Accanella, Radicipes*), LG: large gorgonian and antipatharian (see Table 1).

| Fish species | | 0–200 m | 200–400 m |
|---|---|---|---|
| | | n = 743 | n = 443 |
| *Gadus morhua* | N | NC > SC > SP = LG, P = 0.08 | SC > NC > SP = AR = LG* |
| | Wt | NC>SC > SP = LG* | NS |
| *Coryphaenoides rupestris* | N | NS | SP > NC = SC = AR = LG*** |
| | Wt | NS | SP > NC = SC = AR = LG*** |
| *Macrourus berglax* | N | NS | LG > SC = AR > NC = SP*** |
| | Wt | NS | LG > SC = AR > NC =SP*** |
| *Sebastes marinus* | N | NS | NS |
| | Wt | NS | NS |
| *Sebastes mentella* | N | SP > LG = NC > SC** | AR > SP > NC = SC= LG* |
| | Wt | SP > LG = NC > SC** | AR = SP > NC> SC = LG** |
| *Reinhardtius hippoglossoides* | N | LG > SP = SC > NC*** | LG > SC > AR = NC > SP* |
| | Wt | LG > SP = SC > NC*** | LG > SC > AR = NC > SP, P = 0.052 |
| *Cyclopterus lumpus* | N | NS | SC > NC > SP = AR = LG** |
| | Wt | NS | SC > NC > SP = AR = LG** |
| *Anarhichas* spp. | N | NS | AR > LG > SC > NC > SP** |
| | Wt | NS | AR > LG > SC > NC > SP** |
| *Lycodes* spp. | N | SC > NC = LG > SP* | LG > SC > NC > AR > SP** |
| | Wt | SC > NC = LG > SP* | LG = SC > NC > R > SP** |
| *Glyptocephalus cynoglossus* | N | SP > NC = LG > SC*** | SP > AR = NC > SC = LG*** |
| | Wt | SP > NC = LG > SC*** | SP > AR = NC > SC = LG*** |
| *Pandalus borealis* and *Pandalus montagui* | N | SC = LG > NC = SP**** | LG > NC = SP = AR > SC* |
| | Wt | SC = LG > NC = SP**** | SC > NC = LG > SP = AR*** |
| *Chionecetes opilio* | N | SC > NC > SP = LG*** | SP > SC = NC > AR = LG* |
| | Wt | SC > NC > SP = LG*** | SP > SC = NC > AR > LG*** |

Table 5. Continued.

| Fish species | 400–600 m<br>n = 219 | 600–1,000 m<br>n = 113 | > 1,000 m<br>n = 103 |
|---|---|---|---|
| *Gadus morhua* | NS<br>NS | NS<br>NS | NS<br>NS |
| *Coryphaenoides rupestris* | AR > LG > NC = SC > SP***<br>AR > LG > NC = SC > SP*** | NC = SP = AR = LG > SC*<br>LG = AR > NC = SP > SC* | NS |
| *Macrourus berglax* | SC > AR > LG = NC > SP***<br>SC > AR > LG = NC > SP*** | SC > LG = NC > AR > SP**<br>SC > LG = NC > AR > SP** | LG = SC > NC > SP > AR***<br>LG = SC > NC > SP > AR* |
| *Sebastes marinus* | NS<br>NS | NS<br>NS | NS<br>NS |
| *Sebastes mentella* | AR > SP > NC = SP = LG, P = 0.08<br>AR > SC = SP > LG = NC* | | NS |
| *Reinhardtius hippoglossoides* | NC = SC > AR > SP > LG**<br>NC = SC > AR = SP > LG** | NS<br>NS | NS<br>NS |
| *Cyclopterus lumpus* | NS<br>NS | NS<br>NS | SC > NC = SP = AR = LG, P = 0.079<br>SC > NC = SP = AR = LG, P = 0.079 |
| *Anarhichas* spp. | AR > SC > NC = SP = LG, P = 0.057<br>AR > SC > NC = SP = LG, P = 0.053 | SC = SP > NC > AR > LG, P = 0.059<br>SC = SP > NC > AR > LG, P = 0.073 | NS<br>NS |
| *Lycodes* spp. | SC > AR > NC = SP = LG, P = 0.071<br>SC > AR > NC = SP = LG* | SC > NC > AR = LG > SP*<br>SC > NC > AR = LG > SP* | NC > SC = SP = AR = LG**<br>NC > SC = SP = AR = LG** |
| *Glyptocephalus cynoglossus* | NS<br>NS | SP = SC = AR > NC = LG, P = 0.079 | SP = AR > NC = SC = LG*<br>SP = AR > NC = SC = LG** |
| *Pandalus borealis* and *Pandalus montagui* | NS<br>NS | NC > LG > SC = SP = AR*** | NC > SP = LG > SC = AR***<br>AR = SC > SP > NC > LG* |
| *Chionecetes opilio* | NS<br>NS | NS<br>NS | NC > SP = LG > SC = AR**<br>NC > LG > SC = SP = AR* |

## Macrourus berglax

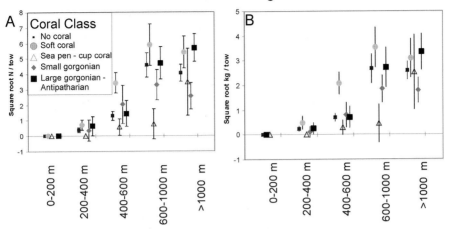

Figure 6. *Macrourus berglax* average number (A) and wet weight (B) per tow, square-root-transformed, stratified by depth class and coral class in 2003–2005 scientific survey data. Error bars: 95% confidence intervals.

## Sebastes mentella

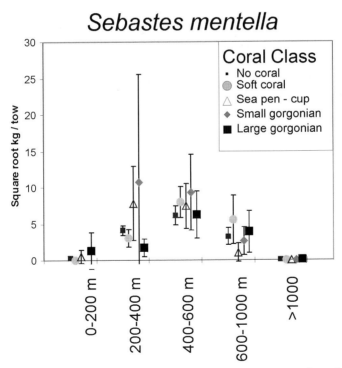

Figure 7. *Sebastes mentella* average wet weight per tow, square-root-transformed, stratified by depth class and coral class in 2003–2005 scientific survey data in 2003–2005 scientific survey data. Error bars: 95% confidence intervals.

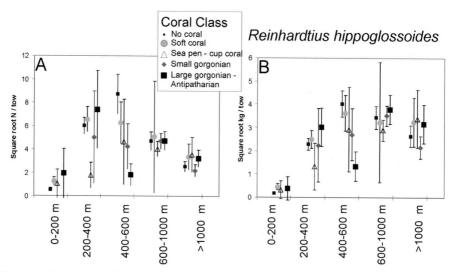

Figure 8. *Reinhardtius hippoglossoides* (A) number and (B) wet weight per tow, square-root transformed, stratified by depth class and coral class in 2003–2005 scientific survey data. Error bars: 95% confidence intervals.

ment dominated environment in which sea pens were the most common coral observed (Wareham and Edinger, 2007). The greater numbers of witch flounder in deeper waters and greater weights in shallower waters are consistent with the flounder's life history; this species recruits as juveniles in deep water and moves to shallower water after reaching maturity (Scott and Scott, 1988).

Our results highlight the potential importance of soft corals and sea pens as potentially habitat-forming organisms, particularly for juvenile fish and invertebrates such as shrimp and crab. Groundfish and commercial invertebrate abundances on the shallow (< 200 m) or deep (200–400 m) shelf were often greatest in soft coral sets. Although many of the soft corals collected in shallow water were attached to gravel, nephtheid soft corals are able to survive on sandy or even muddy bottoms, unlike the large gorgonian

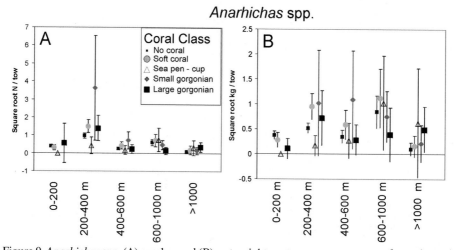

Figure 9. *Anarhichas* spp. (A) number and (B) wet weight per tow, square-root transformed, stratified by depth class and coral class in 2003–2005 scientific survey data. Error bars: 95% confidence intervals.

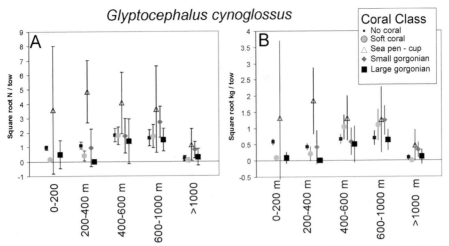

Figure 10. *Glyptocephalus cynoglossus* number (A) and wet weight (B) per tow in 2003–2005 scientific survey data, square-root transformed, stratified by depth class and coral class. Error bars: 95% confidence intervals.

corals that typically require exposed hard substrates. The contrasting patterns between numbers and weights per tow, as observed in wolffish, shrimp, and snow crab, imply that juveniles may occupy one depth range and coral habitat; and adults another. Both shrimp and crab recruit in shallow gravelly habitats, and then migrate to soft bottom habitats with maturity (Bergström, 2000). Thus corals, or physical habitats that support corals, may be more important as juvenile fish and invertebrate habitat than as habitat for adults, which may benefit from a size refuge from predation. Habitat segregation by age has been well-documented in juvenile Atlantic cod and other groundfish species (e.g., Gregory and Anderson, 1997). Distinguishing biological structuring of habitat from coincidental occurrence of corals and fish in the same environment will require in situ observations (e.g., Costello et al., 2005), but may still be ambiguous (cf. Auster et al., 2005).

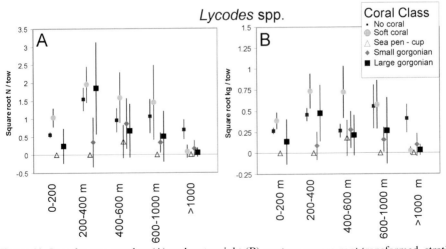

Figure 11. *Lycodes* spp. number (A) and wet weight (B) per tow, square-root transformed, stratified by depth class and coral class in 2003–2005 scientific survey data. Error bars: 95% confidence intervals.

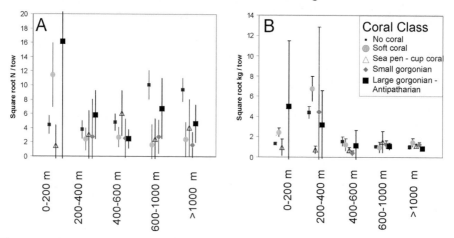

Figure 12. Shrimp (*Pandalus borealis* and *Pandalus montagui*) number (A) and wet weight (B) per tow, square-root transformed, stratified by depth class and coral class in 2003–2005 scientific survey data. Error bars: 95% confidence intervals.

be most strongly associated with deep-sea corals (Mortensen et al., 1995; Husebø et al., 2002; Costello et al., 2005; Mortensen et al., 2005). Previous coral-redfish associations have been described for *Lophelia pertusa* (Linnaeus) reefs (Husebø et al., 2002; Costello et al., 2005), large skeletal gorgonians near the Aleutian Islands (Kreiger and Wing, 2002; Stone, 2006) and the Scotian margin (Mortensen et al., 2005). Similarly, the lack of a clear association between corals and *R. hippoglossoides* weight in deep water was surprising, because coral by-catch rates in the *R. hippoglossoides* fishery far exceed those of most other fisheries, as recorded in the fisheries observer data (Wareham and Edinger, 2007, and unpubl. data). *Reinhardtius hippoglossoides* in the Northwest Atlantic are fished in waters as deep as 1800 m with trawls, gillnets, and longlines. We caution that our results are preliminary, and are based on a very limited dataset for coincident

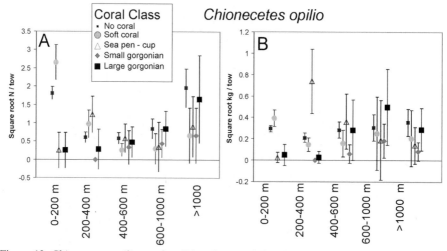

Figure 13. *Chionecetes opilio* number (A) and wet weight (B) per tow, square-root transformed, stratified by depth class and coral class in 2003–2005 scientific survey data. Error bars: 95% confidence intervals.

Table 6. Results of Spearman rank correlation analysis between groundfish abundance and total weight of all corals, coral species richness, coral class, depth, and bottom temperature in 2003–2005 Scientific Survey data. All depths combined. * P < 0.05; ** P < 0.01. *** P < 0.001. Refer to Table 1 for definition of coral classes.

| Fish species | | Coral biomass n = 1,621 | Coral species richness n = 1,621 | Coral class (0–4) n = 1,621 | Depth n = 1,621 | Bottom temperature n = 1,389 |
|---|---|---|---|---|---|---|
| Gadus morhua | N | –0.099(**) | –0.101(**) | –0.107(**) | –0.371(**) | –0.152(**) |
| | Wt | –0.109(**) | –0.111(**) | –0.115(**) | –0.356(**) | –0.125(**) |
| Coryphaenoides rupestris | N | 0.234(**) | 0.257(**) | 0.260(**) | 0.569(**) | 0.284(**) |
| | Wt | 0.237(**) | 0.260(**) | 0.262(**) | 0.569(**) | 0.284(**) |
| Macrourus berglax | N | 0.277(**) | 0.285(**) | 0.284(**) | 0.683(**) | 0.372(**) |
| | Wt | 0.278(**) | 0.286(**) | 0.284(**) | 0.682(**) | 0.366(**) |
| Sebastes marinus | N | 0.00 | 0.01 | 0.01 | 0.092(**) | 0.083(**) |
| | Wt | 0.00 | 0.01 | 0.008 | 0.092(**) | 0.083(**) |
| Sebastes mentella | N | 0.02 | 0.00 | 0.013 | 0.444(**) | 0.570(**) |
| | Wt | 0.04 | 0.02 | 0.033 | 0.476(**) | 0.599(**) |
| Reinhardtius hippoglossoides | N | 0.187(**) | 0.187(**) | 0.188(*) | 0.661(**) | 0.346(**) |
| | Wt | 0.224(**) | 0.228(**) | 0.231(*) | 0.750(**) | 0.420(**) |
| Cyclopterus lumpus | N | –0.01 | –0.01 | –0.015 | –0.110(**) | –0.214(**) |
| | Wt | –0.01 | –0.01 | –0.015 | –0.114(**) | –0.215(**) |
| Anarhichas spp. | N | 0.05 | 0.05 | 0.037 | 0.076(**) | 0.137(**) |
| | Wt | 0.062(*) | 0.058(*) | 0.050(*) | 0.090(**) | 0.155(**) |
| Lycodes spp. | N | 0.04 | 0.03 | 0.024 | 0.085(**) | –0.082(**) |
| | Wt | 0.03 | 0.02 | 0.012 | 0.071(**) | –0.099(**) |
| Glyptocephalus cynoglossus | N | 0.00 | –0.01 | 0.003 | 0.151(**) | 0.402(**) |
| | Wt | 0.00 | –0.01 | 0.003 | 0.129(**) | 0.375(**) |
| Pandalus borealis and Pandalus montagui | N | 0.03 | 0.01 | 0.029 | 0.266(**) | 0.055(*) |
| | Wt | 0.120(**) | 0.117(**) | –0.111(**) | 0.146(**) | –0.147(**) |
| Chionecetes opilio | N | –0.04 | –0.04 | –0.045 | –0.218(**) | –0.229(**) |
| | Wt | –0.04 | –0.04 | –0.043 | –0.129(**) | –0.071(**) |

coral and fish records, with significant sampling gaps in areas likely to contain both corals and important groundfish concentrations.

**Do corals matter to fish?**—The lack of strong associations between large skeletal corals and most groundfish species observed in our data differs from the results of studies concerning scleractinian deep-sea reefs in the Northeast Atlantic (Husebø et al., 2002; Costello et al., 2005), and the Gulf of Mexico (Sulak et al., 2007; S. L. Harter and A. N. Shepard, unpubl. data), and gorgonian coral habitat in Alaska (Kreiger and Wing, 2002; Stone, 2006). Scleractinian deep reefs, in which the corals build a geological structure, may be much more likely to host endemic or specialized fish fauna than gorgonian coral habitats, in which the geological features of the habitat are formed from bedrock or cobble-boulder fields, and not by living organisms (cf. Auster et al., 2005; Tissot et al., 2006; Auster, 2007).

The contrast between our results and previous studies may also stem from the scale difference between trawl survey data and direct observations using gillnet fishing in reef and nearby off-reef settings (< 200 m scale, Husebø et al., 2002) or video observations

The contrast between our results and previous studies may also stem from the scale difference between trawl survey data and direct observations using gillnet fishing in reef and nearby off-reef settings (< 200 m scale, Husebø et al., 2002) or video observations (Auster et al., 2005; Costello et al., 2005; Mortensen et al., 2005). In the trawl surveys used here, each point represents the aggregate organism occurrence within a trawl path 1.4 km long and 17 m wide, further reduced according to the catchability of the various species (cf. Schneider et al., 1987). The different results and difference in scale highlight the importance of in situ observations and collections, especially for studying juvenile fish, crustaceans, and other fauna potentially associated with corals (cf. Buhl-Mortensen and Mortensen, 2005). Other important patterns, such as a quantification of coral bio-mass per unit area, can only be achieved through in situ observations.

Although there were no dramatic relationships between corals and abundance of the 10 groundfish species studied, there was a weak but statistically significant positive cor-relation between coral species richness and fish species richness, suggesting that habitats that support diverse corals are also likely to support diverse assemblages of fishes. The correlation between coral species richness and fish species richness may be coincidence, may reflect greater abiotic structural variation in coral-rich areas, or may reflect diffuse effects of corals on fish habitat and fish communities similar to those observed on shal-low water reefs (cf. Sale et al., 1994). For example, most tropical reef fish do not have an obligate relationship with individual coral species, but depend on the reef environment (e.g., Sale, 2004). By increasing the spatial and hydrodynamic complexity of habitats, deep-sea corals may provide important, but not obligate, habitat for a wide variety of fishes. Diffuse effects of deep-sea corals on fish habitat and communities could include higher prey abundance, greater water turbulence, and resting places for a wide variety of fish size classes (Auster et al., 2005; Costello et al., 2005). Similar patterns have been well-documented in shallow water tropical reef fish communities using in-situ studies, but have yet to be studied in detail for deep-sea corals.

Similarly, coral-rich habitats support a wide diversity of invertebrates, especially epibionts on exposed corals skeletons (Jensen and Frederiksen, 1992; Buhl-Mortensen and Mortensen, 2005; Wareham and Edinger, 2007). In this study, the calcareous gorgo-nian *K. ornata* had a particularly diverse suite of epibionts. The biodiversity of continen-tal slope habitats around the world are poorly known in general, but appear to support a wide diversity of fishes, including rare and/or endemic species (e.g., Williams et al., 2001; Last and Yearsley, 2002). Continental slope environments in the northwest Atlan-tic appear to have lower diversity and lower endemism than similar depth environments elsewhere (e.g., Williams et al., 2001), and in particular, lower diversity and endemism than seamount environments (e.g., Koslow et al., 2001).

***Fishing history and current coral biomass per tow.***—The average biomass and fre-quency of coral bycatch in all coral classes reported in this study were two to three orders of magnitude lower than those reported from previously untrawled areas in the Gulf of Alaska (Kreiger, 2001). We caution that coral biomass per tow cannot be directly compared with data from other regions using different fishing gear, and measurements of coral biomass per unit seafloor area will require in situ observations.

The relatively low average coral biomass per tow observed in this study may result from habitat limitations on coral density, low coral catchability, or past damage to cor-al populations from several decades of intensive bottom trawling (Watling and Norse, 1998; Kulka and Pitcher, 2002; Anderson and Clarke, 2003; Stone, 2006). Of these, trawling damage may be the most likely explanation. The highest wet-weight-per-tow values were recorded on the margins of the northern sampling gap in the Davis Strait. This gap was a deliberate omission due to net damage from rough bottoms, and this area will likely remain unsampled by the Northern Shrimp Survey (Wareham and Edinger, 2007). Previously untrawled sites in this area sampled in August 2006 recovered up to 500 kg of *P. resedaeformis*, *P. arborea,* and other corals in a single tow (Wareham and Edinger, unpubl. data). The Davis Strait sampling gap enjoys no formal protection, is close to areas heavily fished by other trawl fisheries, and is likely to suffer damage from trawling activity, particularly that directed at *P. borealis* or at *R. hippoglossoides* spawning aggregations (Scott and Scott, 1988).

Fisheries observer reports suggest that the largest corals being recovered currently are far smaller than the largest corals caught 10 yrs ago aboard commercial trawlers in the Davis Strait (R. Beazley, SeaWatch, pers. comm.). These observations are consistent with exponential decline in size and wet weight of corals recovered in trawl fisheries through time, as documented on seamounts affected by the orange roughy, *Hoplostethus atlanticus* (Collett, 1889), fishery (Anderson and Clarke, 2003). The general lack of strong patterns of association between corals and fish may also result from a reduced abundance of corals caused by trawling damage (cf. Koslow et al., 2000, 2001). Although coral distributions are unlikely to have changed due to trawling damage, coral abundance, and particularly the abundance of large corals, is likely to have been diminished (cf. Fosså et al., 2001; Gass and Willison, 2005; Mortensen et al., 2005). If large, long-lived corals have the greatest importance in structuring habitat, and have been greatly reduced in frequency due to fisheries impacts, patterns of association between fishes and corals are likely to have been obscured.

**Corals and deep-sea conservation strategies.**—Corals may provide potential habitat for a variety of marine fish species in the Northwest Atlantic, although the relationships between fish and coral are not obligate, and may be coincidental. Species for which coral habitat may be important include endangered species and commercially important species, many of which have experienced dramatic declines in recent years. *Anarhichas* spp., already protected under Canada's Species at Risk Act (SARA), have experienced a dramatic range contraction. These once widespread fishes now have their greatest (albeit reduced) abundance along the shelf edge (O'Dea and Haedrich, 2002), and, according to our results, in habitats that also support small gorgonian corals. Both *C. rupestris* and *M. berglax* meet IUCN criteria for listing as endangered (Devine et al., 2006). Our results suggest that soft coral habitats may be important to juvenile *P. borealis*, *P. montagui*, and *C. opilio*. Corals provide important hard-substrate habitat for many encrusting invertebrates and mobile coral commensals, (e.g., Buhl-Mortensen and Mortensen, 2004, 2005), thus coral-rich environments constitute local biodiversity peaks in the deep-sea (Roberts et al., 2006).

Establishment of large no-take zones (NTZ) in locations including coral diversity hotspots may conserve populations of those fish species effectively, whether or not any direct biological interaction between the corals and the fish species exists. Indeed, the broad patterns we observed in corals (see also Wareham and Edinger, 2007) are similar in many respects to the fish species richness hotspot patterns that are found in demersal fishes across this same large region (Baker, 2003). Thus while coral conservation efforts should focus on corals for their own sake, and for their general importance to biodiversity (Auster, 2005), doing so in the Newfoundland region could also bring benefits for conservation of fishes. Since definitive ocean science often lags behind reality (Haedrich et al., 2001), lack of present knowledge should not be used as an excuse to delay action on conservation.

## Acknowledgments

We thank J. Firth, D. Pittman, and P. Vitch (DFO) for swiftly and efficiently transferring large fish catch and position datasets from the DFO trawl survey and Northern Shrimp Survey programs. Special thanks to the crew and scientific staff of CCGS TELEOST and CCGS WILFRID TEMPLEMAN, and to fisheries observers, especially D. Benson and R. Beazley. R. Devillers gave extensive assistance with cartography, and K. Larade (WWF-Canada) provided bathymetric data. K. Gilkinson, J. Rendell, J. Simms, P. Snelgrove, and two anonymous reviewers provided feedback on earlier drafts. Supported by Fisheries and Oceans Canada, NSERC Discovery grants to E. E. and R. L. H., and Memorial University.

## Literature Cited

Anderson, O. F. and M. R. Clark, 2003. Analysis of bycatch in the fishery for orange roughy, *Hoplostethus atlanticus*, on the South Tasman Rise. Mar. Freshwat. Res. 54: 643–652.

Auster, P. J. 2005. Are deep-water corals important habitats for fishes? Pages 747–760 *in* A. Freiwald and J. M. Roberts, eds. Cold-water corals and ecosystems. Springer-Verlag, Heidelberg.

_____. 2007. Linking deep-water corals and fish populations. Pages 93–99 *in* R. Y. George and S. D. Cairns, eds. Conservation and adaptive management of seamount and deep-sea coral ecosystems. Rosenstiel School of Marine and Atmospheric Science, University of Miami. Miami. 324 p.

_____, J. Moore, K. B. Heinonen, and L. Watling. 2005. A habitat classification scheme for seamount landscapes: assessing the functional role of deep-water corals as fish habitat. Pages 761–769 *in* A. Freiwald and J. M. Roberts, eds. Cold-water corals and ecosystems. Springer-Verlag, Heidelberg.

Baker, K. D. 2003. Potential sites for marine protected areas in Canada's North Atlantic. MS Thesis. Memorial University of Newfoundland. 73 p.

Bergström, I. 2000. The biology of *Pandalus*. Adv. Mar. Biol. 38: 55–244.

Buhl-Mortensen, L. and P. B. Mortensen. 2004. Crustaceans associated with the deep-water gorgonian corals *Paragorgia arborea* (L., 1758) and *Primnoa resedaeformis* (Gunnerus, 1763). J. Nat. Hist. 38: 1233–1248.

_____ and _____. 2005. Distribution and diversity of species associated with deep-sea gorgonian corals off Atlantic Canada. Pages 849–879 *in* A. Freiwald and J. M. Roberts, eds. Cold-water corals and ecosystems. Springer-Verlag, Heidelberg.

COSEWIC. 2005. Canadian species-at-risk, August 2005. Committee on the Status of Endangered Wildlife in Canada, Environment Canada. Website: http://www.cosewic.gc.ca

Costello, M. J., M. McCrea, A. Freiwald, T. Lundalv, L. Jonsson, B. J. Bet, T. C. E. van Weering, H. de Haas, J. M. Roberts, and D. Allen. 2005. Role of cold-water *Lophelia pertusa* coral reefs as fish habitat in the NE Atlantic. Pages 771–805 *in* A. Freiwald and J. M. Roberts, eds. Cold-water corals and ecosystems. Springer-Verlag, Heidelberg.

Devine, J. A., K. D. Baker, and R. L. Haedrich. 2006. Deep-sea fishes qualify as endangered. Nature 429: 29.

Fosså, J. H., P. B. Mortensen, and D. M. Furevik. 2001. The deep-water coral *Lophelia pertusa* in Norwegian waters: distribution and fishery impacts. Hydrobiologia 471: 1–12.

Gass, S. E. and J. H. M. Willison. 2005. An assessment of the distribution of deep-sea corals in Atlantic Canada by using both scientific and local forms of knowledge. Pages 223–245 *in* A. Freiwald and J. M. Roberts, eds. Cold-water corals and ecosystems. Springer-Verlag, Heidelberg.

Husebø, Å., L. Nøttestad, J. H. Fosså, D. M. Furevik, and S. B. Jørgensen. 2002. Distribution and abundance of fish in deep-sea coral habitats. Hydrobiologia 471: 91–99.

Jensen, A. and R. Frederiksen. 1992. The fauna associated with the bank-forming deepwater coral *Lophelia pertusa* (Scleractinia) on the Faroe shelf. Sarsia 77: 53–69.

Koslow, J. A., G. W. Boehlert, J. D. M. Gordon, R. L. Haedrich, P. Lorance, and N. Parin. 2000. Continental slope and deep-sea fisheries: implications for a fragile ecosystem. ICES J. Mar. Sci. 57: 548–557.

_____, K. Gowlett-Holmes, J. K. Lowry, T. O'Hara, G. C. B. Poore, and A. Williams. 2001. Seamount benthic macrofauna off southern Tasmania: community structure and impacts of trawling. Mar. Ecol. Prog. Ser. 213: 111–125.

Kreiger, K. J. 2001. Coral (*Primnoa*) impacted by fishing gear in the Gulf of Alaska. Pages 106–116 *in* J. H. M. Willison, J. Hall, S. E. Gass, E. L. R. Kenchington, M. Butler, and P. Doherty, eds. Proc. First Int. Symp. on Deep-Sea Corals. Ecology Action Centre and Nova Scotia Museum, Halifax, Nova Scotia.

_____ and B. L. Wing, 2002. Megafauna associations with deepwater corals (*Primnoa* spp.) in the Gulf of Alaska. Hydrobiologia 471: 83–90.

Kulka, D. W. and D. A. Pitcher. 2002. Spatial and temporal patterns in trawling activity in the Canadian Atlantic and Pacific. ICES CM 2001/R:02. 57 p.

Last, P. R. and G. K. Yearsley. 2002. Zoogeography and relationships of Australasian skates (Chondrichthyes: Rajidae). J. Biogeogr. 29: 1627–1641.

Mortensen, P. B., M. Hovland, T. Brattegard, and R. Farestveit. 1995. Deep-water bioherms of the scleractinian coral *Lophelia pertusa* (L.) at 64° N on the Norwegian shelf: structure and associated megafauna. Sarsia 80: 145–158.

_____, L. Buhl-Mortensen, D. C. Gordon, Jr., G. B. J. Fader, D. L. McKeown, and D. G. Fenton. 2005. Effects of fisheries on deepwater gorgonian corals in the Northeast Channel, Nova Scotia. Am. Fish. Soc. Symp. 41: 369–382.

O'Dea, N. R. and R. L. Haedrich. 2002. A review of the status of the Atlantic Wolffish, *Anarhichas lupus*, in Canada. Can. Field-Nat. 116: 423–432.

Reed, J. K., A. N. Shepard, C. C. Koenig, K. M. Scanlon, and R. G. Gilmore, Jr. 2005. Mapping, habitat characterization, and fish surveys of the deep-water *Oculina* coral reef Marine Protected Area: a review of historical and current research. Pages 443–465 *in* A. Freiwald and J. M. Roberts, eds. Cold-water corals and ecosystems. Springer-Verlag, Heidelberg.

Roberts, J. M., A. J. Wheeler, and A. Freiwald. 2006. Reefs of the deep: the biology and geology of cold-water coral ecosystems. Science 312: 543–547.

Sale, P. F. 2004. Connectivity, recruitment variation, and the structure of reef fish communities. Integr. Comp. Biol. 44: 390–399.

_____, G. E. Forrester, and P. S. Levin. 1994. The fishes of coral reefs: ecology and management. Res. Explor. 10: 224–235.

Schneider, D. C., J-M. Gagnon, and K. D. Gilkinson. 1987. Patchiness of epibenthic megafauna on the outer Grand Banks of Newfoundland. Mar. Ecol. Prog. Ser. 39: 1–13.

Scott, W. B. and M. G. Scott. 1988. Atlantic fishes of Canada. Can. Bull. Fish. Aquat. Sci. 219: 731 p.

Stone, R. P. 2006. Coral habitat in the Aleutian Islands of Alaska: depth distribution, fine-scale species associations, and fisheries interactions. Coral Reefs 25: 229–238.

Sulak, K. J., R. A. Brooks, K. E. Luke, A. D. Norem, M. Randall, A. J. Quaid, G. E. Yeargin, J. M. Miller, W. M. Harden, J. H. Caruso, and S. W. Ross. 2007. Demersal fishes associated with *Lophelia pertusa* coral and hard-substrate biotopes on the continental slope, northern Gulf of Mexico. Pages 65–92 *in* R. Y. George and S. D. Cairns, eds. Conservation and adaptive management of seamount and deep-sea coral ecosystems. Rosenstiel School of Marine and Atmospheric Science, University of Miami. Miami. 324 p.

Tissot, B. N., M. M. Yoklavich, M. S. Love, K. York, and M. Amend. 2006. Benthic invertebrates that form habitat on deep banks off southern California, with special reference to deep sea coral. Fish. Bull. 104: 167–181.

management of seamount and deep-sea coral ecosystems. Rosenstiel School of Marine and Atmospheric Science, University of Miami. Miami. 324 p.

Tissot, B. N., M. M. Yoklavich, M. S. Love, K. York, and M. Amend. 2006. Benthic invertebrates that form habitat on deep banks off southern California, with special reference to deep sea coral. Fish. Bull. 104: 167–181.

Wareham, V. E. and E. N. Edinger. 2007. Distributions of deep-sea corals in the Newfoundland and Labrador region, Northwest Atlantic Ocean. Pages 289–313 *in* R. Y. George and S. D. Cairns, eds. Conservation and adaptive management of seamount and deep-sea coral ecosystems. Rosenstiel School of Marine and Atmospheric Science, University of Miami. Miami. 324 p.

Watling, L. and E. A. Norse. 1998. Disturbance of the seabed by mobile fishing gear: A comparison to forest clearcutting. Conserv. Biol. 12: 1180–1197.

Williams, A., J. A. Koslow, and P. R. Last. 2001. Diversity, density and community structure of the demersal fish fauna of the continental slope off western Australia (20 to 35°S). Mar. Ecol. Prog. Ser. 212: 247–263.

ADDRESSES: (E.E.) *Geography Dept. and Biology Dept., Memorial University, St. John's, Newfoundland and Labrador, Canada.* (V.E.W.) *Environmental Science Program, Memorial University, St. John's, Newfoundland and Labrador, Canada; and Fisheries and Oceans Canada, St. John's, Newfoundland and Labrador, Canada.* (R.L.H) *Biology Dept., Memorial University, St. John's, Newfoundland and Labrador, Canada.* CORRESPONDING AUTHOR: (E.E.) *E-mail:* <eedinger@mun.ca>.

# Designing management measures to protect cold-water corals off Nova Scotia, Canada

HEATHER BREEZE AND DEREK G. FENTON

## Abstract

In 2002, Department of Fisheries and Oceans (DFO) implemented its first fisheries closure to protect cold-water corals. The Northeast Channel Coral Conservation Area, southwest of Nova Scotia, was put in place to protect concentrations of gorgonian corals. Since then, DFO has established the Gully Marine Protected Area, which includes cold-water coral habitats, and implemented another fisheries closure (the Lophelia Coral Conservation Area) to protect a small, damaged *Lophelia pertusa* (Linnaeus, 1758) reef complex. The design criteria and the management measures used in each area were different and the activities that are permitted vary. These differences reflect the circumstances particular to each area, as well as evolving knowledge of protecting coral habitats. The lessons learned in establishing the marine protected area and coral conservation areas have been applied to the development of a coral conservation plan for the region. Experience to date suggests that protecting cold-water coral areas may require a variety of approaches, even within a single jurisdiction.

In the waters off Nova Scotia, a long peninsula on the east coast of North America (Fig. 1), Canada's Department of Fisheries and Oceans (DFO) is the government agency primarily responsible for managing fisheries and conserving ocean biodiversity. In the last 5 yrs, DFO established two fisheries closures to protect cold-water corals in this region. In addition to the fisheries closures, a marine protected area was established to protect a diversity of species and habitats, including cold-water corals. The purpose of this paper is to provide a description of these management measures and highlight key considerations in their design, with the aim of assisting others involved in planning and designing cold-water coral conservation measures.

Although corals were documented off Nova Scotia's coasts in the nineteenth century, most conservation and research efforts began in the late 1990s. Off Nova Scotia, these species are largely found in deep channels, submarine canyons, and steep slope areas (Breeze et al., 1997; Gass and Willison, 2005; Mortensen et al., 2006). Twenty-three species of cold-water coral have been confirmed in the region (Mortensen et al., 2006). Freiwald et al. (2004) and Butler (2005) provided a historical perspective on coral conservation in the region, including a brief overview of the management measures in place at the time of publication. Conservation efforts have focused on large gorgonian corals and on the stony coral *Lophelia pertusa* (Linnaeus, 1758).

DFO conducted opportunistic coral research in this area starting in 1997, with photographic and video observations in areas reported to have high coral densities and examination of bycatch records from DFO groundfish surveys. A dedicated coral research program took place in 2001–2003, with the goal of collecting information on the distribution, abundance, biology, and condition of corals in order to make informed manage-

George, R. Y. and S. D. Cairns, eds. 2007. Conservation and adaptive management of seamount and deep-sea coral ecosystems. Rosenstiel School of Marine and Atmospheric Science, University of Miami.

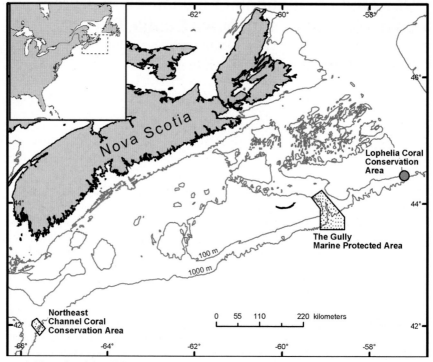

Figure 1. Nova Scotia's offshore and locations of areas with restrictions to protect corals. See Figure 4 for a detailed map of the *Lophelia* Coral Conservation Area.

ment decisions (Mortensen et al., 2006). The research findings led to the development of management measures to protect corals.

Freiwald et al. (2004) documented known and potential threats to cold-water corals worldwide; of those threats, the activities that occur off Nova Scotia include bottom fishing, petroleum exploration and development, disposal of materials at sea, placement of cables and pipelines, and scientific research (DFO, 2005c). Off Nova Scotia, fishing is the most widespread of these activities and overlaps with known cold-water coral distributions. Increasing levels of carbon dioxide in the atmosphere may also impact cold-water corals and other calcifying marine animals (Orr et al., 2005) and is a concern for management of these species.

## Coral conservation in Canada

Canada has no national policy or program specifically directed at the conservation of cold-water corals. However, several federal policy documents guide Canada's current and future actions to protect cold-water corals. For example, the 1986 "Policy for the Management of Fish Habitat" supports "taking direct action to establish sanctuaries for the preservation of living marine resources and associated habitats, consistent with fisheries management objectives [...]" (DFO, 1986). Canada signed and ratified the Convention on Biological Diversity in 1992, committing itself to the conservation and sustainable use of biological diversity (Convention, 1992). Most recently (2005), Canada's Oceans Action Plan was released in which the government of Canada committed to protecting important, productive and biologically diverse areas, and vulnerable species (Canada, 2005).

Globally, several other countries have implemented measures to protect cold-water corals, including the United Kingdom, Norway, and the United States. Freiwald et al. (2004) summarized international actions taken to date. These have mainly been activity-specific closures, such as the fisheries closures put in place off the United Kingdom (Darwin Mounds), Norway (several *L. pertusa* reefs), and Florida (*Oculina* Bank Habitat Area of Particular Concern).

In Canada, the Fisheries Act, Oceans Act, and National Marine Conservation Areas Act can be used to create area-based restrictions in marine areas (Butler, 2005). To date, two of those laws have been used to protect corals: the Fisheries Act, which can establish fisheries closures, and the Oceans Act, which can restrict or put limitations on many types of activities through the designation of a marine protected area (MPA). In Canada, MPA specifically refers to an area designated by regulation under the Oceans Act.

DFO has six administrative regions. Measures to protect marine areas and species, including cold-water corals, have been developed and implemented on a regional basis. The waters off Nova Scotia are within the Maritimes Region of DFO and to date, this is the only region that has implemented measures to protect corals. Cold-water corals are found in other parts of Canada, such as off Newfoundland and Labrador (Wareham and Edinger, 2007) and on Canada's Pacific Coast (Jamieson et al., 2006). A cold-water coral and sponge conservation plan is under development in the Pacific Region (K. Conley, Fisheries and Oceans Canada, pers. comm.), and a research program is in progress in the Newfoundland and Labrador Region (J. Simms, Fisheries and Oceans Canada, pers. comm.).

**Conservation approaches and areas closed to protect corals off Nova Scotia.**—In the Maritimes Region, DFO and other government regulators have taken three general approaches to coral conservation: integration with existing regulatory process, inclusion in regional management planning, and activity restrictions in specific areas. Coral conservation has been integrated into existing regulatory processes, such as federal environmental assessments of proposed oil and gas exploration activities. Corals have also been included in broader regional management planning mandated under Canada's Oceans Act, such as the recent development of a draft integrated management plan for the eastern part of the Scotian Shelf (DFO, 2005a).

There are currently three areas with management measures to protect corals, one marine protected area and two fisheries closures (Fig. 1): (1) The Gully Marine Protected Area, (2) Northeast Channel Coral Conservation Area, and (3) *Lophelia* Coral Conservation Area. These areas were set up using different legislation and for different purposes, as discussed below.

**The Gully Marine Protected Area.**—The Gully Marine Protected Area was implemented under the Oceans Act to protect a diversity of habitats and species associated with a large submarine canyon, including cold-water corals. In total, 19 taxa of corals (Alcyonacea, Gorgonacea, and Scleractinia) have been documented in the general area of the canyon (Breeze and Davis, 1998; MacIsaac et al., 2001; Mortensen and Buhl-Mortensen, 2005). The 2364-km$^2$ MPA is divided into three management zones (Fig. 2), with differing levels of activity allowed in each zone. The distribution of coral species was one of the considerations in designing the zones. Most coral species are located in

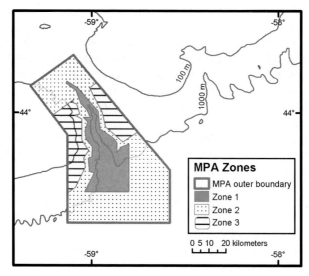

Figure 2. The Gully Marine Protected Area (2364 km²) off Nova Scotia showing the location of its three zones. Zone 1 has the highest level of protection while some fishing activities are permitted in Zones 2 and 3. Other activities may also be permitted in Zone 3.

zones 1 and 2 of the MPA. Zone 1 has the greatest restrictions on activities that directly impact the ocean bottom while longline fisheries are permitted in Zone 2 (Canada Gazette, 2004).

Interim protection for the canyon was announced in 1998, with final designation in 2004 as the first MPA on Canada's east coast (DFO, 2004a). In keeping with the National Framework for Establishing and Managing Marine Protected Areas (DFO, 1999), various assessments of the canyon and the potential impact of the MPA on human activities were conducted prior to designation (see Canada Gazette, 2004 for a summary of these assessments). Historically, several activities occurred directly in the area or nearby, such as fishing, petroleum exploration and development, shipping, and scientific research. An advisory group provided advice on the design of the MPA and continues to provide guidance on its management. Its members include representatives of government agencies, fishing organizations, the oil and gas industry, non-government organizations, and researchers.

General regulations for the MPA prohibit activities that disturb, damage, destroy or remove any living marine organism, its habitat or the seabed, as well as activities that deposit, discharge or dump any substance (Canada Gazette, 2004). Certain activities are identified in the regulations as exceptions to these prohibitions, e.g., longline fisheries. Video and photographs of the ocean bottom have shown few signs of fisheries impacts on corals in the area and researchers have speculated that the rugged topography of the Gully provides protection from fishing gear (Mortensen et al., 2006). Other activities, such as research activities, may be approved by submitting a plan to DFO for review.

**Northeast Channel Coral Conservation Area.**—The Northeast Channel area was initially identified as an important area for corals based on reports from fishers (Breeze et al., 1997; Willison et al., 2002). Conservation interest began in the late 1990s and there was a proposal from an environmental non-government organization (ENGO) for a protected area (Willison et al., 2002). Visual surveys by scientists in 2000 and 2001 confirmed high abundances of the gorgonian corals *Primnoa resedaeformis* (Gunnerus, 1763) and *Paragorgia arborea* (Linnaeus, 1758) (Mortensen and Buhl-Mortensen, 2004), as well as damage from fishing activities (Mortensen et al., 2005). Following the surveys, DFO focused management actions on fishing activity because the channel was a highly-used fishing area and few other activities occurred or were planned for the area. With this in mind, it was decided that a fisheries closure using the Fisheries Act would be used rather than another option such as a marine protected area under the Oceans Act. DFO representatives met with different fishing organizations, and in addition, a DFO-industry working group was formed specifically to advise on the design of a closure for this area. Fishing activity in the area was examined in detail to identify potential socio-economic impacts on users of the area.

Early in the process, several boundary proposals were submitted by fishing organizations and an NGO (Fig. 3A). From these proposals and through examining the scientific survey and fishing activity data, a closure boundary was developed that captured the highest concentrations of intact corals observed in the channel. It was agreed to extend the boundary to depths below 500 m, where there were no scientific data at that time, as a precautionary measure. (Recent research has shown that significant coral populations do exist below 500 m, with some species found only in the deeper water [E. Kenchington, Fisheries and Oceans Canada, pers. comm.]). In the spring of 2002, this boundary proposal was presented at a community meeting. Concerns were raised about the initial boundary, particularly from longline fishers who actively fished in a portion of the area.

Following additional meetings with individual fishers and the working group, the final proposal put forth a zoning scheme. In the largest part of the area, the "restricted fisheries zone" (Fig. 3B), fishing with bottom-contact gear would not be permitted. In a small part of the area, the "limited fisheries zone," fishing for groundfish using longline or handline gear would be permitted if the vessel carried an at-sea observer. Monitoring of this zone would be carried out to better understand the effects of the fishery. Under Canada's Fisheries Act, variation orders can be issued that vary where and when specific fleets can fish. The 424-km$^2$ closure was implemented using a variation order and added to fishing license conditions. In 2003, there was a minor adjustment to the boundaries of the limited fisheries zone to simplify monitoring and management.

***Lophelia* Coral Conservation Area.**—In 2003, DFO researchers found a damaged *L. pertusa* reef complex at the Stone Fence off eastern Nova Scotia (Fig. 1) (Mortensen et al., 2006). The general location of the reef had long been suspected from fishers' reports (Collins, 1884; Breeze et al., 1997; Gass and Willison, 2005). The reef complex was very small (about 1.5 km$^2$) and the main activity in the area was groundfish fishing. As this was the only known location with living *L. pertusa* off Canada's east coast, DFO sought to protect the reef and allow for recovery. This objective was made clear with fishing associations from the outset, with the aim of developing a closure for all bottom impacting fisheries, beginning with the 2004 fishing season. Lessons learned from the Northeast Channel and Gully experiences were applied to the development of the *Lophelia* Coral Conservation Area and several new issues also emerged.

The DFO-fishing industry working group was brought back together, with the addition of fishers who used that particular area. Inclusion of non-fishing interests in the group was considered at this phase but was not welcomed by the fishing industry. A separate meeting was held with other interested parties, where participants included academ-

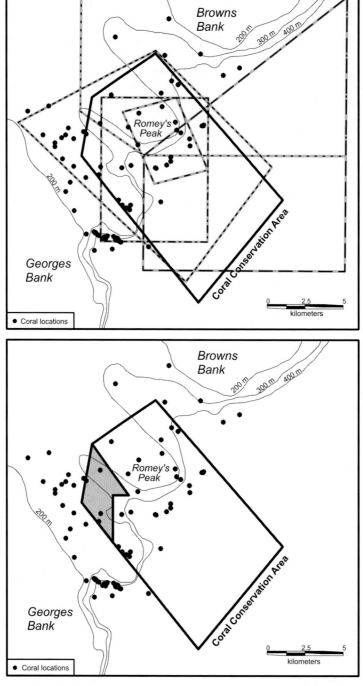

Figure 3. (A) The Northeast Channel area off Nova Scotia, showing the boundary proposals considered by DFO and the fishing industry working group (dashed and dotted lines) and the boundary proposal that went forward to community meetings with fishermen (solid line). (B) The final zoning scheme for the Northeast Channel Coral Conservation area (424 km$^2$). The limited fisheries zone is shaded.

ics and representatives from government departments outside DFO and environmental groups.

Fishing activity in a 150-km$^2$ area surrounding the reef complex—an area many times the size of the reef—was examined in detail to identify users and potential socio-economic impacts. Much of the information was obtained from the commercial landings and observer databases maintained by DFO. The two main fisheries that used bottom gear were a halibut fishery carried out by longline and a redfish fishery that used otter trawl.

Many different considerations went into the type of management measure chosen for this area and the design of the closure. Voluntary measures were initially considered but were deemed insufficient for protecting this unique area. A closure within an overall plan for a fishing fleet (known as a conservation harvesting plan) was considered. Breaking the rules of these plans may result in fleet-wide implications and not just penalties for the individual who breaks the rules. This was the less preferred option for most of the fishers consulted. As a result, a closure under Fisheries Act regulations was the management measure selected.

Discussions with DFO enforcement staff on the effectiveness of fisheries closures resulted in several design considerations. Support from fishers was considered essential to the success of the closure. The area has few fisheries, is far offshore, and thus has less surveillance activity than other parts of Nova Scotia's offshore. The use of prominent coordinates was recommended to prevent confusion among both fishers and enforcement officers.

A buffer around this small reef complex was needed both for enforcement purposes and to account for navigational accuracy and gear movement. This is a high current area and thus had some potential for gear to unintentionally drift from set positions. In developing the boundaries, various buffers were drawn around the reef complex at set distances. As in the Northeast Channel, several boundaries were considered (Fig. 4). During discussions of the initial boundary proposals, halibut fishers pointed out that access to the main halibut fishing ground was severely restricted, as they would have to "buffer the buffer" to make sure their gear did not enter the area. Zoning was considered to address the needs of both groundfish fleets, but would require a larger overall closure in order to make the separate zone boundaries enforceable.

One boundary proposal included the addition of a large area to the south of the reef complex as a precautionary measure, as was done in the Northeast Channel. This southern area had not been fully explored by scientists. Those involved in the research felt that this entire reef complex was mapped (D. Gordon, Fisheries and Oceans Canada, pers. comm.). They concluded that this reef complex did not extend to the area of the proposed precautionary boundary. However, it is possible that other, separate reefs may occur to the south of the known reef complex.

The final boundary configuration was more acceptable to fishers who used the area. It did not include the deeper waters and the design provided greater access to the halibut fishing ground while still allowing a buffer around the reef. At 15 km$^2$, it was 10 times the size of the reef complex. The closure was implemented in June 2004 through a variation order and was later added to license conditions under the Fisheries Act.

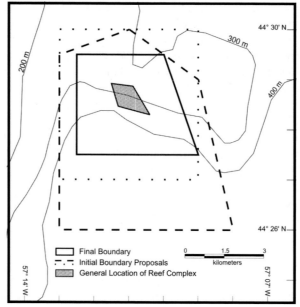

Figure 4. Initial boundary considerations for *Lophelia* Coral Conservation Area, off Nova Scotia. The final closure (15 km²) is indicated by the solid line and other proposed boundaries by the dashed and dotted lines.

## Post-establishment activities

Once established, fisheries closures and marine protected areas require ongoing management activities to ensure protection measures are fully implemented. Since these areas were put in place, there have been regular overflights and ongoing vessel-based fisheries patrols. Data from fisheries logbooks, at-sea observers, vessel monitoring systems, and surveillance overflights are regularly analyzed to determine the type and level of activity within and around the conservation areas. In 2004, DFO successfully prosecuted a longline fisher who did not have a fisheries observer onboard while fishing in the limited fisheries zone of the Northeast Channel (DFO, 2004b).

Following the establishment of the Northeast Channel Conservation Area in 2002, there was a DFO groundfish research trawl survey in the area which received widespread and negative attention (see e.g., CBC News, 2002). This led to the development of a research protocol for DFO trawl surveys to avoid placing trawl survey stations within the Conservation Areas. To prevent impacts from activities such as oil and gas exploration and submarine cable placement, DFO has made arrangements to protect these areas with relevant government agencies.

Further research in the Northeast Channel has been conducted. A visual benthic survey in the region in June 2005 and 2006 included stations within and near the conservation area with the goal of evaluating the design of closure. Data from previous research carried out in the Gully is being re-examined with the goal of better understanding the benthic species found there (V. Kostylev, Natural Resources Canada, pers. comm.).

The establishment of the conservation measures occurred without a clear overall vision for coral conservation and management activities in the region. Many stakeholders felt that DFO should develop an overall vision for coral conservation and place these efforts in the context of other management activities. Following a multi-stakeholder workshop to identify issues and priority activities, a Coral Conservation Plan for the Maritimes Region (DFO, 2006) was developed as part of integrated management efforts. It describes where more work is needed and prioritizes activities by DFO and others for the next 5 yrs, including a list of research needs. The plan also describes existing management actions and provides an improved process for evaluating additional areas that may require conservation action.

## Lessons learned in planning and design

Based on our experiences, we conclude with what was learned and provide several suggestions for other managers to consider in designing conservation measures related to corals.

**Match the management tool to the situation.**—We suggest that it is important to match the appropriate tool to the objective for the area and existing and potential threats. In Canada, marine protected area designation is a long process that may result in a comprehensive approach to conservation. On the other hand, single activity closures can be implemented relatively quickly to deal with immediate concerns but may not address all potential threats. Other jurisdictions considering area-based closures should evaluate whether activity-specific closures or broader, MPA-style closures will better meet management objectives. Further, the use of zoning schemes in either MPAs or fisheries closures may address socio-economic concerns and result in greater community support while still meeting conservation objectives.

**Include the actual users of an area.**—While industry representatives and working groups can provide much information, it is necessary to identify and meet with individual fishers who use the area to gain a full picture of potential economic costs and operational impacts. Commercial fishing data are often not at a scale for detailed socio-economic analysis of particular areas and fishing gear may be set or used differently in different areas. Those who use a particular area are the best equipped to know how a closure will affect them. In addition, including the actual users fosters greater support among the people who will be most impacted by changes in management.

**Feedback from non-resource users.**—Conservation organizations, researchers, and many members of the general public have an interest in deep-sea corals and other marine conservation issues. It is important for these groups to have an opportunity to provide feedback on proposed management measures prior to their establishment.

**Collaborate on boundary design.**—In our experience it is unlikely that a preliminary boundary proposal, even one developed after much consultation, will be acceptable to all parties. Boundaries may gain greater acceptance if they are developed jointly and transparently with options for adjustments.

**The path may be bumpy.**—The best intentions and processes may run into roadblocks. Some people may feel that they are being unfairly targeted by conservation activities and will not wish to provide input on further management measures that will affect them. In addition, fishing and other groups are expected to participate in many government advisory processes and may not have the capacity to take part in them all. A dedicated effort is needed to obtain information from all parties and to fully understand different points of view.

**Corals may be a useful starting point for benthic conservation.**—In Nova Scotia, the implementation of coral conservation measures has been a useful avenue for opening discussion on broader conservation issues. For example, since these measures have been implemented there has been interest in developing appropriate research protocols for sensitive areas in general. As well, a benthic classification scheme that will help in understanding the sensitivity of all marine benthic communities has been developed (DFO, 2005b).

**Need for an overall conservation plan.**—The development of an overall coral conservation plan is an opportunity to set out clear objectives and build support for coral conservation among a diverse group of organizations. By clearly stating objectives and priorities for action, management and research activities can be focused where they are most needed.

## Acknowledgments

The views expressed in this article are those of the authors and not of Fisheries and Oceans Canada. Many people contributed to developing and implementing the coral conservation measures that are currently in place in the Maritimes Region of DFO. The authors would like to thank D. Gordon, J. Hansen, P. Mortensen, the fishers and industry representatives who have participated in the Scotia-Fundy Fishing Industry Roundtable Corals Working Group, and the reviewers of this manuscript.

## Literature Cited

Breeze, H. and D. S. Davis. 1998. Deep sea corals. Pages 113–120 *in* W. G. Harrison and D. G. Fenton, eds. The Gully: a scientific review of its environment and ecosystem. Can. Stock Asses. Sec. Res. Doc. 98/83.

_____, D. S. Davis, M. Butler, and V. Kostylev. 1997. Distribution and status of deep sea corals off Nova Scotia. Mar. Issues Comm. Spec. Publ. 1. Ecology Action Centre, Halifax, NS. 58 p.

Butler, M. 2005. Conserving corals in Atlantic Canada: a historical perspective. Pages 1199–1209 *in* A. Freiwald and J. M. Roberts, eds. Cold-water corals and ecosystems. Springer, Berlin.

Canada. 2005. Canada's Oceans Action Plan. Fisheries and Oceans Canada, Ottawa. 20 p.

Canada Gazette. 2004. The Gully Marine Protected Area Regulations and regulatory impact analysis statement. Can. Gaz. Part II. [serial online]; 138 (10). Available from: http://gazettedu-canada.gc.ca/partII/2004/20040519/html/sor112-e.html.

CBC News. 2002. Rare coral destroyed. With Katherine Morris. CCAN-TV. September 4, 2002.

Collins, J. W. 1884. On the occurrence of corals on the Grand Banks. Bull. U.S. Fish Comm. 4: 237.

Convention on Biological Diversity. 1992. Secretariat of the Convention on Biological Diversity, Text of the Convention. Available from: http://www.biodiv.org/convention/convention.shtml Accessed 21 August 2005

DFO. 1986. Policy for the management of fish habitat. Fisheries and Oceans Canada, Ottawa. 28 p.

_____. 1999. National framework for establishing and managing marine protected areas – a working document. Fisheries and Oceans Canada, Ottawa. 21 p.

_____. 2004a. The Gully Marine Protected Area [Internet]. DFO Backgrounder. Available from: http://www.dfo-mpo.gc.ca/media/backgrou/2004/hq-ac61a_e.htm Accessed 14 May 2004

_____. 2004b. Fisher pleads guilty for fishing in Coral Conservation Area. DFO Maritimes Region News Release. Available from: http://www.mar.dfo-mpo.gc.ca/communications/maritimes/news04e/NR-MAR-04-02E.html Accessed 14 May 2004

_____. 2005a. Eastern Scotian Shelf integrated ocean management plan (2006–2011): draft for discussion. Oceans and Coastal Management Report, 2005-02. Maritimes Region, Fisheries and Oceans Canada, Dartmouth. 73 p.

_____. 2005b. Framework for classification and characterization of Scotia-Fundy benthic habitats. Can. Sci. Advis. Sec. Sci. Advis. Rep. 2005/071. 14 p.

_____. 2005c. The Scotian Shelf: an atlas of human activities. Maritimes Region, Fisheries and Oceans Canada, Dartmouth. 113 p.

_____. 2006. Coral Conservation Plan. Maritimes Region (2006–2010). Oceans and Coastal Management Report, 2005-02. Maritimes Region, Fisheries and Oceans Canada, Dartmouth. 59 p.

Freiwald, A., J. H. Fosså, A. Grehan, T. Koslow, and J. M. Roberts. 2004. Cold-water coral reefs. United Nations Environment Programme-World Conservation Monitoring Centre, Cambridge. 84 p.

Gass, S. E. and J. H. M. Willison. 2005. An assessment of the distribution of deep sea corals in Atlantic Canada by using both scientific and local forms of knowledge. Pages 223–245 *in* A. Freiwald and J. M. Roberts, eds. Cold-water corals and ecosystems. Springer, Berlin.

Jamieson, G. S., N. Pellegrin, and S. Jessen. 2006. Taxonomy and zoogeography of cold water corals in explored areas of coastal British Columbia. Can. Sci. Advis. Sec. Res. Doc. 2006/062. 45 p.

Mortensen, P. B. and L. Buhl-Mortensen. 2005. Deep-water corals and their habitats in The Gully, a submarine canyon off Atlantic Canada. Pages 247–277 *in* A. Freiwald and J. M. Roberts, eds. Cold-water corals and ecosystems. Springer, Berlin.

_____, _____, D. C. Gordon, Jr., G. B. J. Fader, D. L. McKeown, and D. G. Fenton. 2005. Effects of fisheries on deep-water gorgonian corals in the Northeast Channel, Nova Scotia. Pages 369–382 *in* P. W. Barnes and J. P. Thomas, eds. Benthic habitats and the effects of fishing. Am. Fish. Soc. Symp. 41. American Fisheries Society, Bethesda, Maryland.

_____, _____, S. E. Gass, D. C. Gordon, Jr., E. L. R. Kenchington, C. Bourbonnais, and K. G. MacIsaac. 2006. Deep-water corals in Atlantic Canada: a summary of ESRF-funded research (2001–2003). Environmental Studies Research Funds Report 143. ESRF, Calgary. 83 p.

Orr, J. C., V. J. Fabry, O. Aumont, L. Bopp, S. C. Doney, R. A. Feely, A. Gnanadesikan, N. Gruber, A. Ishida, F. Joos, et al. 2005. Anthropogenic ocean acidification over the twenty-first century and its impact on calcifying organisms. Nature 437: 681–686.

Wareham, V and E. Edinger. 2007. Distribution of deep-sea coral in the Newfoundland and Labrador region, Northwest Atlantic Ocean. Pages 289–313 *in* R. Y. George and S. D. Cairns, eds. Conservation and adaptive management of seamount and deep-sea coral ecosystems. Rosenstiel School of Marine and Atmospheric Science, University of Miami. Miami. 324 p.

Willison, J. H. M., D. P. Jones, and S. Atwood. 2002. Deep sea corals and marine protected areas in Nova Scotia. Pages 1157–1163 *in* S. Bondrup-Neilsen, N. W. P. Munro, G. Nelson, J. H. M. Willison, T. B. Herman, and P. Eagles, eds. Managing protected areas in a changing world. SAMPAA, Wolfville.

ADDRESSES: (H.B., D.G.F.) *Oceans and Coastal Management Division, Oceans and Habitat Branch, Maritimes Region, Fisheries and Oceans Canada, P.O. Box 1006, Dartmouth, NS, B2Y 4A2, Canada.* CORRESPONDING AUTHOR: (H.B.) *E-mail <breezeh@mar.dfo-mpo.gc.ca>.*

# Deep-sea coral distribution on seamounts, oceanic islands, and continental slopes in the Northeast Atlantic

Jason Hall-Spencer, Alex Rogers,
Jaime Davies, and Andy Foggo

## Abstract

A database of deep-water (> 200 m) antipatharians, scleractinians, and gorgonians has been assembled for the NE Atlantic to determine what their distribution and diversity was before coral habitats became heavily impacted by bottom fishing gear. Benthic sampling expeditions from 1868–1985 have provided 2547 records showing the deep-water distribution of 22 species of antipatharians, 68 species of scleractinians, and 83 species of gorgonians with the majority of records found from seamounts, oceanic islands, and the continental slope of the warm temperate region. Too little is known about the coral biota of boreal and tropical seamounts to assess their levels of endemism, but on seamounts in the warm temperate region of the NE Atlantic the level endemism in antipatharian, scleractinian and gorgonian corals is low (< 3%). Many of the species found on seamounts are characteristic of oceanic islands in this region and the oceanic islands have a significantly different coral fauna to that recorded at the same depths on the continental slope. Given the key role that corals can play in structuring deep-sea habitats it is hoped that our database will help inform the development of a network of marine protected areas to provide long-term protection for the differing communities found on continental slopes and isolated offshore habitats.

A surge of international research into seamount ecology (Malakoff, 2003; Pitcher et al., 2007) was recently boosted by work on South Pacific seamounts reporting that > 30% of the species found were new to science and potentially endemic to seamounts (Parin et al., 1997; Richer de Forges et al., 2000). Surveys to date show that deep-water coral communities often characterize seamount habitats and that these communities have been heavily impacted by trawling (Koslow et al., 2001; Rogers et al., 2007). The North Atlantic has the longest and most intensive history of biogeographic research, yet even there most seamounts remain unsampled (Stocks et al., 2004) and coral reefs several km in extent have only recently been discovered (Fosså et al., 2005). At the same time as these amazing discoveries are being made, benthic surveys are revealing the increasing extent to which bottom-trawling is altering the habitats of deep-sea corals worldwide (Hall-Spencer et al., 2002; Clark and O'Driscol, 2003). This is of particular concern considering that deep-water corals are amongst the longest lived and slowest growing organisms on Earth (Roark et al., 2006). In most parts of the world offshore expansion of the bottom-trawling industry began before any habitat assessment had taken place (Gordon, 2003). Fortunately, there is a wealth of historical deep-water coral data for the NE Atlantic providing detailed records dating back to the expeditions of the vessels LIGHTENING, PORCUPINE, and JOSEPHINE (1868–1869). Regular sampling expedi-

George, R. Y. and S. D. Cairns, eds. 2007. Conservation and adaptive management of seamount and deep-sea coral ecosystems. Rosenstiel School of Marine and Atmospheric Science, University of Miami.

tions such as those funded by Prince Albert I of Monaco from 1886–1915 provide a rich source of information on the coral fauna (e.g., Studer, 1901; Thomson, 1927).

Deep-water bottom-trawling was pioneered in the North Atlantic by the Russians in the 1970s targeting *Coryphaenoides rupestris* Gunnerus, 1765 (roundnose grenadier) south of Iceland and on the Reykjanes Ridge, but it was not until the late 1980s that deep-water bottom-trawling began to increase rapidly along the European continental slope with the development of new markets for *Molva dypterygia* (Pennant, 1784) (blue ling), *Aphanopus carbo* Lowe, 1839 (black scabbardfish), and *Hoplostethus atlanticus* Collett, 1889 (orange roughy) (Gordon, 2001, 2003). Deep-water trawling is now common, with trawl tracks clearly visible on acoustic images and seabed photographs of the European continental slope (Hall-Spencer et al., 2002). Damage continues partly because coral-rich areas are poorly mapped, so trawlers do not know the key sites to avoid, and partly because only very small areas are closed to bottom-trawling (Butler, 2005; Wheeler et al., 2005).

We aim to build on the approach adopted by Cairns and Chapman (2001) and Watling and Auster (2005), drawing attention to valuable baseline information on the diversity and distribution of deep-water scleractinian, gorgonian and antipatharian corals before bottom-trawling became widespread. We have collated a "pre-1985" database of deep-water corals of the NE Atlantic and use it to compare coral faunas on seamounts, continental slopes, and oceanic islands of the warm temperate region.

## Methods

We restricted our analyses to scleractinian, antipatharian, and gorgonian corals because these are the most commonly recorded coral groups in the NE Atlantic, providing the richest source of biogeographic data for our comparisons of seamount, continental slope, and oceanic island habitats. The following publications were used as sources of historical data on the distributions of NE Atlantic scleractinian, antiptharian, and gorgonian corals: Grasshoff (1972, 1973, 1977, 1981a,b,c, 1985a,b, 1986, 1989), Arnaud and Zibrowius (1973), Zibrowius (1973, 1980, 1985), Grasshoff and Zibrowius (1983), Keller (1985), Pasternak (1985), Tendal (1992), Tyler and Zibrowius (1992), Rogers (1999), and Molodtsova (2006), in addition to the original cruise reports of the Prince of Monaco expeditions 1886–1915 held at the United Kingdom's National Marine Biological Library in Plymouth. We updated historical synonyms following the high-level taxonomic nomenclature used for octocorals by Williams and Cairns (http://www.calacademy.org/research/izg/OCTOCLASS.htm, accessed September 2006) and the species-level nomenclature given by the UNESCO-IOC register of marine organisms (http://annual.sp2000.org/2006/search.php, accessed September 2006) together with Molodtsova (2006) for the most recently described antipatharians. Our database has been made available through Seamounts Online (http://seamounts.sdsc.edu), an open-access portal for deep-sea data.

Coral data (taxon, position, depth, date collected, cruise details, and other notes) were entered into a Microsoft Access database—some scleractinians were recorded as dead at the time of collection and these were excluded from further analyses because they could have been fossil occurrences of species no longer living at the sites in question. Those records from > 200 m depth which provided a latitude and longitude position were plotted using ArcView GIS to show the distribution of historical scleractinian, antipatharian, and gorgonian records on a General Bathymetric Chart of the Oceans (GEBCO) showing known seamounts, continental slope areas, and oceanic islands in an area from 12°N–62.5°N and 10°E–40°W. We classified seamounts as topographic rises with limited extent across the summit which are > 1000 m elevation (derived from the December 2005 GEBCO seamount list: http://www.ngdc.noaa.gov/mgg/gebco/gazet_dec2005.xls for the area north of 35.9568°N and OASIS (2003) for seamounts south of 35.9568°N). Coral records from 200–2000 m depth around islands which rise from abyssal depths (the Azores, Madeira, and the Cape Verde Islands) were classified as oceanic island data whereas records from 200–2000 m depth on the shelf and along the edge of the European continent (e.g., the Canaries and Rockall) were classified as continental slope data. Some records could not be included in the GIS database because they did not provide latitude and longitude, but we were able to use these data in analyses of the faunas of seamounts, continental slopes, and oceanic islands when they provided enough information to be able to attribute those records to a particular province (e.g., "240 m depth between Pico and Sao Jorge in the Azores" was attributed to the Azores).

The statistical package PRIMER-E (Clarke and Warwick, 2001a) was used to analyse differences in coral communities among seamounts, continental slope, and oceanic islands in the warm temperate NE Atlantic north of Mauritania and south of Cape Finisterre, as this region proved to be the most data-rich for deep-water corals. A distance matrix was produced of between-site similarities based on presence/absence data for all species at all sites, employing the Bray-Curtis similarity coefficient and a dendrogram of site similarities was produced using group-averaged clustering in PRIMER-E. The distance matrix was used to produce an ordination plot of between-site similarities using non-metric multidimensional scaling (n-MDS). Reliability of the n-MDS plot was assessed by calculating Kruskal's stress value (Kruskal, 1964), a measure of goodness-of-fit of the plotted sample distances in the n-MDS and the calculated distances in the parent matrix. Stress values $\leq 0.05$ give an excellent representation of the actual distances between points with no prospect of misinterpretation (Clarke and Warwick, 2001a). To further define the biodiversity at the different sites we calculated values of average taxonomic distinctness, $\Delta^+$ ("taxonomic breadth" sensu Clarke and Warwick, 2001a) and variation in taxonomic distinctness $\Lambda^+$ (Clarke and Warwick, 2001b) for the coral assemblages at each site. $\Delta^+$ and $\Lambda^+$ are calculated using the overall species list for all sites analyzed (see Clarke and Warwick, 2001a). $\Delta^+$ of a species list is a measure of the average taxonomic distance apart of all its pairs of species, whereas $\Lambda^+$ is a measure of variance of the taxonomic distances between pairs of species at a site around the site $\Delta^+$ value. We then tested whether these values deviated from the expected values for a coral assemblage of known species richness drawn from the same regional fauna, using the TAXDTEST routine in PRIMER-E.

## Results

A database of 2547 pre-1985 records of deep-water corals (> 200 m) was assembled for the NE Atlantic area shown in Figure 1, which included 22 species of antipatharians, 83 species of gorgonians, and 68 species of scleractinians. Most records of corals collected prior to 1985 came from the campaigns of a few research vessels such as the R/Vs METEOR, JEAN CHARCOT, and THALLASSA (Table 1). Latitude and longitude data were available for 1808 of the deep-water coral records and are shown on the bathymetric map of the area in Figure 1. This map shows that most of historical coral records are clustered along the continental slope and around the oceanic islands of the Azores, Madeira, and the Cape Verde Islands. Well-sampled seamount chains to the south of the Azores and to the northeast of Madeira were also hotspots of historical coral records.

Because of a lack of sampling, very little is known about the pre-trawling coral fauna of seamounts in the cold temperate or tropical regions of the NE Atlantic. However, the warm temperate region north of Mauritania and south of Cape Finisterre has been intensively sampled and by combining data from 1869–1985 we were able to examine the overall biogeography and potential degree of endemism on seamounts in this region. Ordination showed that the composition of historical records of seamount corals was very different from records at the same depths (200–2000 m) around the islands of the Azores and Madeira and the continental slope (Fig. 2). Some of this dissimilarity is a result of apparent endemism; in our dataset of 18 species of antipatharians, 58 species of gorgonians, and 60 species of scleractinians for the warm temperate region of the

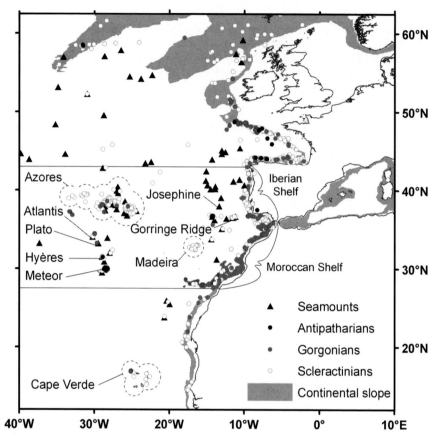

Figure 1. Georeferenced records of deep-water (> 200 m) antipatharian (n = 65), scleractinian (n = 505), and gorgonian (n = 1264) corals for the NE Atlantic compiled from pre-1985 reports. Seamounts > 1000 m in height, oceanic islands, Iberian, and Moroccan continental slopes (200–2000 m) are shown. Horizontal lines delimit a data-rich area north from Mauritania to Cape Finisterre used for comparison of seamount/oceanic island/continental shelf faunas. Named seamounts had sufficient pre-1985 coral data for inclusion in our analyses.

NE Atlantic, the scleractinian *Paracyathus arcuatus* Lindström, 1877 has only been recorded on seamounts and the gorgonians *Dentomuricea meteor* Grasshoff, 1977 and *Tubigorgia cylindrica* Pasternak, 1985 to date, have only been recorded on Great Meteor seamount (Grasshoff, 1985a; Pasternak, 1985). Thus, < 3% of the antipatharian, scleractinian, and gorgonian corals recorded in the warm temperate NE Atlantic are potentially endemic to seamounts.

Cluster analysis indicated that the deep coral faunas recorded around the Azores and Madeira were 61.4% similar in composition and had strong affinities with those recorded from the warm temperate continental slope (Fig. 3). Seamounts were on average only 29% similar to the neighboring continental slope. This partly reflects the reduced sampling of seamounts but there is a suite of scleractinian species (e.g., *Caryophyllia alberti* Zibrowius, 1980, *Caryophyllia foresti* Zibrowius, 1980, *Leptopsammia formosa* Gravier, 1915, *P. arcuatus*) that have not been recorded along the continental slope but have been recorded frequently in open oceanic conditions. Conversely, species such as *Caryophyllia seguenzae* Duncan, 1873 have been found all along the continental slope

Table 1. Expeditions used as major sources of deep-water (> 200 m) data on corals for mapping NE Atlantic coral distribution in the area 12°N–62.5°N and 10°E–40°W. Total numbers of records are shown with georeferenced records in parentheses. "Other" were obtained from expeditions with < 20 coral records and from literature cited in the methods section.

| Expeditions | Date | Antipatharians | Gorgonians | Scleractinians |
|---|---|---|---|---|
| RV JOSEPHINE | 1869 | 0 | 2 (1) | 19 (8) |
| RV PORCUPINE | 1869–1870 | 0 | 0 | 32 (30) |
| PRINCE OF MONACO | 1886–1915 | 40 (40) | 169 (126) | 142 (94) |
| RV CHALLENGER | 1873 | 0 | 1 (1) | 30 (21) |
| RV TRAVAILLEUR | 1881–1883 | 0 | 20 (17) | 18 (16) |
| RV TALISMAN | 1883 | 0 | 50 (45) | 51 (44) |
| RV CALYPSO | 1958–1959 | 0 | 0 | 36 (18) |
| RV SARSIA | 1958–1974 | 0 | 1 (1) | 53 (41) |
| RV JEAN CHARCOT | 1966–1976 | 5 (1) | 66 (20) | 300 (174) |
| RV METEOR | 1967–1970 | 9 (4) | 118 (81) | 98 (90) |
| RV THALASSA | 1967–1973 | 2 (2) | 39 (36) | 139 (123) |
| RV BARTLETT | 1975 | 0 | 19 (15) | 17 (14) |
| RV CRYOS | 1984 | 0 | 36 (36) | 0 |
| Other | 1868–1985 | 78 (18) | 250 (126) | 707 (591) |
| Totals | | 134 (65) | 771 (505) | 1,642 (1,264) |

from Senegal to Scotland but appear not to have colonized oceanic islands or open oceanic seamounts north of Cape Verde (Fig. 4).

Our analysis of taxonomic patterns indicated that Madeira had greater variation in taxonomic distinctness ($\Lambda^+$) than expected for the size of its species list (P = 0.012); all other sites fell within the expected bounds for this measure. Greater deviations from expectation were clear for taxonomic breadth (Fig. 5): the oceanic islands had less taxonomic breadth than expected for their species list sizes (Madeira P = 0.004; Azores P = 0.044) as did two of the seamounts, Gorringe Ridge (P = 0.002) and Hyères (P = 0.016). Thus, these sites either had lower high-level (e.g., families, orders) taxonomic variation, or more low-level (species) taxonomic variation than expected given the composition of the database and their relative species richnesses. This low taxonomic breadth suggests proliferation of taxa in a few genera, whereas other genera from different higher families or orders remain relatively species poor. Seamounts emerge as impoverished vs the island sites in terms of species richness, but only Hyères and Gorringe Ridge had a coral fauna that fell outside limits of taxonomic breadth predicted from the taxonomic composition of the regional fauna.

## Discussion

The NE Atlantic has the best available historical data on deep-water corals, dating back to the first descriptions of life in the deep-sea (Duncan, 1873), but the records are dispersed in specialist publications, written in a variety of languages, and differ in their use of taxonomic nomenclature. The UNESCO-IOC register of marine organisms (http://annual.sp2000.org/2006/search.php) allowed us to use historical reports to construct a database of > 2500 pre-1985 records of antipatharians, gorgononians, and scleractinians recorded in the NE Atlantic. This database provides valuable insights into the biogeography of these corals before the impacts of deep-water trawling became widespread. Although the geographic precision of the records vary, from detailed latitude/longitude

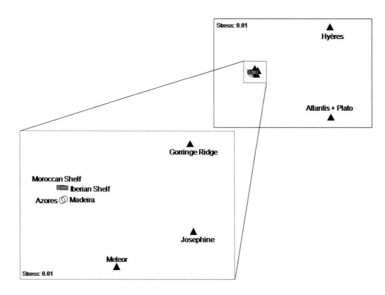

Figure 2. n-MDS ordination of coral assemblages in the warm temperate NE Atlantic north of Mauritania and south of Cape Finisterre. Ordination based upon Bray-Curtis similarity values calculated using species presence/absence. ▲ = seamounts, ■ = continental shelf slope, ○ = oceanic islands. An expanded view of the relationships amongst sites contained within the inset box is provided to facilitate interpretation.

positions to more vague information in the oldest cruise reports such as "Bay of Biscay" or "NW Flores, Azores", the data nevertheless allow comparisons of the distribution of the species present.

Our GIS map of georeferenced data indicates that the vast majority of deep-sea coral records come from steeply-sloping seabed types around seamounts and oceanic islands and along the continental slope. This is to be expected since high relief topography provides the hard substrata that antipatharian, scleractinian, and gorgonian corals require for attachment. Habitats with abrupt changes in topography also accelerate the flow of oceanic water masses, increasing food supply to filter-feeding communities and preventing their burial in silt (Rogers, 1994, 1999; Rogers et al., in press). The Mid-Atlantic Ridge appears to be a strong biogeographic boundary between corals that characterize the American boreal continental slope and those that live in the north-flowing warmer waters on the European continental slope. Cairns and Chapman (2001) and Watling and Auster (2005) point out striking dissimilarities between the deep-water coral faunas of the eastern and western Atlantic and work by Schröder-Ritzrau et al. (2005) indicates that the palaeoceanographic history of the North Atlantic may be key to the high levels of coral biodiversity recorded in deep-water in the warm temperate NE Atlantic. They found that seamounts off NW Africa, the low-latitude Mid-Atlantic Ridge and the Azores have supported coral growth for the past 50,000 yrs whereas the last glacial maximum was associated with conditions that were unsuitable for coral growth in more northern parts of the Atlantic.

It is useful to assess the degree of endemism on seamounts in this region, given that recent data on South Pacific seamounts indicates that > 30% of the species found are potentially endemic to seamounts (Parin et al., 1997; Richer de Forges et al., 2000). As with the majority of seamounts world-wide (Stocks et al., 2004), those in the cold tem-

## Similarity

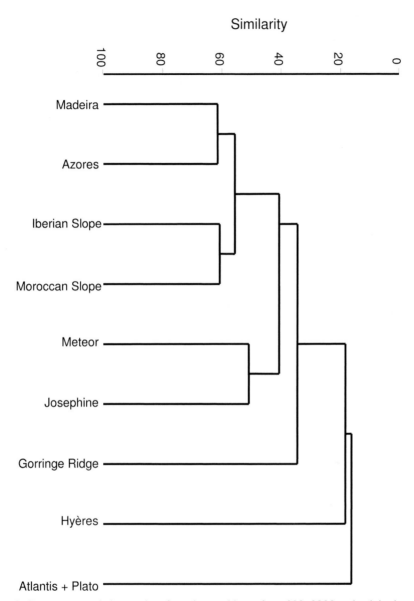

Figure 3. Group-averaged cluster plot of coral assemblages from 200–2000 m depth in the warm temperate NE Atlantic, based upon Bray-Curtis similarity values calculated from species presence/absence data.

It is useful to assess the degree of endemism on seamounts in this region, given that recent data on South Pacific seamounts indicates that > 30% of the species found are potentially endemic to seamounts (Parin et al., 1997; Richer de Forges et al., 2000). As with the majority of seamounts world-wide (Stocks et al., 2004), those in the cold temperate and tropical regions of the NE Atlantic have been poorly studied and have few historical coral records. However, GIS mapping revealed that seamount chains south of the Azores and NE of Madeira have detailed baseline information on coral distributions allowing a comparison with the faunas found elsewhere in the warm temperate region of the NE Atlantic. We found that < 3% of the 136 species of antipatharians, gorgonians,

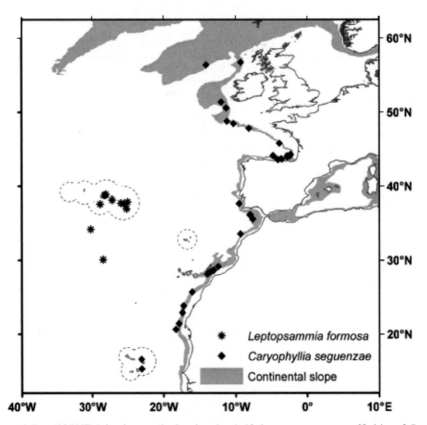

Figure 4. Pre-1985 NE Atlantic records showing the shelf-slope vs open ocean affinities of *Caryo-phyllia seguenzae* vs *Leptopsammia formosa*.

Meteor is part of a seamount chain that would be expected to facilitate the dispersal of corals. In the light of these results it is possible that the degree of endemism on South Pacific seamounts may be far lower than the 30% estimate of Parin et al. (1997) and Richer de Forges et al. (2000) and that these estimates reflect a lack of sampling in the deep-sea of the South Pacific region.

Despite the low levels of endemism recorded on NE Atlantic seamounts, our analyses reveal that the coral communities present are significantly different from those recorded on the continental slope. For example, the offshore islands and seamounts have a suite of species that do not occur along the continental slope. The coral fauna of seamounts in the region appears to be depauperate compared with those found on oceanic islands and the continental slope which both present larger "targets" for colonization by the planktonic phases in the coral life history. Most of the scleractinian corals of the warm temperate region are widespread (Cairns and Chapman, 2001) and are found on oceanic islands, seamounts, and shelf slope sites, although there are exceptions such as *L. formosa* which appears to be restricted to open ocean areas, away from the continental slope. Molodtsova (2006) notes that several NE Atlantic antipatharians appear to be restricted to open ocean areas, with *Antipathes erinaceus* (Roule, 1905), *Distichopathes* sp., *Phanopathes* sp., and *Stauropathes punctata* (Roule, 1905) only recorded on Josephine seamount, the Azores and Cape Verde Islands.

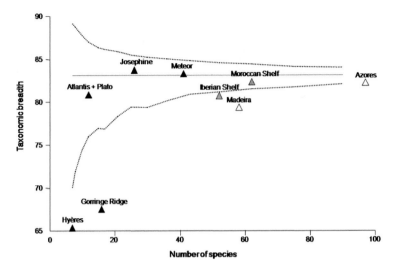

Figure 5. Funnel plot of taxonomic breadth (Δ⁺) for deep-water coral assemblages at sites in the warm temperate NE Atlantic. Dotted lines indicate the 95% confidence intervals for taxonomic breadth at any given level of species richness (x-axis); sites lying below these bounds have lower taxonomic breadth than expected at their richness level given the taxonomic composition of the pool of species from which they are drawn. Seamounts–black symbols, Continental Shelf–grey symbols, Oceanic Islands–open symbols.

Large, deep-water coral reefs are afforded a degree of physical protection from bottom trawling due to the risks they pose to fishing gear (Hall-Spencer et al., 2002) which helps explain why live reefs can still be found along the coasts of Ireland (Grehan et al., 2004), Scotland (Roberts et al., 2005), and Norway (Fosså et al., 2005). However, the vast majority of deep-water coral species (e.g., many scleractinians and all stylasterids, antipatharians, zooanthids, and gorgonians) form more isolated colonies that are highly vulnerable to damage from rockhopper gear and reinforced sweeper nets. Unlike scleractinian reefs, these corals do not leave clear evidence of trawling damage so it is not possible to determine their historical distribution and abundance based on post-fishing surveys. The accumulation of historical information on deep-water habitats is time-consuming but can clearly provide a wealth of information to help our understanding of the ecology of the deep-sea biota and to inform policy influencing the management of remaining resources. Given that deep-water trawling is the major cause of damage to deep-water habitats, decisions over where to curb these activities must be underpinned by pooling data on the distributions of sensitive benthic species with data on the distribution of deep-water trawling to highlight areas where pristine habitats are likely to still be found.

## Acknowledgments

We would like to thank T. Molodtsova, S. Cairns, and M. Grasshoff for friendly advice on antipatharians, scleractinians, and gorgonians, respectively, and to G. Gonzales-Mirelis for producing Figure 1. This project was supported by a Royal Society University Research Fellowship to J.H.-S. with additional grant aid from the Esmée Fairbairn Foundation and the CenSeam project. A.R. would like to thank G. Mace for provision of facilities at the Institute of Zoology, Zoological Society of London.

## Acknowledgments

We would like to thank T. Molodtsova, S. Cairns, and M. Grasshoff for friendly advice on antipatharians, scleractinians, and gorgonians, respectively, and to G. Gonzales-Mirelis for producing Figure 1. This project was supported by a Royal Society University Research Fellowship to J.H.-S. with additional grant aid from the Esmée Fairbairn Foundation and the CenSeam project. A.R. would like to thank G. Mace for provision of facilities at the Institute of Zoology, Zoological Society of London.

## Literature Cited

Arnaud, P. M. and H. Zibrowius 1973. Capsules ovigères de gastéropodes Turridae et corrosion du squelette des scléractiniaires bathyaux des Açores. Rev. Fac. Ciênc. Lisb. Sér C. 17: 581–597.

Butler, M. 2005. Conserving corals in Atlantic Canada: a historical perspective. Pages 1199–1209 *in* A. Freiwald and J. M. Roberts, eds. Cold-water corals and ecosystems. Springer-Verlag Berlin Heidelberg. 1243 p.

Cairns, S. D. and R. E. Chapman. 2001. Biogeographic affinities of the North Atlantic deep-water Scleractinia. Pages 30–57 *in* J. H. W. Willison et al., eds. Proc. 1st Int. Symp. Deep-sea Corals. Ecology Action Centre, Halifax. 231 p.

Clark, M. and R. O'Driscoll. 2003. Deep-water fisheries and aspects of their impact on seamount habitat in New Zealand. J. Northwest Atl. Fish. Sci. 31: 441–458.

Clarke, K. R. and R. M. Warwick. 2001a. Change in marine communities: an approach to statistical analysis and interpretation. 2nd ed. Plymouth Marine Laboratory, Plymouth, UK. 169 p.

_____ and _____. 2001b. A further biodiversity index applicable to species lists: variation in taxonomic distinctness. Mar. Ecol. Prog. Ser. 216: 265–278.

Duncan, P. M. 1873. A description of the Madreporaria dredged up during the expeditions of H.M.S. Porcupine in 1869 and 1870. Trans. Zool. Soc. Lond. 8: 303–344.

Fosså, J. H., B. Lindberg, O. Christensen, T. Lundälv, I. Svellingen, P. Mortensen, and J. Alvsvåg. 2005. Mapping of *Lophelia* reefs in Norway: experiences and survey methods. Pages 259–392 *in* A. Freiwald and J. M. Roberts eds. Cold-water corals and ecosystems. Springer-Verlag Berlin Heidelberg. 1243 p.

Gordon, J. D. M. 2001. Deep-water fisheries at the Atlantic Frontier. Cont. Shelf Res. 21: 987–1003.

_____. 2003. The Rockall Trough, northeast Atlantic: the cradle of deep-sea biological oceanography that is now being subjected to unsustainable fishing activity. J. Northwest Atl. Fish. Sci. 31: 57–83.

Grasshoff, M. 1972. The Gorgonia of the eastern North Atlantic and Mediterranean. I. Family Ellisellidae (Cnidaria: Anthozoa). Results of the Atlantic Seamount Cruises 1967 with RV Meteor. Meteor Forsch. Ergebnisse 10: 73–87.

_____. 1973. Die Gorgonaria des östlichen Nordatlantik und des Mittelmeeres II. Die Gattung Acanthogorgia (Cnidaria: Anthozoa) Auswertung der Atlantischen Kuppenfahrten 1967 von FS Meteor. Meteor Forsch. Ergebnisse 13: 1–10.

_____. 1977. Die Gorgonarien des östlichen Nordatlantik und des Mittelmeeres. III Die Familie Paramuriceidae (Cnidaria, Anthozoa). Meteor Forsch. Ergebnisse 27: 5–76.

_____. 1981a. Gorgonaria und Pennatularia (Cnidaria: Anthozoa) vom Mittelatlantischen Rucken SW der Azoren. Steenstrupia 7: 213–230.

_____. 1981b. Die Gorgonaria, Pennatularia und Antipatharia des tiefwassers der Biskaya (Cnidaria, Anthozoa). I Allgemeiner Teil. Bull. Mus. Nat. Hist. Nat. Paris 3: 731–766.

_____. 1981c. Die Gorgonaria, Pennatularia und Antipatharia des tiefwassers der Biskaya (Cnidaria, Anthozoa). II Taxonomischer Teil. Bull. Mus. Nat. Hist. Nat. Paris 3: 941–978.

_____. 1985a. Die Gorgonaria und Antipatharia der Großen Meteor-Bank und der Josephine-Bank. (Cnidaria: Anthozoa). Senckenb. Marit. 17: 65–87.

_____ and H. Zibrowius 1983. Kalkkrusten auf achsen von hornkorallen, rezent und fossil. Senckenb. Marit. 15: 111–145.

Grehan, A., V. Unnithan, A. Wheeler, X. Monteys, T. Beck, M. Wilson, J. Guinan, J. M. Hall-Spencer, A. Foubert, M. Klages, and J. Thiede. 2004. Evidence of major fisheries impact on cold-water corals in the deep-waters off the Porcupine Bank, west coast of Ireland: are interim management measures required? ICES Annual Science Symp., Vigo, Spain, 2004, CM 2004/AA:07.

Hall-Spencer J. M., V. Allain, and J. H. Fosså. 2002. Trawling damage to Northeast Atlantic ancient coral reefs. Proc. R. Soc. Lond., Ser. B: Biol. Sci. 269: 507–511.

Keller, N. B. 1985. The Madreporarian corals of the Reykjanes Ridge and the Plato submarine mountains (the north part of the Atlantic Ocean). Trudy Inst. Okeanol. 120: 39–51.

Koslow, J. A., K. Gowlett-Holmes, J. K. Lowry, T. O'Hara, G. C. B. Poore, and A. Williams. 2001. Seamount benthic macrofauna off southern Tasmania: community structure and impacts of trawling. Mar. Ecol. Prog. Ser. 213: 111–125.

Kruskal, J. B. 1964. Multidimensional scaling by optimizing goodness of fit to a nonmetric hypothesis. Psychometrica 29: 1–27.

Malakoff, D. 2003. Deep-sea mountaineering. Science 301: 1034–1037.

Molodtsova, T. N. 2006. Black corals (Antipatharia:Anthozoa:Cnidaria) of the north-east Atlantic. Pages 141–151 in A. N. Mironov, A. V. Gebruk, and A. J. Southward, eds. Biogeography of the north Atlantic seamounts. KMK Press, Moscow.

OASIS. 2003. Seamounts of the North-East Atlantic. OASIS, Hamburg and WWF Germany, Frankfurt am Main, November 2003. 38 p.

Parin, N. V., A. N. Mironov, and K. N. Nesis. 1997. Biology of the Nazca and Sala y Gomez submarine ridges, an outpost of the Indo-West Pacific fauna in the eastern Pacific Ocean: composition and distribution of the fauna, its communities and history. Adv. Mar. Biol. 32: 145–242.

Pasternak, F. A. 1985. Gorgonians and antipatharians of the seamounts Rockaway, Atlantis, Plato, Great-Meteor and Josephine (Atlantic Ocean). Trudy Inst. Okeanol. 120: 21–38.

Pitcher T. J., T. Morato, P. J. B. Hart, M. Clark, N. Haggan, and R. Santos, eds. 2007. Seamounts: ecology, fisheries and conservation. Fish and aquatic resources series, Blackwell, Oxford. 527 p.

Roark E. B., T. P. Guilderson, R. B. Dunbar, and B. L. Ingram. 2006. Radiocarbon-based ages and growth rates of Hawaiian deep-sea corals. Mar. Ecol. Prog. Ser. 327: 1–14.

Richer de Forges B., J. A. Koslow, and G. C. B. Poore. 2000. Diversity and endemism of the benthic seamount fauna in the southwest Pacific. Nature 405: 944–947.

Roberts, J. M., C. J. Brown, D. Long, and C. R. Bates. 2005. Acoustic mapping using a multibeam echosounder reveal cold-water coral reefs and surrounding habitats. Coral Reefs 24: 654–669.

Rogers A. D. 1994. The biology of seamounts. Adv. Mar. Biol. 30: 305–350.

_____. 1999. The biology of *Lophelia pertusa* (Linnaeus 1758) and other deep-water reef-forming corals and impacts from human activities. Int. Rev. Hydrobiol. 84: 315–406.

_____, A. Baco, H. Griffiths, and J. M. Hall-Spencer. 2007. Corals on seamounts. Pages 141–169 in T. J. Pitcher et al., eds. Seamounts: ecology fisheries and conservation. Blackwell Fisheries and Aquatic Resources Series, Blackwell Scientific, Oxford.

Schröder-Ritzrau, A., A. Freiwald, and A. Mangini. 2005. U/Th-dating of deep-water corals from the eastern North Atlantic and western Mediterranean. Pages 157–172 in A. Freiwald and J. M. Roberts, eds. Cold-water corals and ecosystems. Springer-Verlag Berlin Heidelberg. 1243 p.

Stocks, K. I., G. W. Boehlert, and J. F. Dower. 2004. Towards an international field programme on seamounts within the Census of Marine Life. Arch. Fish. Mar. Res. 51: 320–327.

Studer, T. 1901. Alcyonaires provenant des campagnes de l'Hirondelle (1886–1888). Résult. Camp. scient. Prince Albert I de Monaco 20: 1–46.

Tendal, O. S. 1992. The north Atlantic distribution of the octocoral *Paragorgia arborea* (L., 1758) (Cnidaria, Anthozoa). Sarsia 77: 213–217.

Schröder-Ritzrau, A., A. Freiwald, and A. Mangini. 2005. U/Th-dating of deep-water corals from the eastern North Atlantic and western Mediterranean. Pages 157–172 *in* A. Freiwald and J. M. Roberts, eds. Cold-water corals and ecosystems. Springer-Verlag Berlin Heidelberg. 1243 p.

Stocks, K. I., G. W. Boehlert, and J. F. Dower. 2004. Towards an international field programme on seamounts within the Census of Marine Life. Arch. Fish. Mar. Res. 51: 320–327.

Studer, T. 1901. Alcyonaires provenant des campagnes de l'Hirondelle (1886–1888). Résult. Camp. scient. Prince Albert I de Monaco 20: 1–46.

Tendal, O. S. 1992. The north Atlantic distribution of the octocoral *Paragorgia arborea* (L., 1758) (Cnidaria, Anthozoa). Sarsia 77: 213–217.

Thomson, J. A. 1927. Alcyonaires provenant des campagnes scientifiques du Prince Albert I de Monaco. Résult. Camp. scient. Prince Albert I de Monaco 73: 1–77.

Tyler, P. A. and H. Zibrowius. 1992. Submersible observations of the invertebrate fauna on the continental slope southwest of Ireland (NE Atlantic Ocean). Oceanol. Acta 15: 211–226.

Watling, L. and P. J. Auster 2005. Distribution of deep-sea Alcyonacea off the northeast coast of the United States. Pages 279–296 *in* A. Freiwald and J. M. Roberts, eds. Cold-water corals and ecosystems. Springer-Verlag Berlin Heidelberg.

Wheeler, A. J., B. J. Bett, D. S. M. Billet, D. G. Masson, and D. Mayor. 2005. The impact of demersal trawling on northeast Atlantic deep-water coral habitats: the case of the Darwin Mounds, United Kingdom. Am. Fish. Soc. Symp. 41: 807–817.

Zibrowius, H. 1973. Revision des espèces actuelles du genre *Enallopsammia* Michelotti, 1871, et description de *E. marenzelleri*, nouvelle espèce bathyale à large distribution : Océan Indien et Atlantique Central (Madreporaria, Dendrophylliidae). Beufortia Misc. Publ. 21: 37–54.

_____. 1980. Les scléractiniaires de la Méditerranée et de l'Atlantique nord-oriental. Mem. Inst. Oceanogr. 11 : 1–227.

_____. 1985. Scléractiniaires bathyaux et abyssaux de l'Atlantique nord-oriental: campagnes BIOGAS (POLYGAS) et INCAL. Pages 311–324 *in* L. Laubier and C. Monniot, eds. Deep-sea fauna from the Gulf of Biscay: BIOGAS campaign. IFREMER, Brest. 367 p.

ADDRESSES: (J.H.-S., J.D., A.F) *Marine Biology and Ecology Research Centre, Biological Sciences, University of Plymouth, Plymouth, United Kingdom, PL4 8AA.* (A.R.) *Institute of Zoology, Zoological Society of London, Regent's Park, London, United Kingdom, NW1 4RY.* CORRESPONDING AUTHOR: (J.H.-S.) *Email: <jhall-spencer@plymouth.ac.uk>.*

# The Darwin Mounds: from undiscovered coral to the development of an offshore marine protected area regime

ELIZABETH M. DE SANTO AND PETER J. S. JONES

## Abstract

The first offshore Marine Protected Area (MPA) in the United Kingdom (UK) is the Darwin Mounds, an area of *Lophelia pertusa* (Linnaeus, 1758) discovered only in 1998. At the time of its discovery, this was considered to be an exceptional example of *L. pertusa*, growing on a sand base, rather than hard substratum, and exhibiting a distinctive "tail" structure not yet seen elsewhere. Damage to the area caused by deep-water trawling has been observed and in 2003, at the UK's request, the European Commission imposed a ban on trawling in a 1380 km² area surrounding the Mounds, which became permanent in 2004. This move was made possible by the revised Common Fisheries Policy (CFP) and represents the first EC example of an offshore fisheries closure for nature conservation (rather than fish stocks). Eventually a network of offshore MPAs will be designated throughout the EU's marine waters, including around the UK. Drawing on a detailed legal and policy analysis and a program of semi-structured interviews with stakeholders, regulators and specialists in the field, this paper explores the unique circumstances and sequence of events that led to the protection of the Darwin Mounds.

*Lophelia pertusa* (Linnaeus, 1758) has been known to scientists and fishermen for hundreds of years. It is a stony coral (Scleratinia, family Caryophylliidae) found globally, except in polar regions (Fosså, 2002). The rich concentration of biodiversity associated with coral reefs and the slow growth of these ecosystems is well-known (Rogers, 1999; Husebø et al., 2002; Roberts et al., 2003, 2006). Traditionally deep-water reefs were considered good sites for net and long-line fishing. With the advent of bottom-trawling, however, substantial damage to *L. pertusa* reefs has been documented (Fosså et al., 2002; Hall-Spencer et al., 2002; Wheeler et al., 2005), increasing with the development of larger vessels, more powerful trawls and gear specially adapted to reaching areas that were previously inaccessible. A recent study on *L. pertusa* found in Norwegian waters determined that between 30%–50% of their reefs had been damaged or impacted by trawling, with an associated decline in fishing success according to local fishermen (Fosså et al., 2002). Until recently, deep-sea species in European Union waters were off-quota and their exploitation came largely from vessels that had surpassed or did not hold quotas for other commercial species.

The Darwin Mounds area of *L. pertusa* was discovered in May 1998 during a seabed survey conducted for the Atlantic Frontier Environment Network (AFEN), a partnership between the oil industry and UK government agencies including the Department of Trade and Industry (DTI), the Joint Nature Conservation Committee (JNCC) and the Scottish Office Agriculture and Fisheries Department (now the Fisheries Research Services (FRS) agency of the Scottish Executive) (AFEN, 2000; Bett, 2001). AFEN was

George, R. Y. and S. D. Cairns, eds. 2007. Conservation and adaptive management of seamount and deep-sea coral ecosystems. Rosenstiel School of Marine and Atmospheric Science, University of Miami.

formed in 1995, a period when survey activity north and west of Scotland surged following the discovery of the Foinhaven and Schiehallion oil fields in the early 1990s, with the objective of gaining an environmental baseline for the areas being licensed.

The Mounds lie approximately 185 km to the northwest of Scotland at a depth of around 1000 m, scattered across approximately 1500 km$^2$ and supporting significant amounts of *L. pertusa* and associated biodiversity, including sessile or hemi-sessile invertebrates and giant protozoan xenophyophores (*Syringammina fragilissima* Brady, 1883) (Bett, 2001). The hundreds of mounds present in the area are approximately 100 m in diameter and 5 m in height (Bett, 2001). Figures 1A and 1B illustrate their geographical location and distribution. The Darwin Mounds were further investigated in June 1998, August 1999 and twice during the summer of 2000, when evidence of damage from trawling was visible over half of the eastern fields (Wheeler et al., 2001, 2005).

**Legal framework.**—A primary legal instrument for nature conservation in the European Union (EU) is the 1992 Directive on the Conservation of Natural Habitats and of Wild Fauna and Flora, herein referred to as the Habitats Directive (Council Directive 92/43/EEC). In November 1999, following a Greenpeace Atlantic Frontier campaign to halt offshore oil exploration in UK waters, the English High Court ruled that the European territory to which the Directive applies includes areas over which Member States exercise sovereign rights beyond territorial waters (i.e., beyond 12 nmi). This ruling, commonly referred to as the Greenpeace Judgment, requires the UK to apply the Habitats Directive to the 200 nmi limit of its Exclusive Fishing Zone (EFZ) including the water column and seabed, and other Member States are following suit, designating protected areas under the Habitats Directive [referred to as Special Areas of Conservation (SACs)] in their offshore waters. In addition to its EFZ, the UK also claims jurisdiction over its Continental Shelf, extending up to 340 nmi from the baseline, but covering only the seabed.

As a result of the Greenpeace Judgment, the UK Government is revising its existing national legislation, the 1994 Conservation (Natural Habitats, etc.) Regulations, in order to transpose the Habitats Directive (and its predecessor the 1979 Birds Directive) into UK law in its offshore waters, including not only its EFZ, as stipulated in the Greenpeace Judgment, but the entire UK Continental Shelf. This process has been prolonged but the new regulations are due to come into effect in 2007. Given the total land area of the UK is 244,101 km$^2$ and that of its territorial sea is approximately 161,200 km$^2$, this extension over the UK Continental Shelf would add an additional 706,200 km$^2$, resulting in a total extent of UK area (territorial and offshore waters, and land area) subject to protection of 1,111,501 km$^2$ or a 2.74 fold increase in area protected by the UK implementation of the Habitats Directive. See Figure 2 for a map of the UK offshore area, outlining its territorial waters, EFZ and Continental Shelf limits.

**The revised Common Fisheries Policy (CFP).**—While the Habitats Directive provides an important framework for protecting habitats and species in Europe, the primary mechanism for enforcing areas closed to fishing in the marine environment lies in the revised Common Fisheries Policy (CFP). A complete overview of the process that went into the revision of the CFP is beyond the scope of this paper, however, some key issues can be highlighted in the context of situations where nature conservation and fisheries management overlap. The most recent reform of the CFP began in 1998, resulting in Council Regulation (EC) 2371/2002 (from this point referred to as the Basic Regulation),

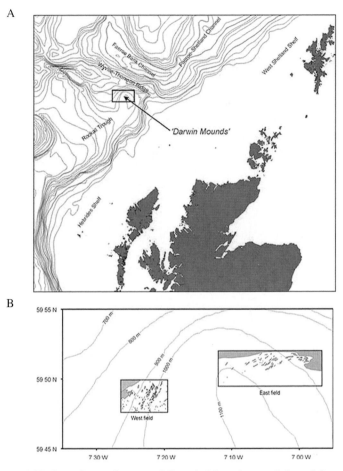

Figures 1A and 1B. Location and extent of Darwin Mounds area (adapted from Johnston and Tasker, 2002. Figures courtesy of Brian Bett, National Oceanography Centre, Southampton).

which came into effect on 1 January 2003. The Basic Regulation emphasizes that: "the Community shall apply the precautionary approach in taking measures designed to protect and conserve living aquatic resources, to provide for their sustainable exploitation and to minimize the impact of fishing activities on marine ecosystems. It shall aim at a progressive implementation of an ecosystem-based approach to fisheries management" (Article 2, para. 1).

In order to implement these approaches, the Basic Regulation outlines specific technical measures including recovery and management plans and the establishment of emergency closures. In particular, Article 7 allows for the Commission to apply emergency measures "if there is evidence of a serious threat to the conservation of living aquatic resources, or to the marine eco-system resulting from fishing activities and requiring immediate action". It was this mechanism that allowed for the initial protection of the Darwin Mounds area.

**The Darwin Mounds closure.**—The sequence of events that led to the closure of the Darwin Mounds area to bottom-trawling are outlined in Table 1. Following their discovery and the outcome of the Greenpeace Judgment, the Secretary of State for the

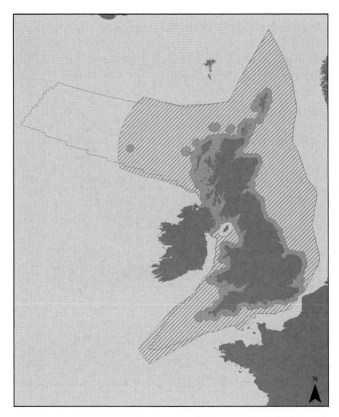

Figure 2. The UK offshore area showing the territorial sea, EFZ, and continental shelf Boundaries. (Adapted from and reproduced with permission from the JNCC).
Legend: ■ Land, ▨ UK Territorial Seas, □ UK Continental Shelf area, ▨ British fishery limit extent.

Environment and Rural Affairs made a commitment in early 2001 to protect the Darwin Mounds as a SAC under the Habitats Directive. From 1999 to 2001, the UK's Joint Nature Conservation Committee (JNCC) undertook a 2-yr research project to identify offshore marine sites for protection under the Habitats and Birds Directives, resulting in the completion of a comprehensive report on implementing the Directives in UK offshore waters (Johnston et al., 2001). Another relevant report was released by WWF in May 2002, suggesting a management framework for the Darwin Mounds as the UK's first offshore SAC (Gubbay et al., 2002). In addition to the draft Regulations mentioned earlier, in late summer 2003, the Department for the Environment, Food and Rural Affairs (DEFRA) released a consultation document proposing the Darwin Mounds as a candidate Special Area of Conservation under the Habitats Directive (DEFRA, 2003). DEFRA subsequently informed the European Commission that the site will become the UK's first offshore SAC under the Habitats Directive.

While at first glance the closure of the Darwin Mounds area may appear to have been a somewhat quick and straightforward process, taking a relatively short period of time to move from a temporary to permanent ban on bottom-trawling in the area (i.e., 7 mo), it required a careful, step-wise approach on the UK's part with a certain degree of compromise. In October 2002, the UK made its first approach to the European Commis-

Table 1. Timeline of Darwin Mounds MPA Designation.

| Date | Action | Outcome |
|------|--------|---------|
| 1998 May | Discovery of Darwin Mounds by AFEN survey. | |
| 1999 and 2000 | Darwin Mounds revisited, damage visible. | |
| 1999 November | Greenpeace Judgment. | UK required to extend Habitats Directive offshore. |
| 1999–2001 | JNCC process established by Defra to identify offshore Natura 2000 sites. | JNCC Report 325: Implementing Natura 2000 in UK Offshore Waters. |
| 2000 July | European Commission requested ICES to provide advice on cold-water corals. | Reports in 2001, 2002, and 2003 on *Lophelia* in ICES waters. |
| 2001 October | Secretary of State Beckett Announcement. | Publicity. |
| 2002 May | WWF-UK Report on Darwin Mounds SAC. | |
| 2002 October | UK first approached European Commission (EC) regarding protecting area. | Positive indications from Commission. |
| 2002 December | Commission agreed on TACs for deep-sea species in 2003 and 2004. | |
| 2003 January | Revised CFP Regulation 2371/2002 came into effect. | Provided mechanism for emergency closure. |
| 2003 March | UK held informal discussions with EC and other Member States | Compromise on degree and extent of closure. |
| 2003 June | UK made formal approach (in writing) to EC for action under CFP Regulation 2371/2002. | Positive response from Commission. |
| 2003 July | UK made formal request for closure of Darwin Mounds area. | Accepted. |
| 2003 August | Emergency closure (Regulation 1475/2003). | |
| 2003 September | Proposal for permanent Regulation submitted. | |
| 2004 February | Emergency closure extended a further six months (Regulation 263/2004). | |
| 2004 March | Closure made permanent (Regulation 602/2004). | |

sion, alerting them to the site and indicating a need for action to be taken, although no mechanism yet existed for implementing a protected area in offshore waters. With the advent of the reformed CFP in 2003, however, a mechanism became available and the UK began informal discussions with the Commission about whether and how to use the emergency closure provisions. As this would be the first use of the mechanism, and as it was the first closure proposed for nature conservation objectives, care was taken by both the UK and the European Commission in order to ensure that the proposal was properly assessed and that no poor precedents would be set, with the UK wanting to be certain the Commission was on board. Consequently, it aimed to provide the most solid case possible for closure based on the best evidence available and recommendations from the JNCC and the International Council for the Exploration of the Sea (ICES) Advisory Committee on Ecosystems (ACE).

The ICES advisory process had begun a bit earlier, in July 2000, when the European Commission made a request for urgent advice "to identify areas where cold-water corals may be affected by fishing" (ICES, 2001). Subsequently, ICES established a Study

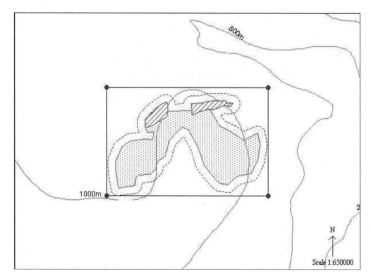

Figure 3. Darwin Mounds closure as recommended by the ICES Advisory Committee on Ecosystems. (Adapted from and reproduced with permission from ICES). Legend: [░░░] Extent of mounds region mapped by AFEN, [▨▨▨] limits of interpreted side-scan of Darwin Mounds East and West fields, [☐] possible site boundary generated using simple point coordinates, - - - - - 2.2 km margin from region of mounds, ──── bathymetric contour ©GEBCO Digital Atlas, British Oceanographic Data Centre on behalf of IOC and IHO 1994 and 1997

Group on Mapping the Occurrence of Cold Water Corals (SGCOR) which compiled maps identifying cold-water coral areas in the North-East Atlantic. These maps were then circulated to ACE and a selection of working group chairs for comment, in order to enable ICES (through ACE) to provide advice to the European Community (ICES 2001, 2002, 2003).

Concurrent with its aforementioned dialogue with the European Commission in 2003, the UK also pursued informal discussions with other Member States, targeting those with fishing interests in the area (primarily France and Spain) and others supportive of a closure in the area (Ireland and the Netherlands). Following pressure from France, a compromise was made regarding the extent of the area closed: the borders of the original square-shaped ICES ACE recommendation for a closed area around the Darwin Mounds were altered, with the North East and North West corners removed, resulting in a hexagon-shape. Figure 3 shows the original ICES proposal for a closure, which was modified to what is shown in Figure 4, the final area surrounding the Darwin Mounds permanently closed to bottom-trawling. The closed area is slightly larger than the extent of the feature to allow for a "buffer zone" such that trawls cannot accidentally cross the Mounds at the end of their 1.5–2 km long trawl warps. In addition, the area is closed only to bottom-trawling methods of fishing, as there was Spanish interest to keep the area open to pelagic fishing and a complete closure would not have been politically feasible.

The UK continued its step-wise approach to the Commission in subsequent months, with a formal letter expressing their intention to pursue an emergency closure in June 2003 before actually making the formal request on 24 July 2003. No objections were received from other Member States during the 5-d comment period, and a 6-mo emergency closure went forward, under Regulation 1475/2003 of 27 August 2003. This temporary

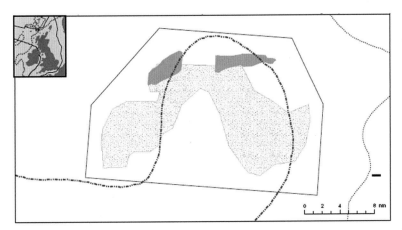

Figure 4. Permanent Darwin Mounds closure. (Adapted from and reproduced with permission from the JNCC).
Legend: ▨ East and West fields of dense mounds, ▨ extent of mounds, ☐ bottom trawling exclusion area, ▬ ▪ ▬ ▪ ▬ 1000 m isobath. GESCO bathymetry ©NERC 1994, 1997. Darwin Mounds East and West fields and mounds extent courtesy of Dr. Brian Bett, Southampton Oceanography Centre.

closure was extended for a further 6 mo under Regulation 263/2004 of 14 February 2004, during which the UK prepared a proposal for a permanent closure of the area, which involved amending Regulation 850/98 of 30 March 1998 on the conservation of fishery resources through technical measures for the protection of juveniles of marine organisms. The permanent ban on bottom-trawling in the Darwin Mounds area came into effect as Regulation 602/2004 on 22 March 2004, adding the geographical location of the Darwin Mounds area to Article 30 of Regulation 850/98 in its section on restrictions on the uses of demersal towed gears.

**Why a success?**—In addition to the stepwise approach made by the UK and the compromise made on the degree and extent of the closure during the negotiation process, other external factors provided incentives for the closure to succeed. First, it should be clarified that from a legal standpoint, a bifurcation between nature conservation and fisheries management exists in the European Union. While the former remains the remit of Member States, the European Commission retains exclusive legislative jurisdiction over fisheries. Consequently, when a Member State is faced with a nature conservation issue that results from fishing activity, before the provisions of the Basic Regulation came into effect there was no mechanism available to handle such a situation. As the first use of the revised CFP's emergency closure mechanism, there was an incentive for the Commission to make certain that the Darwin Mounds closure went through. For the UK's part, it was imperative that the most robust case for closure possible be made, and the role played by the JNCC report and ICES ACE recommendations to the European Commission during the negotiations process should not be overlooked. This irrefutability is of particular importance given that Article 7 of the Basic Regulation requires "evidence of a serious threat to the conservation of living aquatic resources" for the Commission to act. The fact that these corals had been revisited and damage from trawling had been clearly visible made a strong case for an immediate closure.

An important factor in their protection was the "uniqueness" of the Darwin Mounds, with their "tail-like" shapes, associated fauna and the fact that the corals had colonized sandy rather than hard substrate. (At the time of their discovery, the Darwin Mounds were the only example of these characteristics, however since then, similar (though not identical) situations have been found for *L. pertusa* in other areas). In addition, the area under question was relatively small, covering < 1500 km², and lacked the intensive fishing history of other nearby areas containing *L. pertusa*, such as the Rockall Bank. Nevertheless, there was concern on the UK side that fishing in the area could increase in the summer of 2003 following the first allotment of Total Allowable Catch (TAC) quotas for deep-sea fish species released in 2002 (Council Regulation 2340/2002). This threat added further impetus to the UK's efforts to secure a closure as quickly as possible. In addition, momentum was maintained at both the national and European level by the environmental NGO community, notably the WWF with its aforementioned 2002 report on the Darwin Mounds as a potential SAC (Gubbay et al., 2002).

**Conclusion: is it a success?**—While the closure of the Darwin Mounds area can be viewed as a political success, there are several outstanding issues that need to be addressed. The current method of enforcement relies on Vessel Monitoring Systems (VMS), satellite transmitters that relay a fishing vessel's location back to shore via a Global Positioning System (GPS) satellite network. While theoretically an efficient means of tracking fishing activity, it is only recently that UK fishermen have been required to use VMS boxes without "off" switches, and this requirement is not extended throughout the EU's fishing fleets. In addition, the current system relies on data sent every 2 hrs, a rate that may not be sufficient to detect bottom-trawling activity on the edge of a closed area. For the Darwin Mounds, VMS data is supplemented with aerial surveys by the Scottish Fisheries Protection Agency (SFPA) and enforced by SFPA patrol vessels. From an enforcement point of view, however, the most easily and efficiently protected area is one that is closed to all forms of fishing (Guénette et al., 1998). Given that the closure only applies to bottom trawling, the SFPA have to prove "beyond reasonable doubt" that fishing vessels, whether observed by VMS or by air patrols, were actually trawling the seabed, as pelagic trawling is allowed. The burden of evidence in this respect can be problematic, making successful prosecutions very difficult. Boarding by a fisheries patrol vessel may be the only way to secure successful prosecutions, and this is expensive, dangerous, and logistically challenging. This also calls into question the assumption that VMS will provide for the enforcement of offshore fisheries closures. The enforcement of such protected areas is thus likely to continue to pose major challenges (Jones, 2006). With Member States designating further offshore protected areas in coming years, the question of enforcement must be taken into careful consideration as resources are stretched to meet the difficult requirement of policing areas that are spread over wide areas far from shore. To further complicate enforcement matters, in addition to the anticipated extension of the Habitats Directive over the EFZs of Member States, in 2003 the OSPAR and Helsinki Commissions called for the establishment of an "ecologically coherent network" of MPAs by 2010 in the OSPAR and HELCOM areas (i.e., the Northeast Atlantic and Baltic Sea). This network will include sites already designated as SACs under the Habitats Directive, and will also incorporate marine habitats and species not listed in the Annexes to the Directive. The JNCC released a report (JNCC, 2004) on this initiative, exploring the concept of an "ecologically coherent network" of MPAs, as this concept is not formally defined, and providing several recommendations regarding the design of such a network. From a jurisdictional perspective, the overlap between the OSPAR network and that of Natura 2000 in offshore waters may pose some tensions with regard to enforcement, as Member States will be required to monitor those areas comprising Natura 2000 under their obligations stemming from the Habitats Directive, while the European Community will be responsible for OSPAR areas under its commitment to the OSPAR Convention. From our research on this area of institutional overlap (De Santo and Jones, 2007), it does not yet seem clear how the latter goal will be achieved. Nevertheless, there is currently a Marine Strategy Directive in development, as well as a Maritime Policy for the European Union, which may harmonize matters—this remains to be seen.

A key issue raised by the case of the Darwin Mounds is the role of the precautionary principle in the CFP. Whereas this principle, in its simplest form, calls for actions to be taken in the face of uncertainty, the Basic Regulation articles on emergency closures require a degree of scientific certainty that may not be available in all situations. With regard to the Darwin Mounds, irrefutable proof of damage from bottom-trawling was a cornerstone in the argument to close the area to fishing. This may not be the case for other areas in need of protection, and one can also argue that such an approach is counter-productive; if evidence of damage to an area of deep-water coral is required to close it to fishing, then what method is available for protecting pristine areas that are at risk of being damaged? Although the closure of the Darwin Mounds is a success in many respects, it also highlights the division between marine nature conservation and fisheries management in the European Union, a legal and political issue that will require resolution in the near future.

## Acknowledgments

The authors gratefully acknowledge the participation of representatives from relevant regulatory authorities, industry/users and the epistemic/scientific community in informal, off-the record interviews, and thank them for contributing their personal stories, insights and opinions to this study. Special thanks to M. Tasker and an anonymous reviewer for comments on an earlier draft and to the Joint Nature Conservation Committee for the use of their maps of the Darwin Mounds.

## Literature Cited

AFEN (Atlantic Frontier Environmental Network). 2000. The UK Atlantic Margin environment: towards a better understanding. Orcadian Limited, Orkney. 78 p.

Bett, B. J. 2001. UK Atlantic Margin environmental survey: introduction and overview of bathyal benthic ecology. Cont. Shelf Res. 21: 917–956.

DEFRA (Department for the Environment, Food and Rural Affairs). 2003. European Habitats Directive 92/43/EEC: Darwin Mounds – candidate Special Area of Conservation. Consultation document Hab/1/1222. 14 p.

De Santo, E. M. and P. J. S. Jones. 2007. Offshore marine conservation in the North East Atlantic: emerging challenges and opportunities. Mar. Policy 31: 336–347.

Fosså, J. H., P. B. Mortensen, and D. M. Furevik. 2002. The deep-water coral Lophelia pertusa in Norwegian waters: distribution and fishery impacts. Hydrobiologia 471: 1–12.

Gubbay, S., C. M. Baker, and B. J. Bett. 2002. The Darwin Mounds and the Dogger Bank, case studies of the management of two potential Special Areas of Conservation in the offshore environment. WWF-UK, Surrey. 72 p.

Guénette, S., T. Lauck, and C. Clark. 1998. Marine reserves: from Beverton and Holt to the present. Rev. Fish. Biol. Fish. 8: 251–272.

Hall-Spencer, J., V. Allain, and J. H. Fosså. 2002. Trawling damage to Northeast Atlantic ancient coral reefs. Proc. R. Soc. Lond., B. 269: 507–511.

Husebø, Å., L. Nøttestad, J. H. Fosså, D. M. Furevik, and S. B. Jørgensen. 2002. Distribution and abundance of fish in deep-sea coral habitats. Hyrdobiologia 471: 91–99.

ICES (International Council for the Exploration of the Sea). 2001. Initial report of the study group on cold water corals in relation to fishing. ICES, Copenhagen. 22 p.

____. 2002. Report of the study group on the mapping of cold water coral, Report CM 2002/ACE:05. ICES, Copenhagen. 21 p.

____. 2003. Report of the study group on cold water corals, Report CM 2003/ACE:02. ICES, Copenhagen. 23 p.

Johnston, C. M. and M. Tasker. 2002. Darwin Mounds proposed Special Area of Conservation, Report 02 P10. JNCC, Peterborough. 8 p.

_____, C. G. Turnbull, and M. L. Tasker. 2001. Natura 2000 in UK offshore waters: advice to support the implementation of the EC Habitats and Birds Directives in UK offshore waters, Report 325. JNCC, Peterborough. 162 p.

Jones, P. J. S. 2006. Collective action problems posed by no-take zones. Mar. Policy 30: 43–156.

JNCC (Joint Nature Conservation Committee). 2004. Developing the concept of an ecologically coherent network of OSPAR marine protected areas, Report 02 No 08. JNCC, Peterborough. 23 p.

Roberts, J. M., A. J. Wheeler, and A. Freiwald. 2006. Reefs of the deep: the biology and geology of cold-water coral ecosystems. Science 312: 543–547.

_____, D. Long, J. B. Wilson, P. B. Mortensen, and J. D. Gage. 2003. The cold-water coral *Lophelia pertusa* (Scleractinia) and enigmatic seabed mounts along the north-east Atlantic margin: are they related? Mar. Pollut. Bull. 46: 7–20.

Rogers, A. D. 1999. The biology of *Lophelia pertusa* (Linnaeus 1758) and other deep-water reef-forming corals and impacts from human activities. Int. Rev. Hydrobiol. 84: 315–406.

Wheeler, A. J., B. J. Bett, D. G. Masson, and D. Mayor. 2005. The impact of demersal trawling on North East Atlantic deepwater coral habitats: the case of the Darwin Mounds, United Kingdom. Pages 807–817 *in* P. W. Barnes and J. P. Thomas, eds. Benthic habitats and the effects of fishing: American fisheries society, symposium 41, Bethesda. 890 p.

_____, D. S. M. Billett, D. G. Masson, and A. J. Grehan. 2001. The impact of benthic trawling on NE Atlantic coral ecosystems with particular reference to the northern Rockall Trough, Report CM 2001/R:11. ICES, Copenhagen. 2 p.

ADDRESSES: (E.M.D.S., P.J.S.J.) *Environment and Society Research Unit, Department of Geography, University College London, Pearson Building, Gower Street, London WC1E 6BT, United Kingdom.* CORRESPONDING AUTHOR: (E.M.D.S.) *Phone: +44-(0)207-679-0531, Fax: +44-(0)207-679-0565, E-mail: <e.santo@ucl.ac.uk>.*

# Spatial identification of closures to reduce the by-catch of corals and sponges in the groundfish trawl fishery, British Columbia, Canada

JEFF A. ARDRON, GLEN S. JAMIESON, AND DORTHEA HANGAARD

## Abstract

From 1996 to 2004, approximately 322 t of cold-water corals and sponges were observed as by-catch in British Columbia's (BC) groundfish bottom trawl fishery. We explore an efficient spatial establishment of closures in BC to significantly reduce by-catch and destruction of habitat-forming corals and sponges. Density analyses of by-catch locations indicate twelve areas of high coral/sponge species concentration, representing about 7.5% of BC's continental shelf and slope, but about 97% of all coral/sponge by-catch by weight. These twelve areas represent the diversity of corals and sponges identified in the observer data, though site-specific verification is required due to low confidence in species identification in the dataset. These twelve areas are of average economic value to the fishery, however, because the fishery is an individual quota fishery, and due to the mobility of many groundfish species, it is difficult to estimate the potential economic cost of establishing these closures. Closing an area does not necessarily mean that mobile individuals of targeted species would not be caught elsewhere. Overall, the proposed potential closure areas contain about one quarter of historic (1996–2002) trawl sets.

Cold-water corals have been poorly studied in British Columbia (BC) to date (see Jamieson et al., 2007, for known species and occurrences). The following three families in particular are potential widely distributed, habitat-forming corals in BC and Alaskan waters (Etnoyer and Morgan, 2003; Jamieson et al., 2007): (1) Primnoidae ("red tree"): Mostly at depths between 100–500 m attached to boulders, and large specimens > 1 m in height. (2) Paragorgiidae ("bubblegum trees"): Mostly at depths between 0–800 m and species, and large specimens > 2.5 m in height. (3) Isididae ("bamboo coral"): Mostly at depths between 600–1200 m, and are rivalled in size only by paragorgids.

The reef-building hexacoral *Lophelia pertusa* (Linnaeus, 1758) has been found in Juan de Fuca Canyon (Hyland et al., 2004), Alberni Inlet, and the Strait of Georgia (Jamieson et al., 2007), but its overall abundance and distribution in BC is unknown.

As with corals, the ecological significance of habitat-forming sponges has been scantily studied in BC, although the extent of sponge reefs, or bioherms, has been better documented (Conway et al., 1991; Conway, 1999; Jamieson and Chew, 2002; Conway et al., 2005; Krautter et al., 2006). Cook (2005) provides the most comprehensive analysis, while Jamieson and Chew (2002) list trawl by-catch species from the bioherms and nearby locations. The discovery of globally unique hexactinellid (glass) sponge reefs in BC's trawling grounds has resulted in considerable scientific and public interest. They have

George, R. Y. and S. D. Cairns, eds. 2007. Conservation and adaptive management of seamount and deep-sea coral ecosystems. Rosenstiel School of Marine and Atmospheric Science, University of Miami.

been mapped in four general locations in Hecate Strait and Queen Charlotte Sound and cover about 38,604 ha (K. Iwanowska, Natural Resources Canada, March 2006, pers. comm.), and smaller reefs have been found in the Strait of Georgia and Queen Charlotte Strait (K. Conway, Natural Resources Cananda, pers. comm.). Bioherms can increase in size over the course of centuries to > 19 m in height, and can have live sponges up to 1.5 m high (Conway et al., 2001; Krauter et al., 2001). Numerous species of fish and crustaceans have been observed in video transects of the bioherms (Cook, 2005).

This paper seeks to explore on the basis of current data the most efficient use of spatial protection through fishery closures to significantly reduce both coral and sponge trawl by-catches and the destruction of suggested significant, deep-water biogenic habitat.

**The Groundfish Trawl Observer Program.**—Beginning in 1996, 100% observer coverage became mandatory in BC's bottom trawl fishery. Coral and sponge by-catches have been recorded from that time, but as non-commercial species they were not reported in much taxonomic detail, and training of observers on their taxonomic identification was minimal to non-existent. Furthermore, the observer reporting categories were not always easily transferable to a particular taxonomic category. This situation has improved somewhat since 2003, but unfortunately, these more recent data were not made available for the present analysis.

**Coral and sponge trawling closures to date.**—In BC, trawl damage to the globally unique hexactinellid sponge reefs has been detected by both multi-beam acoustic surveys and video transects (Conway, 1999; Conway et al., 2001; Krauter et al., 2001). In 2000, the industry was asked to comply with voluntary closures around the reefs. However, video evidence that voluntary protection of the reefs was not being achieved led to mandatory fisheries closures in July 2002 (Jamieson and Chew, 2002), and these were subsequently expanded on 1 April 2006, on the basis of revised mapping of sponge bioherm locations. No specific closures to protect cold-water corals currently exist in BC.

## Methods

**Groundfish Trawl Observer database, 1996–2002.**—Observed by-catch was collated from the DFO groundfish trawl observer database from 1996 (the year the program began) to 2002. Summarized landings were also provided for 2003 and 2004, but the spatial data were withheld. Although the master DFO database includes start and end points (and more recently recorded mid-points), for reasons of confidentiality the dataset that was provided to us had only calculated midpoints. Data for the months of Jan–Mar 2001 were inadvertently not provided. However, these three missing months of data are unlikely to appreciably alter results as fishing in the winter is limited. There were 1,351,479 separate observations overall (all species), with 1,301,392 recorded as being from bottom trawls. Of these, 3888 observations were of corals and sponges, or 0.30% overall. Catches of corals and sponges were recorded in 2.62% of all bottom trawls, though many of these were incidental amounts (Table 1). Overall, from 1996 to 2004, about 322 t of corals and sponges were visually estimated by observers as by-catch in BC's groundfish bottom trawl fishery (Table 2).

Observations from the following observer recording categories were considered: "stony corals," "soft corals," "gorgonian corals," "calcareous corals," "sea pens," "glass sponges," "bath sponges," and "sponges." These categories were deemed to likely represent habitat-forming and also long-lived organisms sensitive to damage. On closer examination, reporting in many of these categories appeared to be inconsistent. This lack of consistency reflects the difficult taxonomy of these species and the cursory training received by observers during this time. Thus for our analy-

Table 1. Number of coral or sponge (c-s) observations in the trawl fishery of British Columbia by year; the percentage that these represented of all by-catch observations; and the percentage that these constituted of all fishing tows.

| Year | Recorded c-s observations | Proportion of all bottom trawl spp. observations | Proportion of all bottom trawl tows |
|---|---|---|---|
| 1996 | 271 | 0.14% | 1.10% |
| 1997 | 365 | 0.20% | 1.91% |
| 1998 | 509 | 0.27% | 2.08% |
| 1999 | 613 | 0.31% | 2.61% |
| 2000 | 806 | 0.40% | 3.53% |
| 2001 | 611 | 0.42% | 4.20% |
| 2002 | 713 | 0.36% | 3.67% |
| Overall | 3,888 | 0.30% | 2.62% |

ses, the categories were initially grouped together. Areas that emerged containing higher coral and sponge by-catch density were then individually examined, linking back to the observation records to detect possible trends in the reported categories.

**GIS analyses.**—All calculations were performed in the BC Albers Equal Area projection, which largely preserves area (though not shape or direction). Using an unequal area projection such as geographic longitude-latitude (degrees) would have skewed results southward since individual grid cell area decreases as latitude increases. Our calculations used equal area grids of 1 ha per cell (100 m × 100 m).

ArcView™ 3.2 and 8.2 with the Spatial Analyst extensions for each were used for all GIS analyses. To map the density of coral and sponge by-catch, we employed the standard "out of the box" kernel density analyses available in the Spatial Analyst extension. A density analysis moves though each grid cell on the map, summing the values of the points found within that grid cell, and to a lesser extent, all other points found within a specified "search radius" of that cell, weighted inversely by distance. Thus, a region with several moderate landings will show as denser in by-catch than an area with, say, only one larger landing. This approach mitigates results being skewed by single large landings or misreporting, as can happen if only largest values are considered. Also, this approach addresses the issue of grid squares straddling areas of high by-catch, and thereby dividing the results, as can occur in systems that simply bin results into grids or statistical areas, without consideration of neighboring areas.

The steps of the GIS density analysis are outlined below, and then further explained in the Results section, under the sub-section "Spatial distribution of by-catch."

Table 2. Total observed British Columbia groundfish trawl by-catch of corals and sponges. Due to changes in the observer program over time, trends should be interpreted cautiously. Generally, data from more recent years were considered more reliable.

| Year | By-catch (g) | Comments |
|---|---|---|
| 1996 | 7,894 | Observer program begins |
| 1997 | 39,444 | |
| 1998 | 22,178 | |
| 1999 | 21,813 | |
| 2000 | 78,778 | Voluntary sponge closures |
| 2001 | 101,332 | Voluntary sponge closures |
| 2002 | 23,155 | Legal sponge closures |
| 2003 | 17,216 | Legal sponge closures |
| 2004 | 10,570 | Legal sponge closures |
| Total | 322,379 | |

(1) Median tow length: All tow lengths were plotted in a histogram to (a) detect normality or multi-modal distributions, and (b) determine a median distance to be used as the "search radius" in the density analyses (10.0 km);

(2) Density analyses: Using the mid-points of all trawl tows that had coral and/or sponge observations (from the eight categories listed above), density analyses were performed using observed by-catch weight and calculated CPUE, separately;

(3) Normalization: Density values were square root transformed;

(4) Break point: A histogram of the transformed by-catch density was examined alongside the mapped results, classed according to standard deviations, to suggest a break point in the distribution of "high" density areas (0.5 standard deviations);

(5) Demarcating High Density: Boundaries of these areas of high density were hand-drawn following the above-selected break point and visually smoothed;

(6) Coral-sponge by-catch data from points found within these boundaries were compared with points outside, and also data were compared across each of the twelve identified areas of high by-catch.

**Preliminary economic analysis.**—The value of every bottom trawl tow from 1996–2002 was estimated using historical prices for all commercial species noted in the observer records. Because there was not much fluctuation in yearly average prices, 2002 prices were used to standardize the values. Next, the value (2002 dollars) per unit effort (VPUE) was calculated for each tow. A density analysis (not shown) was then conducted to identify areas of higher and lower VPUE, using the same methods as described above for calculating the density of by-catch.

## Results

**Numerical distribution of by-catch.**—From 1996 to 2002, 3915 catches of corals and sponges were recorded in 2.62% of all bottom trawls. The vast majority of these records were of small by-catches, usually just a few kilograms (median = 2.3 kg; i.e., 5.0 lb—the observers noted their observations in pounds). In each year, however, there were also several very large by-catches observed. Overall, 9.4% of recorded coral-sponge by-catch was > 91 kg (200 lb), and 3.9% (i.e., 154 records) were > 454 kg (1000 lb). The largest observed coral-sponge by-catches were very large, with seven catches estimated to be > 4536 kg (10,000 lb), and three > 11,340 kg (25,000 lb). It should be noted that all of these larger landings were visually estimated by the observers and may have been over- or under- estimated to a considerable degree. Nonetheless, taken altogether, the observed by-catch of corals and sponges was extremely steeply skewed. Despite log-transformation, plots noticeably rise to the right, and the heaviest quartile of landings accounted for about 96% of total landings—99% by CPUE. Such curves indicate that while the majority of landings were quite small, large landings heavily influenced the overall statistics. This was compounded by the aggregated taxonomic categories, wherein many small species were lumped together with a few larger ones.

**Spatial distribution of by-catch.**—Mapping the distribution of coral and sponge by-catch initially showed no particular trends; the midpoints of sets occurred throughout the trawling grounds of BC's shelf and slope. Mapping the top quartile of points began to show some concentrations, but these were difficult to discern or interpret. A density analysis of all mid-point coral and sponge data produced much clearer patterns, the details of which are explained below.

Plotting bottom trawl observations in a histogram of calculated tow distance (duration * speed), exhibited a bimodal distribution, with two different fishing behaviors distributed according to location and depth. Most trawling occurred on the relatively narrow continental shelf or at the top edges of the continental slope; however, fishing in waters deeper than 500 m, down the shelf slope, was markedly longer and slower, accounting for the second smaller "hump" of the bimodal distribution. Because most trawling and by-catch of corals and sponges occurred in waters shallower than 500 m, however, this distribution of tows was considered when determining what would constitute a neighboring point in the density analysis ("search radius"). The median length of these bottom trawls in waters < 500 m depth was calculated to be 10.0 km, for the years of 2001 and 2002—the only years in the provided dataset when speed was recorded. This is comparable to the mean tow length of 9.6 km around the sponge reefs found in a separate analysis performed by Jamieson and Chew (2001) that had access to start and end point data.

Using the "kernel" density option, in ArcView Spatial Analyst©, whereby points nearby are weighted more than points farther away (inverse distance weighted), longer search radii (e.g., 20 km) gave "fuzzier", more generalized results, whereas shorter radii (e.g., 5 km) appeared fragmented and somewhat more difficult to interpret. Thus, using the 10.0 km search radius also made visual sense. Nonetheless, the density analyses were robust to such wide variations in search radii, and results grew or contracted in a predictable and consistent fashion.

Once plotted in a histogram, results of the 10.0 km density analyses appeared nonnormally distributed. To aid in the calculation of distribution statistics, these data were square-root transformed. Densities of both transformed observed weights (kg km$^{-2}$)$^{0.5}$ and catch per unit effort (CPUE) (kg hr$^{-1}$ km$^{-2}$)$^{0.5}$ were considered. CPUE was approximated by dividing total by-catch observed by the hours the net was towed (net size data were not available). It was found that CPUE calculations tended to over-emphasize the areas of extremely large by-catch, as well as sets of very short duration. We postulate that this is because in sets of large by-catch, the presence of corals or sponges was likely detected by the fisher and the set was terminated early; or, in the cases of very short duration sets, these may have been terminated for technical reasons. Unfortunately, access to the "success code" of each set was not provided, which might have provided answers. In both cases, a key assumption of CPUE analysis—the normal distribution of fishing effort in all areas—was violated, which appears to have produced spurious results. For this reason, we used only the density of catch weight to inform our final results.

The histogram of transformed by-catch density examined alongside the mapped results (both classed according to standard deviations) suggested a break point in the distribution of "high" density areas at approximately 0.5 standard deviations. Using the 0.5 standard deviations as a guideline, areas of high density were hand-drawn using a GIS, and smoothed visually to identify twelve key by-catch areas (Fig. 1).

While these twelve areas captured 61.5% of all coral-sponge records, they accounted for 97% of all by-catch by weight, and 98.8% by CPUE. Closer examination revealed that the majority of by-catch was captured in the three areas that overlapped the hexactinellid sponge reef locations; i.e., area numbers 4, 6, and 8. These three areas, which had just 16.3% of all coral-sponge records, accounted for 85.0% of by-catch by weight (92.3% by CPUE). This strongly suggests that landings from these three areas represented either larger species, such as the hexactinellid sponges, or a higher density of species. It would also suggest that the large landings from these three areas could be obscuring trends with regard to the capture of other smaller or less dense species, and that the data ought to be re-stratified based on these three areas. When the three areas were removed from the analysis, the remaining nine areas captured 54.0% of remaining

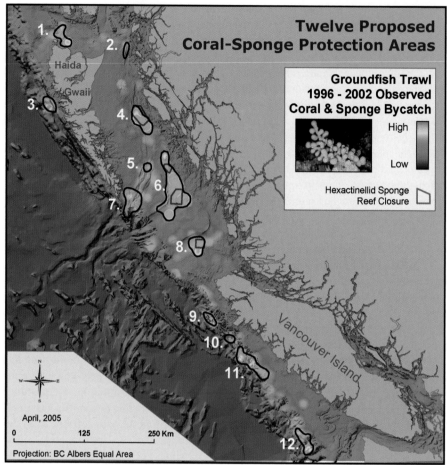

Figure 1. Relative density (square root transformed) of coral and sponge by-catch, where orange areas account for more than one standard deviation, and yellow is approximately 0.5 standard deviations of the overall relative density after square-root transformation. Proposed Coral-sponge protection areas: (1) Learmonth Bank; (2) Bell Passage; (3) Kindakun; (4) McHarg Bank; (5) Mid-Moresby Trough; (6) Mitchell's Trough; (7) South Moresby Gully; (8) Goose Trough; (9) Kwakiutl Canyon; (10) Crowther Canyon; (11) Esperanza Canyon; (12) Barkley Canyon (numbers correspond to Table 3).

coral-sponge records, and 80.9% of by-catch by weight (84.5% by CPUE). While not as large a proportion as the overall values quoted above, which were heavily biased by the three hexactinellid reef areas, these area identifications contain a much greater proportion of records than would be expected from a random species distribution.

**Preliminary biodiversity analysis.**—The proposed twelve protected areas were examined to consider how well by-catch from each coral-sponge category in the observer database would have been minimized with spatial closure. As stated above, although data in these categories are somewhat unreliable, they do provide a first indication of the possible diversity of corals and sponges from that area. From the raw data, subsets from each category were selected that appeared to represent the most reliable observations. For most categories, the most reliable data were from the most recent years; however, for the more readily identifiable "gorgonian" and "sea pen" categories, data from all years were included. Overall, results showed large proportions of each category were in the identified areas—about 80% or higher. However, "stony corals" had 64% inclusion and "sea pens" 47%, due to the fairly widespread distribution of these organisms (Table 3).

Due to the low confidence placed in species identification in the data analyzed, the above values should be interpreted cautiously. At best, this preliminary analysis suggests that a diversity of structural organisms would benefit by minimizing gear impacts in these areas. Detailed non-destructive surveys of the specific sites being considered would clarify the actual spatial occurrences of species in the different groupings.

**Preliminary economic analysis.**—While some of the proposed Coral-Sponge Protection Areas were in higher VPUE areas, many were not. The twelve proposed protected areas accounted for 30.3% of 1996–2002 landings by weight and 30.6% of their calculated value, indicating that they were of average value by overall landed weight. Spatially, they occupy about 7.5% of the BC coast (shelf and slope to 2000 m) and contained 24.1% of the 1996–2002 trawl mid-points used in our analyses.

**Evaluation of the effectiveness of existing closures.**—Most present-day marine protected areas and trawl fishery closures occur in areas that historically had little trawl activity (1.4%–Table 4) and virtually no coral or sponge by-catch. The exceptions are the hexactinellid fishery closures. From 1996–2001, about one third of all coral and sponge by-catch would have been prevented by the hexactinellid sponge reef closures, had they been enacted prior to 2002 (75,126 kg out of 229,469 kg). The remaining two thirds of all coral and sponge by-catch did not occur in the initial sponge closure areas (154,343 kg). Thus, while helpful, the 2002 hexactinellid closures did not by themselves adequately minimize the coral and sponge by-catch issue.

## Discussion

A total of about 322 t of corals and sponges were reported as by-catch in BC's trawl fishery from 1996 to 2004. Because these are non-commercial species, these observations were likely under-reported. Also, it is likely that many of the damaged coral and sponge fragments remained on the sea floor. Thus, this estimate is possibly many times smaller than the actual destruction or damage to these species that occurred. Some fishers, on the other hand, believe that because the by-catch is usually thrown overboard, fragments of previously caught corals could be being repetitively caught in subsequent sets. Regardless, we would suggest this magnitude of damage is very high, particularly since the actual abundance and spatial distribution of corals and sponges in fished areas remains largely unknown.

Sinclair et al. (2005) recently analyzed BC trawl data to describe conditions important for determining fishing locations and areas of high fish density. Their analyses of fishing effort spatial distribution (not shown) were similar to, but more focused than ours. The two approaches differed in objectives and methods: they mapped results binned into 1 km$^2$ grid cells, looking for species-habitat associations; whereas we performed a density analysis to look for spatial trends in by-catch occurrence. We believe each approach, though different, was appropriate to the respective objectives. Our density analysis may make "hotspot" areas perhaps appear somewhat larger than they actually are due to the decay radius of the interpolation. In the context of conservation planning, a slightly larger area automatically provides a buffer against data uncertainty and spatial gaps. In

Table 3. Coral-sponge taxonomic groupings estimated to be hypothetically "protected" by the proposed Coral-Sponge Protection Areas (CSPAs) off British Columbia. The most reliable subsets of observations for each observer category were used as indicator species (see text); subset years are in the column headings. In the rightmost column, main species groups are indicated for each area. Go = gorgonian corals, St = Stony Corals, So = Soft Corals, Sg = Sponges, Gl = Glass Sponges, Ca = Calcareous Corals, Sp = Sea Pens; Hex = in the vicinity of a known hexactinellid sponge reef. CSPA numbers correspond to Figure 1. Table entries are weights in kg.

| Subset (yr) CSPA No. | Go 96–02 | St 2002 | So 2002 | Sp 96–02 | Sg 00–02 | Gl 2001 | Ca. 97, 02 | Main species |
|---|---|---|---|---|---|---|---|---|
| 1 | 2,396 | 0 | 0 | 0 | 2 | 0 | 0 | Go |
| 2 | 0 | 0 | 0 | 517 | 30 | 0 | 0 | Sp |
| 3 | 62 | 3 | 3 | 3 | 2,198 | 2 | 0 | Sg |
| 4 | 24 | 0 | 0 | 1 | 907 | 454 | 0 | Sg, Hex |
| 5 | 471 | 97 | 0 | 36 | 3 | 0 | 0 | Go, St |
| 6 | 196 | 202 | 1,928 | 3 | 86,824 | 2,853 | 206 | St, So, Sg, Gl, Ca, Hex |
| 7 | 623 | 12 | 24 | 132 | 242 | 1 | 0 | Go, Sp |
| 8 | 16 | 0 | 0 | 2 | 1,293 | 0 | 68 | Sg, Hex |
| 9 | 7 | 0 | 0 | 136 | 5,875 | 0 | 9 | Sg, Sp |
| 10 | 59 | 0 | 5 | 68 | 242 | 7 | 3 | Go, Sg, Sp |
| 11 | 30 | 8 | 2 | 262 | 5,549 | 40 | 102 | Sp, Gl, Ca |
| 12 | 3 | 2 | 0 | 42 | 515 | 71 | 0 | Sg, Gl |
| Total (kg) | 3,888 | 324 | 1,959 | 1,202 | 103,682 | 3,428 | 388 | |
| All BC (kg) | 4,906 | 510 | 2,134 | 2,553 | 105,241 | 3,486 | 488 | |
| Potentially "Protected" | 79.20% | 63.50% | 91.80% | 47.10% | 98.50% | 98.30% | 79.50% | |

Table 4. Hind-casting the possible effect that existing closures would have had on historical pre-closure fishing activity off British Columbia. 1996–1997: total trawl tows (sets) = 39,859; total tows with observed coral-sponge (C-S) by-catch = 636. RPA (rockfish protection areas): commercial rockfish capture is illegal; RCA (rockfish conservation area): both commercial and recreational rockfish capture is illegal.

| Closure | Affected tows 96–97 | | Affected c-s tows 96–97 | |
|---|---|---|---|---|
| Thornyhead | 190 | 0.48% | 0 | 0.00% |
| Hex. Sponges | 275 | 0.69% | 28 | 4.40% |
| RPAs | 62 | 0.16% | 0 | 0.00% |
| RCAs | 7 | 0.02% | 0 | 0.00% |
| Other Closures | 17 | 0.04% | 0 | 0.00% |
| Total | 551 | 1.38% | 28 | 4.40% |

this context, we believe the buffering effect is a reasonable and desirable attribute, since it is likely neighboring areas of higher coral and sponge density exist where trawling may not yet have occurred in the vicinity of areas identified here. The proposed areas that have a high density of by-catch are not necessarily evenly populated with corals and sponges and may more likely have patchy distributions. Density analyses link these otherwise disparate zones, thereby identifying spatial trends in by-catch. On the other hand, density analyses can bridge together areas that might be better managed separately, due to topography or other factors. These specific considerations were considered outside the scope of a preliminary analysis, and would be best dealt with on a case by case basis, during the course of delineating specific closure boundaries.

It is difficult to characterize any potential economic loss that would occur if the twelve proposed areas identified here were closed to trawling. At first glance, the proportion of historic value (30.6%) from these areas might appear to be an appropriate statistic, but this would be an implausible "worse case" as it incorrectly assumes that both commercial fishes within the areas would not move outside the identified potential protected areas, where they could then be caught (spill-over effect), and that species compositions would remain constant. Modeling studies suggest that the substantial mobility of many commercial species largely offsets conservation benefits for them from spatial closures smaller in size than the area ranges of individual species (Walters and Bonfil, 1999), indicating that while closures may benefit non-motile species, their effect on mobile species can be quite different. But, as found in protected areas worldwide, protected areas can actually allow for increases in neighboring fisheries through spill-over effects of some fished species (Halpern, 2003; Hastings and Botsford, 1999, 2003). Thus, the economic effects of these potential coral and sponge protected areas may even be positive for some species if recruitment rates are enhanced through greater reproduction of some fished species. In either case, the observer data indicate that the commercial species caught within the twelve proposed areas widely occur outside of them. Because much of the bottom trawl fishing is based on individual quotas, it is most likely that such quotas could be filled elsewhere with a minimum of economic hardship. However, "freezing the footprint" of current trawling would minimize harm to more pristine areas. The inherent complexity of the trawl fishery and its related benthic ecosystems, however, suggests that in situ studies measuring actual effects would be the only true indicator of proposed closure economic costs and ecological benefits.

While it is recognized that hind-casting has limitations, it is the best tool presently available to evaluate possible futures. However, the point inevitably emerges that protecting areas that have shown high coral/sponge by-catch in the past may be a bit like closing the barn door after the animals have escaped, i.e., it is not presently known if and when corals could recover at these sites, or if patches of untrawled areas of coral and/or sponge abundance remain within the overall trawled area. We considered this possibility by analyzing across years to see if there were spatial shifts in coral/sponge landings. Generally, no such trends were evident, possibly due to: (1) the shortness and high noise of the 7-yr time series obscuring any temporal trends; and (2) trawl tows may be interspersed on softer bottoms throughout harder bottom coral-sponge habitats, and are "nibbling" away at them. The continued reporting of corals and sponges as by-catch suggests that it is quite likely that patches containing some living healthy structural organisms, even long-lived ones, still exist in the trawled areas, but if trawling is allowed to continue, in time they will almost certainly be largely destroyed.

## Acknowledgments

Coral and sponge observer data used in this analysis were provided by Fisheries and Oceans Canada (DFO), Pacific Region. We thank reviewers from the DFO Centre for Science Advice, Pacific Region, for their helpful comments; J. Hall-Spencer of the University of Plymouth who reviewed an earlier version of the paper; as well as two other anonymous reviewers. The authors would also like to thank the observers who shared their extensive knowledge. J.A.'s and D.H.'s research was funded by Lazar Foundation, EJLB Foundation, and Mountain Equipment Coop.

## Literature Cited

Conway, K. W. 1999. Hexactinellid sponge reefs on the British Columbia continental shelf: geological and biological structure with a perspective on their role in the shelf ecosystem. Canadian stock assessment secretariat research document. 99/192: 21 p.

————, J. V. Barrie, W. C. Austin, and J. L. Luternauer. 1991. Holocene sponge bioherms on the western Canadian continental shelf. Cont. Shelf. Res. 11: 771–790.

————, M. Krautter, J. V. Barrie, and M. Neuweiler. 2001. Hexactinellid sponge reefs on the Canadian shelf: a unique "living fossil". Geosci. Can. 28: 271–278.

————, J. V. Barrie, W. C. Austin, P. R. Ruff, and M. Krautter. 2005. Deep-water sponge and coral habitats in the coastal waters of British Columbia, Canada: multibeam and ROV survey results (abstract). Presentation at the: third international symposium on deep-sea corals, Miami, Nov 28–Dec 2, 2005: 32.

Cook, S. E. 2005. Ecology of the hexactinellid sponge reefs on the western Canadian continental shelf. M.Sc. thesis, Univ. of Victoria, Victoria, BC. 136 p.

Etnoyer, P. and L. Morgan. 2003. Occurrences of habitat-forming cold water corals in the Northeast Pacific Ocean. A report to NOAA's office of habitat protection. Marine Conservation Biology Institute. 33 p. Available from: http://www.mcbi.org/publications/pub_pdfs/Etnoyer_Morgan_2003.pdf via the internet. Accessed 18 Feb. 2007.

Halpern, B. 2003. The impact of marine reserves: Do reserves work and does size matter? Ecol. Appl. 13: 1 Supp. 117–137.

Hastings, A. and L. Botsford. 1999. Equivalence in yield from marine reserves and traditional fisheries management. Science 284: 1–2.

———— and ————. 2003. Comparing designs of marine reserves for fisheries and for biodiversity. Application of ecological criteria for evaluating candidate sites for marine reserves. Ecol. Appl. 13:1 Supp. 215–228.

Jamieson, G. and L. Chew. 2002. Hexactinellid sponge reefs: areas of interest as marine protected areas in the North and Central Coasts areas. Canadian science advisory secretariat research document 2002/122: 78 p.

_____, N. Pellegrin, and S. Jessen. 2007. Taxonomy and zoogeography of cold water corals in coastal British Columbia. Pages 215–229 *in* R. Y. George and S. D. Cairns, eds. Conservation and adaptive management of seamount and deep-sea coral ecosystems. Rosenstiel School of Marine and Atmospheric Science, University of Miami. Miami. 324 p.

Krautter, M. W. Conway, and J. V. Barrie. 2006. Recent hexactinosidan sponge reefs (silicate mounds) off British Columbia, Canada: frame-building processes. J. Paleontol. 80: 38–48.

_____, _____, _____, and M. Neuweiler. 2001. Discovery of a "living dinosaur": Globally unique modern hexactinellid sponge reefs off British Columbia, Canada. Facies 44: 265–282.

Sinclair, A. F., K. W. Conway, and W. R. Crawford. 2005. Associations between bathymetric, geologic and oceanographic features and the distribution of British Columbia bottom trawl fishery. ICES CM 2005/L: 25 p.

Walters, C. J. and R. Bonfil 1999. Multispecies spatial assessment models for the British Columbia groundfish trawl fishery. Can. J. Fish. Aquat. Sci. 56: 601–628.

ADDRESSES: (J.A.A., D.H.) *Living Oceans Society, 235 First St., Sointula, BC, V0N 3E0, Canada.* CURRENT ADDRESS: (J.A.A.) *German Federal Agency for Nature Conservation, Isle of Vilm, 18581 Putbus, Ruegen, Germany.* (G.S.J) *Fisheries and Oceans Canada, Pacific Biological Station, Nanaimo, BC, V9T 6N7, Canada.* CORRESPONDING AUTHOR: (J.A.A.) *E-mail: < jeff.ardron@ bfn-vilm.de>.*

# U.S. Pacific Coast experiences in achieving deep-sea coral conservation and marine habitat protection

GEOFFREY SHESTER AND JON WARRENCHUK

## Abstract

We constructed comprehensive management proposals to protect deep-sea corals, sponges, and other seafloor habitat while maintaining fishing opportunities. The proposals were largely adopted by the North Pacific and Pacific Fishery Management Councils in recent decisions regarding Essential Fish Habitat in the United States. The proposals were based on an approach we developed following a comprehensive literature review of fish habitat studies, habitat-fishery linkages, life history of habitat-forming invertebrates, and fishing impacts. The approach freezes the existing bottom trawl footprint, closes habitat areas within the footprint that have low fishing effort, closes sensitive habitat such as coral gardens and seamounts, and requires ongoing research and monitoring. Here, we describe the iterative process through which the proposals were considered by decision-makers and discuss the factors that led to their adoption. Overall, the implementation represents a significant step toward ecosystem-based fishery management and lays a framework for a new era of ocean conservation.

**The challenge of seafloor conservation.**—Sustainability is the challenge of our age. Sustainability requires that we develop innovative approaches to using the planet's renewable natural resources while maintaining the natural capital that produces them. The ecological sustainability of fishing is determined by many factors, including the level and distribution of fishing effort with each gear type, population structure of harvested fish stocks, interactions between marine species, and environmental conditions (Pauly et al., 2002). Because an estimated 98% of all marine species live in or on the seafloor (Thurman and Burton, 2001), a foundation of marine biodiversity conservation is seafloor habitat protection, which we define to include living and non-living structures. Pikitch et al. (2004) state that spatial zoning of destructive fishing gears to protect Essential Fish Habitat (EFH) is a critical component of ecosystem-based fishery management. This paper presents a model framework applied to two case studies that highlight the scientific, legal, and practical challenges of seafloor habitat conservation. In a process designed to protect fish habitat, we built a framework to protect vulnerable habitat areas and to incorporate ecosystem-based management principles into fishery management.

Working for the international conservation organization, Oceana, we developed and submitted proposals to the North Pacific and Pacific Fishery Management Councils (NPFMC and PFMC) with the goal to maintain fishing opportunities while protecting habitat and biodiversity. The proposals were incorporated as management alternatives within the context of two federal environmental impact statements (EIS) aimed at designating and minimizing adverse impacts to EFH. Specifically, we focused on minimizing the adverse impacts of bottom trawling since this gear type was known to cause the

George, R. Y. and S. D. Cairns, eds. 2007. Conservation and adaptive management of seamount and deep-sea coral ecosystems. Rosenstiel School of Marine and Atmospheric Science, University of Miami.

169

largest impacts to seafloor habitats (Watling and Norse, 1998; NRC, 2002; Morgan and Chuenpagdee, 2003). Each proposal was based on an approach developed following a comprehensive literature review of fish habitat studies, habitat-fishery linkages, life history of habitat-forming invertebrates, and fishing impacts (Shester and Ayers, 2005).

The approach used in the proposals relied on the criteria of cost-effectiveness, which we defined as achieving conservation objectives while minimizing costs to the fishing industry. The currently available fisheries data, habitat data, and economic data from Alaska and the U.S. West Coast, while not perfect, allowed us to design a precautionary, cost-effective network of protected areas.

**The legal and scientific frameworks for marine habitat protection in the United States.**—In part in response to increasing concern of destruction of marine habitat, the United States Congress in 1996 passed the Sustainable Fisheries Act, which required fishery managers to minimize to the extent practicable adverse effects on EFH caused by fishing. In this statute, "Essential Fish Habitat" is defined as "waters and substrate necessary to fish for spawning, breeding, feeding or growth to maturity" [16 U.S.C. 1802(10)]. However, due to the failure to protect fish habitat on the part of the responsible U.S. government agency, the National Marine Fisheries Service (NMFS) of the National Oceanic and Atmospheric Administration (NOAA), several conservation groups filed a lawsuit resulting in the American Oceans Campaign vs Daley (2000) court decision. As part of that litigation, the court found that NMFS had not adequately examined the effects of fishing on EFH in five U.S fishery management regions and required the agency to complete an Environmental Impact Statement (EIS) for each region to comply with the U.S. National Environmental Policy Act. In each of the five regions NMFS was compelled to consider a reasonable range of alternatives to identify EFH, determine if adverse impacts to EFH were occurring from commercial fisheries, mitigate adverse impacts to EFH, and respond to public comments. These alternatives were vetted through the quasi-regulatory regional Fishery Councils that advise NMFS on all fisheries management issues. This paper describes the alternatives we developed to protect deep-sea coral and other habitats in two fishery management regions and the process that led to their adoption.

**Corals are "Essential Fish Habitat".**—There are many reasons for protecting deep-sea corals, including their associated biodiversity, potential cures for human diseases, value for scientific study, shelter for a variety of marine species, and their uniqueness (Witherell and Coon, 2001; Roberts and Hirshfield, 2004; Stone, 2006). However, NMFS' emphasis on EFH focused on a narrow subset of these services, which are those that contribute to federally-managed commercial fisheries production. Management of fish habitat in the U.S. prior to the EFH mandate was related primarily to the harvest of habitat-forming species such as precious corals and kelp (WPFMC, 1979; Larson and McPeak, 1995). The U.S. Coral Reef Conservation Act (2000) allocated funding resources for research and education, but neither prohibited damaging activities nor applied to deep-sea corals. The process for EFH designation was designed to be risk-averse and based on the level of information about the various life stages of commercially managed fish species. If scientific research linking fish productivity or density to habitat types was not yet available, NMFS' approach for EFH designation was based on the geographical presence of commercial fish species [EFH Final Rule §600.815 (a)(1)(iii) (A)]. While we recognized that one rationale for habitat protection was commercial fish production, we advocated for a precautionary and expansive designation of EFH while compiling literature on the associations and linkages between commercially important fish and biogenic habitats. To do this, we compiled and submitted a literature review documenting that deep-sea corals and sponges provide structural habitat for commercial groundfish, shellfish, and other marine life and are particularly vulnerable to fishing impacts (i.e., Beaulieu, 2001; Brodeur, 2001; Heifetz, 2002; Husebo et al., 2002; Krieger and Wing, 2002; Buhl-Mortensen and Mortensen, 2004; Freiwald et al., 2004; Costello et al., 2005; Malecha et al., 2005; Stone, 2006). We determined that an appropriate approach in the EFH process was to focus attention on the most sensitive and long-lived benthic habitats (corals and sponges) and the most destructive commercial fishing gear (bottom trawls).

**Habitat impacts of bottom trawling.**—Bottom trawling causes reductions in habitat complexity, changes in species composition, and reductions in biodiversity (Engel and Kvitek, 1998; Watling and Norse, 1998; NRC, 2002; Chuenpagdee et al., 2003). The scale of impact is such that bottom trawling is the most widespread cause of reduced habitat complexity along the North American continental shelf and slope (NRC, 2002). The adverse impacts of bottom trawling on deep-sea coral ecosystems and hard bottom habitats are well documented by both fishery observer data and in situ studies (MacDonald et al., 1996; Krieger, 2001; Thrush and Dayton, 2002; Anderson and Clark, 2003; Hyland et al., 2005). Deep-sea corals epitomize some of the most sensitive type of biogenic habitat in that they are long-lived (Risk et al., 2002; Andrews et al., 2005; Roark et al., 2005), typically live in low-disturbance habitat, and are therefore not resilient to anthropogenic disturbance. During the EFH process, new species of deep-sea corals and sponges were described off the coast of California [*Antipathes dendrochristos* (Opresko, 2005)] and the Aleutian Islands [*Alaskagorgia aleutiana* (Sanchez and Cairns, 2004); R. Stone, NOAA, unpubl. data] highlighting how little we know about these seafloor ecosystems.

## Case 1: The Aleutian Islands, Alaska

The Aleutian Islands provided an ideal geographic setting to construct a model for reducing the habitat impacts of a destructive fishery. Industrial bottom trawling for Atka mackerel *Pleurogrammus monopterygius* (Pallas, 1810), Pacific cod *Gadus macrocephalus* (Tilesius, 1810), and Pacific Ocean perch *Sebastes alutus* (Gilbert, 1890) was first started by Russian and Japanese trawlers in the 1960s and was replaced by American vessels after 1976 (NPFMC, 2005). As of 2004, the combined value of these fisheries was approximately $30 million of ex-vessel value per year (Hiatt, 2005). In contrast to other regions, fishing effort had been relatively patchy as the fisheries themselves were relatively recent and trawlers routinely conduct exploratory tows in new, previously untrawled areas.

Federal groundfish observers have sampled the target and incidental catch (including corals, bryozoans, and sponges) of these fisheries since 1990 (AFSC, 2003). The coral, bryozoan, and sponge by-catch provided evidence of damage to the seafloor and consequently was a more direct indicator of damage than models that assumed homogenous distribution of habitat types (i.e., Appendix B in NOAA, 2005a).

**Case 1 methods.**—The approach incorporated economic and ecological elements in the following sequence: (1) Describe and identify locations of sensitive seafloor habitat using trawl survey data, by-catch information, and submersible dive data; (2) Freeze the bottom trawl footprint by examining the spatial extent of recent bottom trawl effort and closing all areas outside the footprint; (3) assess the distribution of fishery revenue within the footprint and catch per unit effort (CPUE); (4) close areas within the trawl footprint containing high coral and sponge by-catch and low target species CPUE; and (5) close seamounts and submersible-documented coral garden habitats.

Bottom trawl fishing effort in the Aleutian Islands was mapped at a resolution of 10 × 10 km. All areas containing at least one trawl tow per year over the period from 1999 to 2001 were included. Rather than maintain all areas open to bottom trawling and proposing specific closures, we began with the premise that everything is closed to trawling except for areas where trawling will have higher catch per unit effort (CPUE) of target species and lower expected damage to living seafloor structures.

Since no comprehensive map of corals and sponges in the Aleutian Islands had been created, we attempted to compile the best available data on their distribution. Initially, the two key available datasets of coral and sponge locations were trawl survey data and observer by-catch data (NMFS, unpubl. data). Information on substrate type, depths, and ecological features in the Aleutian Islands was either absent or extremely limited. During development of this proposal, submersible dives throughout the central Aleutian Islands documented the most diverse and dense deep-sea coral and sponge assemblages discovered to date (Heifetz et al., 2002; Linder, 2003; Stone, 2006). Several sites contained corals and sponges in such density that no underlying substrate was visible and these unique sites became known as coral "gardens" (Stone, 2006).

We next compared the groundfish catch value of each area to the relative abundance of biogenic habitat features in a Geographic Information System (GIS) using the above described datasets (for further details, see Shester and Ayers, 2005). The output was a proposal map that froze the existing bottom trawl footprint in the Aleutian Islands, closed areas within the footprint with little fishing effort and high coral by-catch, and closed unique coral gardens and seamounts.

In December 2002, we presented and submitted the proposal to the NPFMC which included it as Alternative 5B for analysis in NMFS' Draft Alaska Region EFH EIS. The proposal included bottom trawl effort reduction proportional to displaced revenue, hard limits on coral and sponge by-catch, increased monitoring and research, and a framework for adaptive management.

**An iterative process.**—In the draft EFH EIS, NMFS concluded that there were no habitat impacts caused by fishing that were significant enough to require mitigation or new habitat protection measures (NOAA, 2004). The agency based its rationale on the premise that all commercial fish stocks were above their Minimum Stock Size Thresholds [MSST] which is defined to be > 20% of their unfished biomass.

Oceana and an alliance of marine conservation organizations disagreed with NMFS' conclusion, and with support from scientists and the public, presented a literature review of over 200 scientific articles to refute the conclusion. NMFS, faced with a controversial subject and contrasting scientific opinion, sought peer review of their EIS by a panel of scientists from the Center for Independent Experts (CIE) (69 FR 34136). The CIE report found the MSST criteria to be inadequate as an indicator of adverse habitat impacts as it could not detect localized impacts, could not account for the expected lag between habitat loss and productivity decline, and did not consider that stock abundance was affected by many other factors (Drinkwater, 2004). The CIE report highlighted "the destruction of corals and sponges with their long recovery times are of particular concern" (Drinkwater, 2004).

The NPFMC then reconsidered the spatial components of Alternative 5B with some modification of the trawl footprint based on logbooks brought forth by the bottom trawl industry. The bottom trawl industry opposed using the coral and sponge by-catch data collected by observers as a basis for closures, and opposed by-catch limits. Industry cited that the data did not distinguish where along a tow the by-catch actually occurred, total weights were extrapolated from sub-samples, and that trawl nets were not designed to sample corals and sponges. However, several studies to date have used trawl survey and observer by-catch records to assess distribution of these invertebrates and adverse impacts of trawling (Heifetz, 2002; Anderson and Clark, 2003; Heifetz et al., 2005).

**Case 1 results.**—On 11 February, 2005, the NPFMC adopted a modified version of Alternative 5B for the Aleutian Islands which established an open bottom trawl area based on the footprint of historical trawl effort, protected six known Aleutian Island coral gardens from all bottom contact, and closed all areas outside the footprint to bottom trawling (Fig. 1). The total area protected is substantial and occurs over a wide range of depths (Table 1). The decision was formalized through a Record of Decision by NMFS, incorporated as Amendment 56 to the Bering Sea/Aleutian Islands Groundfish Fishery Management Plan, and implemented through regulations on 28 June, 2006. The adopted plan did not reduce fishing effort or establish by-catch limits for corals and sponges, but required all trawl vessels to use vessel monitoring systems (VMS) in the Aleutian Islands and called for additional research.

In contrast to the original Oceana proposal, the NPFMC did not close many of the areas in the Aleutian Islands with the highest coral by-catch. Only 4% of observed coral by-catch records (9% by weight) fall in areas closed to bottom trawling. However, coral by-catch rates in closed areas were higher than in areas remaining open (87 kg per tow in open areas vs 188 kg per tow in closed areas). Still, despite the unchanged total fishing effort, the implementation of this policy is expected to result in less coral and sponge by-catch, less overall damage to the seafloor and the protection of vast areas of unexplored and untrawled seafloor habitat. The explicit consideration of habitat is a step towards ecosystem-based fisheries management.

## Case 2: California, Oregon, and Washington

Until primitive drag nets began being used to capture fish in 1876 off the Pacific Coast, the seafloor of the Pacific Coast had been largely untouched by commercial fishing operations. These early drag nets were mostly limited to low relief, soft substrate habitat (J. Easely, Oregon Trawl Commission, unpubl. data). The subsequent rise in horsepower of commercial fishing vessels, development of hydraulic gear and synthetic nets in the 1950s and roller gear in the 1960s gave fishermen access to previously inaccessible rough-bottom habitat (J. Easely, Oregon Trawl Commission, unpubl. data). Fishermen now target Dover sole *Microstomus pacificus* (Lockington, 1879), sablefish *Anoplopoma fimbria* (Pallas, 1814), shortspine thornyhead *Sebastolobus alascanus* Bean, 1890, longspine thornyhead *Sebastolobus altivelis* Gilbert, 1896, Petrale sole *Eopsetta jordani*

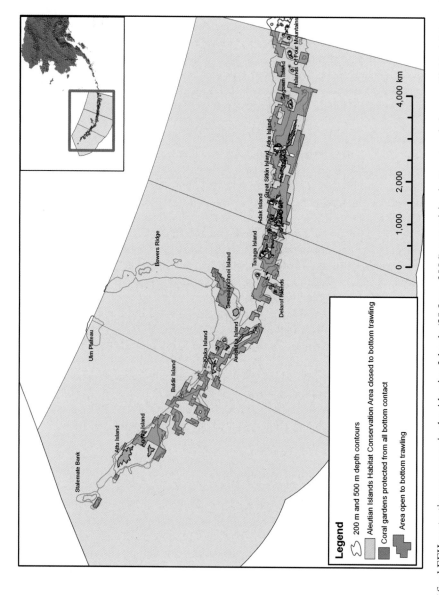

Figure 1. Map of final EFH protection measures in the Aleutian Islands (28 June, 2006) comprising bottom trawl closures (light pink), coral garden no bottom contact zones (bright pink), and remaining open areas to bottom trawling (green).

Table 1. Total area of marine habitat in the Aleutian Islands closed to bottom trawling by depth.

| Depth range | Total area closed to bottom trawling | % of depth range closed to bottom trawling |
|---|---|---|
| 0–200 m | 16,484 of 41,888 km² | 39.4 |
| 200–500 m | 42,378 of 68,802 km² | 61.6 |
| > 500 m | 889,339 of 889,405 km² | 99.9 |

(Lockington, 1879), and Pacific cod (*G. macrocephalus*) with bottom trawls off the Pacific coast for an annual ex-vessel value of approximately $25 million (Daspit, 2004).

We developed a proposal for the EFH EIS under consideration by the PFMC that manages the federal fisheries off Washington, Oregon, and California. Our proposal was one of 23 alternatives to mitigate the effects of commercial fishing on EFH. The proposal was included as Alternative 12, or the "Comprehensive Collaborative Alternative" (Appendix C in NOAA, 2005b).

**Case 2 methods.**—For the U.S. West Coast, we used the approach developed for the Aleutian Islands, and made modifications based on available data and existing management measures. To define the bottom trawl footprint in the PFMC region, we examined bottom trawl records of groundfish catch occurring from 2000–2003 (Daspit, 2004). We selected this time span to include annual variability of trawl activity which incorporates transitions that may have resulted from recent management measures, such as a bottom trawl gear footrope restriction in 2000 (Bellman and Heppell, 2005). Given the constraints placed upon the data by NOAA, a spatial resolution of 10-min blocks was selected to minimize data loss due to confidentiality. The resultant bottom trawl footprint encompassed most of the continental shelf and approximately followed the 1200 m depth contour.

**Areas of sensitive coral and sponge habitat.**—In contrast to the Aleutian Islands, very little observer data on commercial fishery by-catch of corals and sponges was available in the PFMC region due to the infancy of the West Coast Groundfish Observer Program. However, this region had a more extensive database of the substrate and habitat types of the seafloor (PFMC, 2003) which allowed us to adapt the methodology used in the Aleutian Islands to focus on sensitive habitat areas within the bottom trawl footprint. Boundaries of areas of sensitive habitat were identified using the best available datasets based on the following criteria:
• Hard substrate, including rocky ridges and rocky slopes (PFMC, 2003).
• Habitat-forming invertebrates, which included corals, sponges, sea whips, and sea pens (Etnoyer and Morgan, 2003; NMFS, unpubl. data).
• Submarine canyons and gullies (PFMC, 2003).
• Untrawlable areas (trawl hangs and abandoned survey stations) (Zimmerman, 2003).
• Seamounts (PFMC, 2003).
• Highest 20% suitability for overfished groundfish species (PFMC, 2004).

Hard substrates, which include rocky ridges and rocky slopes, are one of the least abundant benthic habitats, yet they are among the most important habitats for fishes (Hixon et al., 1991; NOAA, 2005b). Hard substrates are also the seafloor substrate type most sensitive to bottom trawling and take the longest to recover (NRC, 2002; NOAA, 2005b). We plotted over 10,000 hard substrate polygons from PFMC (2003) in GIS to determine where hard substrate habitat occurred.

We examined an extensive database to identify areas where the presence of corals and sponges were frequently recorded or large samples of these invertebrates occurred. Since sampling effort in the Pacific had not been uniformly distributed across habitat types, our intent in utilizing this data was not to predict the distribution of these animals across unsampled habitats, but rather to focus on regions where repeated samples had occurred. The database included records from NMFS slope and shelf trawl surveys from 1977 to 2003 for a total of 3290 occurrences of corals and sponges comprising 16,765 kg of samples. In addition, Marine Conservation Biology Institute's (MCBI) database, commissioned by NOAA, included 401 additional coral records compiled from various research cruises and scientific collections (Etnoyer and Morgan, 2005).

Point-density analyses were conducted using the ArcView 9.0 Spatial Analyst Point Density Tool (ESRI, 2004) to determine clusters of coral and sponge records. A cell size of 2000 m and a search radius of 10,000 m were used and the point density function computed a mean density per kilometer of coral and sponge records as well as clusters of high survey catches by sampled weight. The proposed closures to bottom trawling under Alternative 12 contained 43% of all occurrences and 66% of the invertebrate biomass that was sampled in NMFS trawl surveys.

Seamounts are sites of enriched biological activity with unique and rare benthic invertebrate communities (Probert et al., 1997; de Forges et al., 2000; Koslow et al., 2001; Stocks, 2004). The location and boundaries of all major seamounts of the U.S. West Coast were identified with a seamount database (PFMC, 2003). An additional seamount, Rodriguez Seamount, was not included in the habitat database, but was included in our analyses. We found that all seamounts were located outside the bottom trawl footprint, indicating that little bottom trawling had occurred there.

Submarine canyons have been shown to contain concentrations of benthic invertebrates (Haedrich et al., 1980; Sarda et al., 1994; Vetter and Dayton, 1999; Brodeur, 2001). Submarine canyons along the Pacific coast include Monterey Canyon, Rogue Canyon, Astoria Canyon, Greys Canyon, Eel River Canyon, Delgada Canyon, and others. The locations and geographic extent of submarine canyon habitat were plotted from a coast-wide marine habitat database (PFMC, 2003).

We also plotted all records from the NOAA West Coast Triennial Trawl Survey where major trawl net hangs and untrawlable survey stations were recorded (Zimmerman, 2003). Since these areas are considered unsuitable for trawling, the assumption is that these records indicate areas of high structural complexity, such as boulders or rock outcrops (Zimmerman, pers. comm., 2004). Trawl hangs have been shown to provide habitat for juvenile fish (Link and Demarest, 2003).

We also ensured that areas identified as the highest 20% habitat suitability for overfished groundfish species (PFMC, 2004) overlapped with our proposed protected areas. However, since the GIS files for habitat suitability were not made publicly available, the extent of overlap was only assessed visually.

With habitat and fishing effort data in the GIS, we constructed the spatial management component of Alternative 12, and proposed areas that would be open or closed to bottom trawling. We attempted to protect as much area meeting our criteria while minimizing displaced revenue, a measure of likely effort shifts. We considered the following factors while drawing boundaries for the open/closed areas to bottom trawling inside the footprint:

Table 2. Habitat features closed to bottom trawling by final EFH regulations off U.S. West Coast.

| Criteria | Closed to bottom trawling in final regulations |
|---|---|
| Hard substrate | 12,813 of 19,548 km$^2$ (65.5%) |
| Coral records | 551 of 2,396 records (23.0%) |
| Sponge records | 206 of 1,294 records (15.9%) |
| Seamounts | 5,151 of 5,151 km$^2$ (100%) |
| Submarine canyons | 8,321 of 15,286 km$^2$ (54.4%) |
| Untrawlable areas/trawl hangs | 398 of 1,847 km$^2$ (21.5%) |
| Total area of seafloor habitat within U.S. West Coast EEZ | 353,500 of 826,680 km$^2$ (42.8%) |

• Avoid highest fishery value areas when considering closed areas;
• When habitat features overlapped high fishery value areas, minimize the overlap of the resultant boundary;
• Distribute open/closed areas equitably throughout different regions of the coastline;
• Draw closed area boundaries that tightly encompass identified habitat features;
• Minimize the number of waypoints (corners) to make the boundaries simple and therefore, easier to enforce (recommendation of NMFS Enforcement and U.S. Coast Guard).

**An iterative process.**—After Alternative 12 was submitted in October 2004, two important sources of information became publicly available; (1) a proposal developed by representatives from the bottom trawl industry indicating habitat areas that they were willing to close, and (2) trawl logbook information from 2003–2004 where the start and end points of bottom trawl tracks were recorded. The trawl track information was far superior to the grossly aggregated 10 × 10 min blocks of fishing effort data as it emphasized the important fishing areas as well as habitat areas that were avoided or lightly fished. The information greatly expanded the ability to reduce economic displacement and we revised the alternative accordingly. The revised alternative reduced the estimated annual displaced revenue by $2,054,950 (38%), reduced the total open area by 19,000 km$^2$ (21%), yet protected an additional 71 records of corals and sponges relative to our initial proposal. On 17 June, 2005 the PFMC adopted a modified version of Alternative 12 that closed 672,175 km$^2$ of seafloor habitat to bottom trawling in the PFMC region.

**Case 2 results.**—The final preferred alternative contained two types of closed areas: a "trawl footprint" closure and ecologically important areas. The footprint closure closed to bottom trawling all seafloor habitat deeper than 700 fathoms (1280 m). Little is known about seafloor habitat at this depth, but since bottom trawling was not occurring there yet, this closure was a sensible precautionary measure. The PFMC voted to close some, but not all, sensitive habitat areas identified in the Oceana proposal. However, the final network of closed areas did protect all known seamounts, and a significant proportion of the known coral and sponge occurrences, submarine canyons and hard substrate habitat (Tables 2 and 3).

In addition to these spatial management measures to restrict bottom trawling, the PFMC established several footrope size restrictions and prohibited dredge and beam trawl gear throughout the region. NMFS deemed the final alternative practicable because it balanced socioeconomic costs and benefits to the fishing industry and communities, impacts to management and enforcement, and protection of habitat and biodiversity (Appendix J in NOAA, 2005b). The final preferred alternative was expected to displace

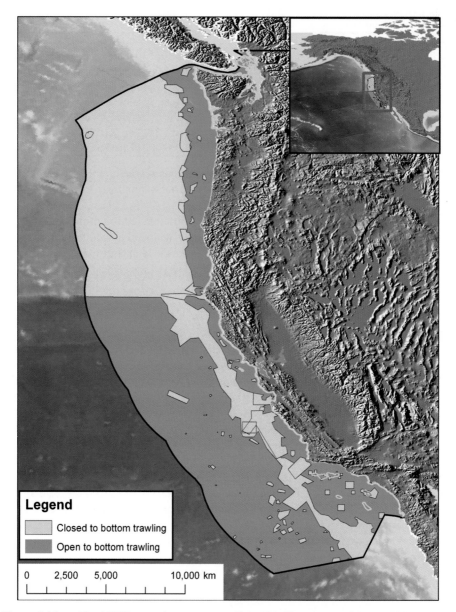

Figure 2. Map of final EFH protection measures off the U.S. West Coast (12 June, 2006) comprising bottom trawl closures (light pink) and remaining open areas to bottom trawling (green).

approximately $2.1 million annually in groundfish bottom trawl revenue, which is expected to be harvested in remaining open areas, so no revenue loss is expected (NOAA, 2005b).

NOAA finalized the action in a Record of Decision as Amendment 19 to the Pacific Coast Groundfish Fishery Management Plan, and established the closures through a regulation that took effect 12 June, 2006 (Fig. 2). However, NOAA only partially approved the PFMC decision regarding the trawl footprint, removing the closure in areas deeper

Table 3. Coral records contained within area closed to bottom trawling off U.S. West Coast.

| Coral records from NMFS trawl surveys (1977–2003) and other institutions (Etnoyer and Morgan, 2005). | Number of records (trawl survey start points and sample locations) | Within final area closed to bottom trawling |
|---|---|---|
| Total | 2,396 | 551 (23.0%) |
| Antipatharians | 199 | 50 (25.1%) |
| Gorgonians | 576 | 231 (40.1%) |
| Pennatulaceans | 1,558 | 248 (15.9%) |
| Scleractinians | 22 | 7 (31.8%) |
| Stylasterids | 41 | 15 (36.6%) |

than 3500 m on the basis that no federally managed species have been documented beyond this depth, and therefore it could not be protected under the EFH regulations. The removal of deep-water habitat from the area closed to bottom trawling reduced the total closure area to 353,500 $km^2$ (Table 2), encompassing 23% of known coral records (Table 3).

## Discussion

**Lessons for deep-sea coral conservation worldwide.**—While legal, political, and enforcement contexts may vary by nation, our approach offers several lessons for coral and sponge protection efforts worldwide. It is unlikely that any nation in the world will have all its seafloor habitat comprehensively mapped in the near future. Until such mapping is complete, freezing the footprint of bottom trawling is a precautionary first step to ensure that trawl effort does not move into areas that may be susceptible to damage. In the context of area closures, it is essential to consider areas where effort will be displaced, as the resulting shifts may offset conservation benefits (Dinmore et al., 2003). Essentially, the lesson of this paper for deep-sea coral conservation is to focus on the most appropriate areas for trawling to continue, in addition to identifying sensitive areas that should be closed. This footprint approach shifts the burden of proof, reduces the perception that fishermen are entitled to fish wherever they want, and provides fishermen an incentive to share information about where they fish.

Once the footprint is established, the question is how to identify significant concentrations of sensitive habitats with different sources of data. We found that for fishery-independent data such as trawl surveys, point-density analysis (by occurrence and by weight) is appropriate for identifying areas of highest occurrence. However, for fishery-dependent data, it is more appropriate to use catch rates to account for the uneven distribution of trawl effort. We found coral and sponge by-catch per unit of target catch or ex-vessel value is more effective for reducing by-catch at the lowest cost, while by-catch per unit effort is more effective for identifying areas of highest concentrations. However, due to skepticism about trawl gear as a means to sample corals, we found that submersible-documented locations of dense coral habitats provided the most compelling evidence to policy-makers despite their limited spatial extent.

Another key lesson for marine conservation efforts is that the process of developing management proposals is iterative. In both case studies, Oceana's original proposals became a starting point. After the proposals began to gain support, additional data came forward from both fisheries agencies and fishermen. Once the proposals became preferred alternatives, the conservation measures were modified with input from stakeholders. We learned that conservation proposals must be reasonable enough to gain traction in a regulatory process, and that iterative adjustments during the process are necessary to ensure implementation.

**Adaptive management.**—Legally, EFH management should be revisited every 5 yrs, which provides the opportunity for adaptive management. Increased use of VMS, on-board observers, and research attention to deep-sea coral ecosystems will likely reveal many new coral beds and other sensitive features of the seafloor in areas remaining open to bottom trawling. While there is a strong tendency within the scientific community and government agencies to withhold these data from the public until they can be formally published, it is imperative that researchers work directly with in-season fishery managers to establish boundaries and precautionary management measures for newly-discovered areas. Management agencies must be prepared to implement emergency closures immediately upon discovery of these new areas in the interim period until comprehensive reviews of EFH measures are conducted. In addition to monitoring compliance, VMS data will provide high resolution data on which areas have higher and lower relative effort, as well as how trawl effort shifts in response to closures. As enforcement and monitoring capabilities improve with technological innovations, the scale of management should decrease so that management can take place at a resolution that better fits the patchiness of the seafloor habitat types and the spatial resolution of fishing effort. As we found in the U.S. West Coast case study, a higher resolution of data allows management measures to simultaneously increase habitat protection and decrease the economic costs of area closures. If coupled with VMS, the approach can be applied in fisheries that have limited enforcement capability on the water, such as "high-seas" trawling in international waters.

In addition, fishery managers should encourage bottom trawl fishermen to convert to less damaging gear types (i.e., bottom longlines, pots, traps) through economic incentives or by changing regulations to allow trawl permit holders to use these gears. Other incentive-based management measures might include hard limits on coral and sponge by-catch resulting in penalties if exceeded, rewarding vessels with lower by-catch with extended fishing seasons or quota, and allocating dedicated access privileges and/or quotas based on conservation performance, rather than catch history, in the event that such privileges and/or quotas are established.

## Conclusion

Rather than viewing recent EFH closures as a final solution to the challenge of seafloor conservation, we view them as a necessary first step in an ongoing, iterative shift toward ecosystem-based fisheries management (see NPFMC, 2006). Oceana's conservation proposals were developed using the best available data to develop a comprehensive approach to protect EFH, as required by law, for the Pacific Coast while maintaining commercial fishing opportunities. In two distinct geographic regions, we successfully constructed a scientific argument that habitat protections were warranted and showed that our management approach can meet conservation objectives in a practical manner. The approach works in highly concentrated industrial trawl fisheries as well as more diffuse industries with a large number of smaller vessels. Already, this model is being proposed to protect coral ecosystems in Canada (see Ardron and Hangaard, 2007). By comparing these case studies, we illustrated that the methodologies of data analysis and mapping must be flexible and responsive to regional differences in data availability, enforcement capacity, and socioeconomics. The implementation of EFH bottom trawl closures represents a significant step toward ecosystem-based fishery management along the U.S. Pacific Coast and provides a model for deep-sea coral conservation worldwide.

## Acknowledgments

The authors would like to thank J. Ayers, who, through his vision and determination, oversaw this project from the concept stage through to completion. L. Morgan, B. Tissot, and two anonymous reviewers provided excellent feedback that greatly improved the content of this paper. We also thank J. Searles, S. Murray, and M. Hirshfield who provided helpful comments, A. Rettinger who provided support throughout the process, and all Oceana staff and supporters who played a role in the success of this effort. Last but not least, we thank all the fishermen, scientists, conservation representatives, and agency staff who are committed to protecting seafloor habitat.

## Literature Cited

Alaska Fisheries Science Center (AFSC). 2003. North pacific groundfish observer manual. North Pacific Groundfish Observer Program. Pagination by section.

Anderson, O. F. and M. R. Clark. 2003. Analysis of bycatch in the fishery for orange roughy, *Hoplostethus atlanticus*, on the South Tasman Rise. Mar. Freshwat. Res. 54: 643–652.

Andrews, A. H., G. M. Cailliet, L. A. Kerr, K. H. Coale, C. Lundstrom, and A. P. DeVogelaere. 2005. Investigations of age and growth for three deep-sea corals from the Davidson Seamount off Central California. Pages 1021–1038 *in* A. Freiwald and J. M. Roberts, eds. Cold-water corals and ecosystems. Springer-Verlag, Berlin Heidelberg.

Ardron, J. A., G. S. Jamieson, and D. Hangaard. 2007. Spatial identification of closures to reduce the bycatch of corals and sponges in the groundfish trawl fishery, British Columbia, Canada. Pages 157–167 *in* R. Y. George and S. D. Cairns, eds. Conservation and adaptive management of seamount and deep-sea coral ecosystems. Rosenstiel School of Marine and Atmospheric Science, University of Miami. Miami. 324 p.

Beaulieu, S. 2001. Life on glass houses: sponge stalk communities in the deep sea. Mar. Biol. 138: 803–817.

Bellman, M. A. and S. A. Heppell. 2005. Evaluation of a US west coast groundfish habitat conservation regulation via analysis of spatial and temporal patterns of trawl fishing effort. Appendix A-19 *in* NOAA, Pacific coast groundfish fishery management plan: essential fish habitat designation and minimization of adverse impacts: draft environmental impact statement. 40 p.

Brodeur, R. 2001. Habitat-specific distribution of Pacific ocean perch (*Sebastes alutus*) in Pribilof Canyon, Bering Sea. Cont. Shelf Res. 21: 207–224.

Buhl-Mortensen, L. and P. B. Mortensen. 2004. Crustaceans associated with the deep-water gorgonian corals *Paragorgia arborea* (L., 1758) and *Primnoa resedaeformis* (Gunn., 1763). J. Nat. Hist. 38: 1233–1247.

Chuenpagdee, R., L. E. Morgan, S. Maxwell, E. A. Norse, and D. Pauly. 2003. Shifting gears: Assessing collateral impacts of fishing methods in the U.S. waters. Front. Ecol. Env. 1: 517–524.

Costello, M. J., M. McCrea, A. Freiwald, T. Lundalv, L. Jonsson, B. J. Bett, T. van Weering, H. de Haas, J. M. Roberts, and D. Allen. 2005. Role of cold-water *Lophelia pertusa* coral reefs as fish habitat in the NE Atlantic. Pages 771–805 *in* A. Freiwald and J. M. Roberts, eds. Cold-water corals and ecosystems. Springer-Verlag, Berlin Heidelberg.

Daspit, W. 2004. 2004 PFMC Groundfish Management Team Reports. Pacific Fisheries Information Network. Available from http://www.psmfc.org/pacfin/pfmc.html. Accessed 23 Oct. 2007.

de Forges, B. R., J. A. Koslow, and G. C. B. Poore. 2000. Diversity and endemism of the benthic seamount fauna in the southwest Pacific. Nature 405: 944–947.

Dinmore, T. A., D. E. Duplisea, B. D. Rackham, D. L. Maxwell, and S. Jennings. 2003. Impact of a large-scale area closure on patterns of fishing disturbance and the consequences for benthic communities. ICES J. Mar. Sci. 60: 371–380.

Drinkwater, K. 2004. Summary report review on evaluation of fishing activities that may adversely affect essential fish habitat (EFH) in Alaska. Center for Independent Experts. 33 p.

Engel, J. and R. Kvitek. 1998. Effects of Otter Trawling on a Benthic Community in Monterey Bay National Marine Sanctuary. Conserv. Biol. 12: 1204–1214.

ESRI. 2004. ESRI ArcMap 9.0. ESRI Inc.

Etnoyer, P. and L. Morgan. 2005. Occurrences of habitat-forming deep-sea corals in the northeast Pacific Ocean. Pages 331–343 in A. Freiwald and J. M. Roberts, eds. Cold-water corals and ecosystems. Springer-Verlag, Berlin Heidelberg.

Freiwald, A., J. H. Fosså, A. Grehan, T. Koslow, and J. M. Roberts. 2004. Cold-water coral reefs. Out of sight — no longer out of mind. UNEP-WCMC, Cambridge. 84 p.

Haedrich, R. L., G. T. Rowe, and P. T. Polloni. 1980. The megabenthic fauna in the deep sea south of New England, USA. Mar. Biol. 57: 165–179.

Heifetz, J. 2002. Coral in Alaska: Distribution, abundance, and species associations. Hydrobiologia 471: 19–28.

_____, B. L. Wing, R. P. Stone, P. W. Malecha, and D. L. Courtney. 2005. Corals of the Aleutian Islands. Fish. Oceanogr. 14: 131–138.

Hiatt, T. 2005. Stock assessment and fishery evaluation report for the groundfish fisheries of the Gulf of Alaska and Bering Sea/Aleutian Islands area: economic status of the groundfish fisheries off Alaska, 2004. 182 p.

Hixon, M. A., B. N. Tissot, and G. W. Percy. 1991. Fish assemblages of rocky banks of the Pacific Northwest. U.S. Department of the Interior, Minerals Management Service, Final Report 91-0052. 410 p.

Husebo, A., L. Nottestad, J. Fossa, D. Furevik, and S. Jorgensen. 2002. Distribution and abundance of fish in deep-sea coral habitats. Hydrobiologia 471: 91–99.

Hyland, J., C. Cooksey, E. Bowlby, and M. Brancato. 2005. A pilot survey of deepwater coral/sponge assemblages and their susceptibility to fishing/harvest impacts at the Olympic Coast National Marine Sanctuary. Cruise report, NOAA ship MacArthur II, AR-04-04: Leg 2. NOAA Technical Memorandum NOS NCCOS 15 June 2005. 13 p.

Koslow, J. A., K. Gowlett-Holmes, J. K. Lowry, T. O'Hara, G. C. B. Poore, and A. Williams. 2001. Seamount benthic macrofauna off southern Tasmania: Community structure and impacts of trawling. Mar. Ecol. Prog. Ser. 213: 111–125.

Krieger, K. J. 2001. Coral (Primnoa) impacted by fishing gear in the Gulf of Alaska. Pages 106–117 in J. Willison, J. Hall, S. Gass, E. Kenchington, M. Butler, and P. Doherty, eds. Proc. First Int. Symp. on Deep-sea Corals. Ecology Action Centre and Nova Scotia Museum, Halifax, Nova Scotia.

_____ and B. L. Wing. 2002. Megafauna associations with deepwater corals (Primnoa spp.) in the Gulf of Alaska. Hydrobiologia 471: 83–90.

Larson, M. L. and R. H. McPeak, eds. 1995. Final Environmental Document, giant and bull kelp commercial and sportfishing regulations: Section 30 and 165, Title 14, California Code of Regulations. The Resources Agency, Sacramento, CA. 150 p.

Lindner, A. 2003. Evolution of shallow-water stylasterid corals (Cnidaria; Hydrozoa; Stylasteridae) from deep-sea ancestors. Integr. Comp. Biol. 43: 1074.

Link, J. S. and C. Demarest. 2003. Trawl hangs, baby fish, and closed areas: a win-win scenario. ICES J. Mar. Sci. 60: 930–938.

MacDonald, D. S., M. Little, N. C. Eno, and K. Hiscock. 1996. Disturbance of benthic species by fishing activities: A sensitivity index. Aquat. Conserv.: Mar. Freshwat. Ecosyst. 6: 257–268.

Malecha, P. W., R. J. Stone, and J. Heifetz. 2005. Living substrate in Alaska: distribution, abundance, and species associations. Pages 289–299 *in* P. Barnes and J. Thomas, eds. Benthic habitats and effects of fishing. American Fisheries Society Symposium 41. Bethesda.

Morgan, M. J. and R. Chuenpagdee. 2003. Shifting gears: addressing the collateral impacts of fishing methods in U.S. waters. Pew science series on conservation and the environment. Washington, D.C. 42 p.

NOAA (National Oceanic and Atmospheric Administration). 2004. Draft Environmental Impact Statement for Essential Fish Habitat Identification and Conservation in Alaska. January 2004. Pagination by section.

_____. 2005a. Final Environmental Impact Statement for Essential Fish Habitat Identification and Conservation in Alaska. April 2005. Pagination by section.

_____. 2005b. Pacific Coast Groundfish Fishery Management Plan, Essential Fish Habitat Designation and Minimization of Adverse Impacts Final Environmental Impact Statement. December 2005. Pagination by section.

National Research Council (NRC). 2002. Effects of Trawling and Dredging on Seafloor Habitat. National Academy of Sciences. Washington, D.C. 126 p.

NPFMC (North Pacific Fishery Management Council). 2005. Stock assessment and fishery evaluation report for the groundfish resources of the Bering Sea/Aleutian Islands regions. Available from: http://www.afsc.noaa.gov/refm/docs/2005/BSAI_Intro.pdf. Accessed 23 Oct. 2007.

_____. 2006. A discussion paper: fishery ecosystem plan for the Aleutian Islands. March 2006. Anchorage, AK. Available from: http://www.fakr.noaa.gov/npfmc/current_issues/ecosystem/FEP306.pdf. Accessed 23 Oct. 2007.

Opresko, D. M. 2005. A new species of antipatharian coral (Cnidaria: Anthozoa: Antipatharia) from the southern California Bight. Zootaxa 852: 1–10.

(PFMC) Pacific Fishery Management Council. 2003. Consolidated GIS data: physical and biological habitat data disk. Volume 1.

_____. 2004. Risk Assessment for the Pacific Groundfish Fishery Management Plan. Available from: http://www.pcouncil.org/habitat/habrisk.html Accessed 5 September 2006.

Pauly, D., V. Christensen, S. Guenette, T. J. Pitcher, U. Rashid Sumaila, C. J. Walters, R. Watson, and D. Zeller. 2002. Towards sustainability in world fisheries. Nature 318: 689–695.

Pikitch, E. K., C. Santora, E. A. Babcock, A. Bakun, R. Bonfil, D. O. Conover, P. Dayton, P. Doukakis, D. Fluharty, B. Heneman, E. D. Houde, J. Link, P. A. Livingston, M. Mangel, M. K. McAllister, J. Pope, and K. J. Sainsbury. 2004. Ecosystem-based fishery management. Science 305: 346–347.

Probert, P. K., D. G. McKnight, and S. L. Grove. 1997. Benthic invertebrate bycatch from a deep-water trawl fishery, Chatham Rise, New Zealand. Aquat. Conserv. Mar. Freshwat. Ecosyst. 7: 27–40.

Risk, M. H., J. M. Heikoop, M. G. Snow, and R. Beukens. 2002. Lifespans and growth patterns of two deep-sea corals: *Primnoa resedaeformis* and *Desmophyllum cristagalli*. Hydrobiologia 471: 125–131.

Roberts, S. and M. Hirshfield. 2004. Deep-sea corals: out of sight, but no longer out of mind. Front. Ecol. Environ. 2: 123–130.

Roark, E. B., T. P. Guilderson, S. Flood-Page, R. B. Dunbar, B. L. Ingram, S. J. Fallon, and M. McCulloch. 2005. Radiocarbon-based ages and growth rates of bamboo corals from the Gulf of Alaska, Geophys. Res. Lett. 32: 5.

Sanchez, J. A. and S. D. Cairns. 2004. An unusual new gorgonian coral (Anthozoa: Octocorallia) from the Aleutian Islands, Alaska. Zool. Meded. Leiden 78: 265–274.

Sarda, F., J. E. Cartes, and J. B. Company. 1994. Spatio-temporal variations in megabenthos abundance in three different habitats of the Catalan deep-sea (Western Mediterranean). Mar. Biol. 120: 211–219.

Shester, G. and J. Ayers. 2005. A cost effective approach to protecting coral and sponge ecosystems with an application to Alaska's Aleutian Islands region. Pages 1151–1169 *in* A. Freiwald and J. M. Roberts, eds. Cold-water corals and ecosystems. Springer-Verlag, Berlin Heidelberg.

Stocks, K. 2004. Seamount invertebrates: composition and vulnerability to fishing. Pages 17–25 *in* T. Morato and D. Pauly, eds. Seamounts: Biodiversity and Fisheries. Fisheries Centre Research Reports 12(5) [with 1 CD-ROM], UBC Vancouver, 78 p.

Stone, R. P. 2006. Coral habitat in the Aleutian Islands of Alaska: depth distribution, fine-scale species associations, and fisheries interactions. Coral Reefs 25: 229–238.

Thurman, H. V. and E. Burton. 2001. Introductory oceanography, Ninth Edition. Prentice Hall. Upper Saddle River. 554 p.

Thrush, S. and P. K. Dayton. 2002. Disturbance to marine benthic habitats by trawling and dredging: Implications for marine biodiversity. Annu. Rev. Ecol. Syst. 33: 449–473.

Vetter, E. and P. Dayton. 1999. Organic enrichment by macrophyte detritus, and abundance patterns of megafaunal populations in submarine canyons. Mar. Ecol. Prog. Ser. 186: 137–148.

Watling, L. and E. A. Norse. 1998. Disturbance of the seabed by mobile fishing gear: a comparison to forest clear-cutting. Conserv. Biol. 12: 1180–1197.

Western Pacific Fishery Management Council (WPFMC). 1979. Fishery management plan for the precious coral fisheries (and associated non-precious corals) of the western Pacific region. Honolulu, Hawaii. 116 p.

Witherell, D. and C. Coon. 2001. Protecting gorgonian corals off Alaska from fishing impacts. Pages 117–125 *in* J. H. M. Willison, J. Hall, S. E. Gass, E. L. R. Kenchington, M. Butler, and P. Doherty, eds. Proc. First Int. Symp. on Deep-Sea Corals. Halifax, Nova Scotia.

Zimmerman, M. 2003. Calculation of untrawlable areas within the boundaries of a bottom trawl survey. Can. J. Fish. Aquat. Sci. 60: 657–669.

ADDRESSES: (G.S.) *Hopkins Marine Station of Stanford University, Oceanview Boulevard, Pacific Grove, California 93950-3094.* (J.W.) *Oceana, Pacific Region, 175 South Franklin St, Ste. 418, Juneau, Alaska 99801.* CORRESPONDING AUTHOR: (J.W.) *E-mail : <jwarrenchuk@oceana. org>.180+90+35*

# Density and habitat of three deep-sea corals in the lower Hawaiian chain

FRANK A. PARRISH

## Abstract

Subphotic contours were surveyed between 350 and 500 m at six deep-sea coral beds in Hawaii. The density and mean height for the deep-sea corals *Corallium secundum* Dana, 1846, *Corallium lauuense* Bayer, 1956, and *Gerardia* sp. at each bed were recorded relative to temperature, substrate, and bottom relief. Species composition and density at the six sites varied; however, the mean size of the coral colonies did not vary except at sites where there was a history of coral harvesting. The *Corallium* species had the highest densities and, at some sites, were found in mono-specific patches. The three coral taxa overlapped in their depth range. Water temperature varied as much as 3–4 °C across the six stations. Multi-year monitoring at two of the sites indicated that temperature differences persisted year-round and exhibited monthly and seasonal fluctuations. All three coral taxa colonized both carbonate and basalt/manganese substrates. The largest patches of *C. secundum* were found on flat exposed bottom, whereas *C. lauuense* encrusted uneven rocky bottom. *Corallium lauuense* was often intermixed with *Gerardia* sp., which colonized cliffs, pinnacles, and the tops of walls. Of the habitat variables, bottom relief best explained the distribution of the three coral taxa. It is hypothesized that the corals favor areas where bottom relief enhances or modifies flow characteristics perhaps improving the colony's feeding success.

Deep-sea coral research in the central Pacific has been focused on describing the geographic range (Grigg, 1974), taxonomic diversity (Grigg and Bayer, 1976), or genetic diversity (Baco and Shank, 2005) of deep-sea or cold-water corals. Few ecological studies have been conducted because of the difficulty of working at greater depths where the corals grow (Genin et al., 1992). Recent analyses of coral distributions and environmental variables at a continental shelf scale indicate there are notable habitat differences between different coral taxa (Gass and Willison, 2005; Leverette and Metaxas, 2005; Watling and Auster, 2005). In the central Pacific, most of what has been learned about deep-sea corals was achieved in conjunction with the Hawaiian precious coral fishery, which uses deep-sea corals as raw materials for the jewelry trade. Three coral species emerged as the basis of this unique small-scale fishery: the short crustose gorgonian fans (Bayer, 1956), *Corallium secundum* Dana, 1846 and *Corallium lauuense* (formally called *Corallium regale*) Bayer, 1956, and the tall flexible hexacoral, *Gerardia* sp. These are commonly referred to as pink, red, and gold coral, respectively. The activity and scale of the fishery is well documented by Grigg (1993, 2002). Coral harvesters initially relied on crude dredges and in 1972 moved to harvesting with small submersibles. For over 30 yrs, harvesting has been sporadic, operationally small, and limited to the coastal coral beds of Makapuu, Oahu, and Keahole, Hawaii.

In recent years, upgrades in research submersibles and support vessels have provided researchers in Hawaii the opportunity to extend surveys to more remote coral beds.

George, R. Y. and S. D. Cairns, eds. 2007. Conservation and adaptive management of seamount and deep-sea coral ecosystems. Rosenstiel School of Marine and Atmospheric Science, University of Miami.

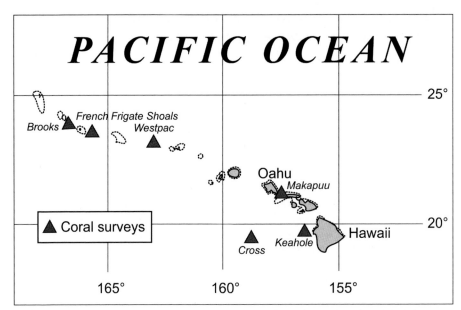

Figure 1. Map of the southern Hawaiian Archipelago with the six coral stations labeled.

Here I report the results of visual surveys at the commercially harvested Makapuu and Keahole coral beds and at four remote coral beds identified in prior studies (Grigg, 1974, Parrish et al., 2002). These surveys systematically assessed density and sizes of *C. secundum, C. lauuense,* and *Gerardia* sp. in relation to depth, water temperature, substrate, and relief types at each of the six sites.

## Methods

All data were collected in a multi-year (1998, 2000, and 2001) series of submersible surveys using the PISCES V, PISCES IV, and RCV-150 at depths ranging from 350 to 500 m. Dive sites included Makapuu, Keahole, Cross Seamount, Brooks Bank, French Frigate Shoals (FFS), and Westpac Bank (Fig. 1). The sites varied widely in their submarine topography. Cross, Westpac, and FFS were located on the summits of oceanic seamounts. The Cross bed was on a peak, the FFS bed on encrusted pinnacles along a ridge, and the Westpac bed was along the crest of the seamount's summit. The Brooks bed is on the slopes of a bank that rises to within 80 m of the surface. The Makapuu and Keahole beds are on the flanks of the large islands of Oahu and Hawaii. Makapuu is a gradual slope and Keahole is a ledge that runs along the top of a seacliff. The sampling at each site was divided into transects loosely focused along the 350 m, 400 m, 450 m, and 500 m contours. As these beds were previously identified, the transects were laid out such that roughly half of the sampling occurred in the bed and half on areas outside of the bed. This maximized the comparability of the geomorphology for areas inside and outside the bed. The sites of Brooks, Westpac, Makapuu, and Keahole had reasonably comparable bottom morphology in and outside of the bed. At Cross Seamount and FFS, the corals grew on bottom peaks and pinnacles that were less prominent (smaller and fewer) outside the coral bed.

The submersibles are three-person vehicles with the pilot located in the center and observers on either side. Each person can view an illuminated bottom area of ~ 55 m² through view ports directed diagonally forward and down. The cumulative view from the three view ports (adjusted for overlap) provides an effective illuminated survey area of ~ 120 m². A video camera directed forward and down was operated continuously during the dive. The RCV-150 is a remote-operated vehicle which surveys a 45 m² illuminated area.

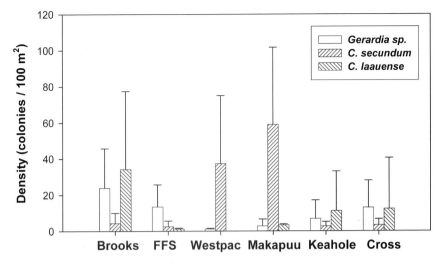

Figure 2. Mean density of three coral types for all pseudo-replicates with coral at the six surveyed coral beds.

Counts and sizes of corals were recorded cumulatively for 5-min segments or "pseudo-replicates" to obtain numerical density and size structure information. The recorded count of the corals was capped at 100 for pseudo-replicates where the density of coral colonies exceeded what observers could count. The coral counts were then normalized by the viewing area of the survey vehicle. This pseudo-replication technique is common in ecological sampling (Oksanen, 2001) and has been used effectively in surveys by the PISCES submersibles (Moffitt and Parrish, 1992; Parrish et al., 2006). A laser reference scale was projected on the bottom within view of the submersible video cameras to assist the observers in estimating coral size.

Surveys were conducted during the fall of each year (September–November). Temperature profiles of water column were recorded at each of the stations and thermographs were deployed at the most northern and southern stations for 2 yrs (2001 and 2002). For each segment, observers logged the primary substrate type and relief scale using three categories. Substrate was divided into categories of sand, carbonate hard bottom, and basalt/manganese. Relief was divided into categories of flat, even bottom termed "hardpan" (< 15 cm relief); uneven bottom "outcrops" (15–90 cm); and steep surfaces such as "pinnacles" or cliffs (> 90 cm) (Fig. 2). Pseudo-replicates with multiple substrate and relief type were coded by whichever substrate and relief type was most abundant during the segment.

The coral count and size data were not normally distributed, and analyses of these data relied on nonparametric techniques. Coral occurrence by substrate type and relief class were assessed using Mann-Whitney (M-W) and Kruskal-Wallis (K-W) tests, respectively. Samples sizes for all analyses were adequate to detect differences at large-effect sizes with alpha at 0.01 and a power of 0.80.

## Results

**Coral surveys.**—Surface tracks of the submersible indicate an average 3.5 km distance covered on the visual surveys. The surveys generated a median of 150 pseudo-replicates for each station. Roughly half of the survey effort at each station focused on areas with corals (in the bed) and the other half on areas immediately adjacent, without corals (Table 1). Coral composition varied among stations. Some stations were primarily *C. secundum* or *Gerardia* sp., and the rest were a mixture of *C. lauuense* and *Gerardia* sp. (Fig. 2).

Table 1. Latitude, morphology, linear survey distance, and number of pseudo-replicates conducted, with the percent with coral in brackets, and the decimal fraction of the substrate and relief categories at each of the six sites.

| | | | | | Psuedo-replicates* | | | | | |
| | | | | | Substrate | | | Relief | | |
| Station (latitude) | Morphology | Dist. (km) | Mean SD | Total (%WC) | snd | carb | bas | hard | out | pin |
|---|---|---|---|---|---|---|---|---|---|---|
| Brooks (23°58.6′) | Bank slope | 3 | 1.2 | 114 (84) | 0.03 | 0.97 | 0.00 | 0.69 | 0.24 | 0.07 |
| FFS (23°55′) | Seamount | 2.5 | 1.07 | 283 (17) | 0.04 | 0.49 | 0.47 | 0.45 | 0.37 | 0.18 |
| Westpac (23°14.2′) | Seamount | 2.8 | 1.13 | 103 (51) | 0.00 | 1.00 | 0.00 | 0.85 | 0.15 | 0.00 |
| Makapuu (21°17.5′) | Island slope | 2.8 | 1.39 | 113 (69) | 0.07 | 0.93 | 0.00 | 0.98 | 0.02 | 0.00 |
| Keahole (19°47.7′) | Island slope | 7.3 | 2.6 | 69 (55) | 0.19 | 0.78 | 0.03 | 0.55 | 0.24 | 0.21 |
| Cross (18°43.3′) | Seamount | 3.2 | 0.73 | 150 (36) | 0.38 | 0.15 | 0.47 | 0.13 | 0.66 | 0.21 |

* %WC = pseudo-replicates with coral, snd = sand substrate, carb = carbonate substrate, bas = basalt/manganese substrate, hard = hardpan relief, out = outcrop relief, pin = pinnacle relief.

The coral beds were generally most dense at the center of the patch with colonies diminishing precipitously around the edges. Coral tallies often captured individual corals on the edge and the high density of the patch center in a single pseudo-replicate, making presence-absence type analyses viable. Of the pseudo-replicates with corals, *C. secundum* had the highest density ($56 \pm 65$ colonies 100 m$^{-2}$) and were found predominantly at Makapuu and Westpac. *Corallium lauuense* exhibited the next highest density ($33 \pm 63$ colonies 100 m$^{-2}$) and then *Gerardia* sp. ($13 \pm 17$ colonies 100 m$^{-2}$). The concave "dish" of the coral colonies in a patch generally oriented in the same direction, presumably perpendicular to the prevailing flow pattern in the area to improve feeding opportunities. Consistency in the orientation of the colonies was most apparent for *C. secundum*, less for *C. lauuense*, and not obvious for *Gerardia* sp. The mean size of *C. secundum* colonies at Makapuu was significantly greater than the mean colony size at Westpac (Table 2). There was no apparent difference for *C. lauuense* coral among Brooks, Keahole, and Cross sites. However, *Gerardia* sp. colonies at Keahole were significantly smaller than at Cross and FFS but did not differ from Brooks (Table 2).

**Habitat variables.**—Overlap occurred in the depth range of the three corals. Plotting coral density by depth showed a tendency for *C. secundum* to be shallower and *C. lauuense* and *Gerardia* sp. to extend deeper. This pattern may not be depth-related; but instead, be an artifact of differing physiography among stations. Corals at Brooks Bank ranged from 350 to 550 m whereas the FFS and Keahole corals clustered over a tighter depth range, encrusting discrete bottom features such as a pinnacle or the top edge of a wall (Fig. 3). Temperature-depth profiles at the six stations showing greater depths did not necessarily signify colder temperatures. The thermograph records indicate temperature varied as

Table 2. Comparison of mean heights of coral colonies among sites.

| Coral taxa | Height (cm)/ mean (SD) | Nonparametric test | df | P | A posterior comparisons (0.05 threshold) |
|---|---|---|---|---|---|
| *Corallium secundum* | 17.9 (4.5) | Mann Whitney: Z = –2.0 | 1 | 0.05 | Westpac < Makapuu |
| *Corallium laauense* | 20.3 (6.4) | Kruskal Wallis: $\chi^2$ = 1.05 | 2 | 0.589 | Brooks, Keahole, Cross; no diff |
| *Gerardia* sp. | 66.5 (30.7) | Kruskal Wallis: $\chi^2$ = 22.6 | 3 | 0.000 | Keahole < FFS, Cross; Brooks no diff |

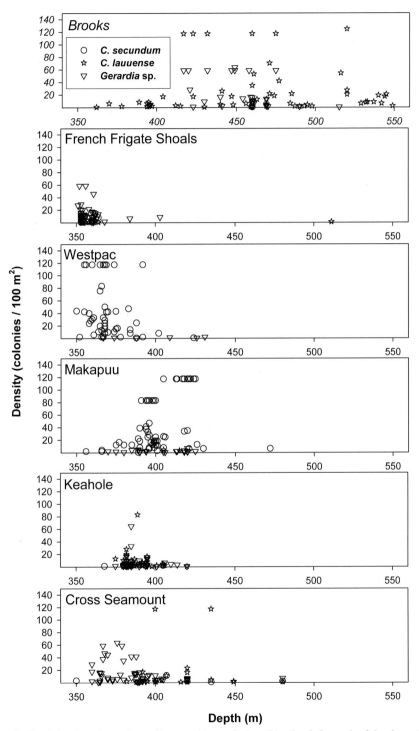

Figure 3. Coral density of pseudo-replicates with coral plotted by depth for each of the three taxa at the six sites.

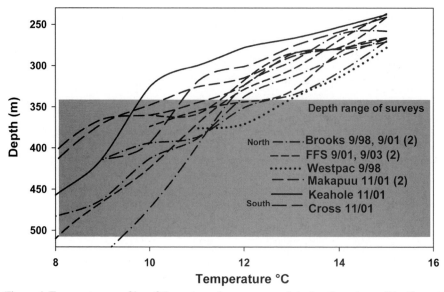

Figure 4. Temperature profiles of the water column measured during the submersible dives at the six coral beds. The number 2 in parentheses following the month/year for each site indicates two dives were made.

much as 4 °C between the latitudinal extremes of the study sites (Fig. 4). These differences persist year-round with monthly, seasonal, and inter-annual variability (Fig. 5).

It was not possible to equally distribute sampling across the three substrate and three relief types for all stations, but all types were well represented in the data (Table 1). The substrate and relief at the six sites ranged from a continuum of one type, such as carbonate hardpan at Makapuu, to a combination of all types at a single site forming a complex matrix, such as the FFS Platform. Sand was the dominant substrate in 11% of pseudo-replicates and, as expected, few corals were found in these segments and the colonies were usually on an isolated rock or high point above the sand scour. Since sand was clearly marginal substrate, pseudo-replicates dominated by sand were eliminated from further analysis. The two *Corallium* species were found in significantly greater density (Mann-Whitney: *C. secundum* Z = −8.3, P < 0.001; *C. lauuense* Z = −1.9, P < 0.04) on carbonate substrate. In contrast, *Gerardia* sp. was found more on manganese/basalt substrate (Mann-Whitney: Z = −6.18, P < 0.01). The patterns seen in the substrate may have been a function of differing scales of bottom relief. *Gerardia* sp. density differed by relief scale (Kruskal-Wallis: $\chi^2$ = 238, df = 2, P < 0.001) and was found encrusting pinnacles and cliff faces more than the hardpan or outcrop relief types. These pinnacles and walls were often of manganese/basalt substrate, suggesting a possible cross correlation between substrate and relief types. Despite the similar morphology of the two *Corallium* species, they exhibited different affinities by relief type. *Corallium secundum* (Kruskal-Wallis: $\chi^2$ = 132, df = 2, P < 0.001) was found in high densities on flat, hardpan bottom (most often carbonate substrate) whereas *C. lauuense* (Kruskal-Wallis: $\chi^2$ = 12.4, df = 2, P = 0.002) was found on outcrops and ledges (usually carbonate but sometimes basalt/manganese substrate).

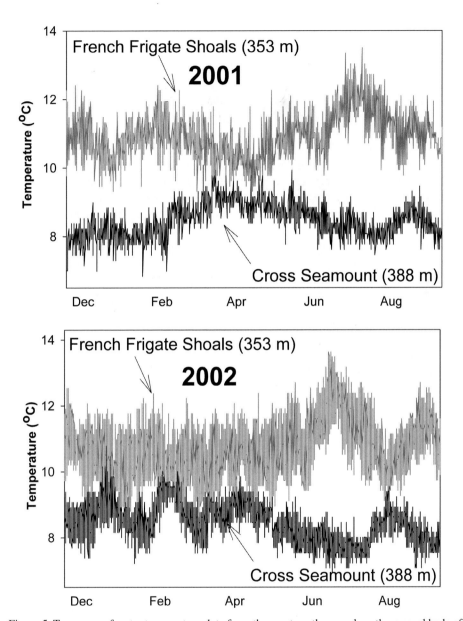

Figure 5. Two years of water temperature data from the most northern and southern coral beds of the six survey sites.

## Discussion

**Coral beds.**—The bed's composition of coral species varied among sites but exhibited surprising within-site homogeneity. The dominance of *C. secundum* at Makapuu and Westpac was the most striking. Colonies of *Gerardia* sp. and *C. lauuense* were seen at these sites, but they were few in number and generally small in size. Makapuu had the highest colony density despite the longest history of coral harvesting (Grigg, 1993). The most recent harvesting activity occurred in 1998–2001. Some areas were encountered where harvesting appeared to have occurred, probably during the most recent phase of the fishery. The bed's large perimeter (Grigg, 2002) and high coral density exceed those of the other coral beds and suggests that habitat and environmental conditions at Makapuu are exceptional for *C. secundum*. Westpac is similar to Makapuu but much more remote and it has never been subject to harvesting. In comparing the size structure of coral colonies between the two sites, it is surprising to see Makapuu having significantly larger mean coral colonies despite a history of coral harvesting. A truncated size structure was expected as a result of earlier harvesting activities, but it is possible recruitment and regrowth (Grigg, 1988) over the intervening decades restored the mean size of coral colonies at the bed (Grigg, 2002). Another explanation would be that the survey dives missed the portions of the coral bed that had been impacted by earlier harvest. Assuming the survey track overlaps with areas where the fishery historically operated, the large mean colony size at Makapuu again would suggest the area has exceptional habitat and environmental conditions for *C. secundum* growth.

*Gerardia* sp. and *C. lauuense* colonies were found mixed to varying degrees across the stations. The FFS site was almost exclusively *Gerardia* sp. whereas Brooks, Keahole, and Cross beds were of somewhat equal proportions of *Gerardia* sp. and *C. lauuense*. *Corallium lauuense* has the stature and morphology of *C. secundum* but colonizes uneven outcrops, including terrain where *Gerardia* sp. occurs in numbers. *Corallium secundum* and *Gerardia* sp. were not seen to coexist in abundance, at any of the stations. Comparing the size structure of colonies at stations where *Gerardia* sp. was found in abundance (Brooks, FFS, Keahole, and Cross) revealed a smaller mean colony size of *Gerardia* sp. at Keahole. It is possible the smaller size is a result of a history of coral harvesters collecting *Gerardia* sp. from the bed. Certainly, recent harvesting (1998–2001) could be the reason, but there could also be lingering effects from earlier harvesting. The high uncertainty of growth rate estimates for *Gerardia* sp. (Grigg, 2002; Roark et al., 2006) could mean recovery to pre-exploitation levels is unlikely even over several decades.

**Temperature.**—Given the overlapping depth range among stations and the fact there was not a single station with a balanced mix of all three coral taxa suggests that depth is not the determining factor for coral distribution within the 350–500 m depth range. Generally, temperatures are cooler with increasing depth. However, the thermal profiles of the water column varied 3–4 °C among the six stations, suggesting bottom topography effects disrupt the water column stratification. Temperature may be an important environmental variable in defining the depth range for these three species to colonize but no obvious patterns emerged within the 350–500 m depth range. The 2-yr data from the thermographs indicate that the southern-most station of Cross Seamount is 3–4 °C colder than the northern-most station of FFS. Some of this difference could be a result of the greater depth (+ 35 m) at the Cross Seamount station. However, a 3–4 °C change was not observed over a 35-m depth change in any of the thermal profiles across the six stations. Based on these limited observations, temperature did not explain occurrence patterns of the deep-sea corals.

**Substrate and relief.**—Obvious differences were seen among substrate, relief, and coral taxa at each of the stations. The detected differences in the substrate type colonized by the different corals appear to be artifacts of bottom relief which was the dominant influence. "Hardpan" relief was mostly carbonate substrate, "outcrops" were a mixture of carbonate ledges and basalt rocks, and "pinnacles" were basalt columns and carbonate cliffs. Using the crude relief scale, it was clear that the carbonate hardpan of the Makapuu station looked the same as that at the Westpac station, and that both supported dense populations of *C. secundum*. The basalt pinnacles on the summits of Cross Seamount and the FFS Platform were similar, and were encrusted with *Gerardia* sp. Brooks and Keahole were a mix of basalt and carbonate outcrops, and both supported *Gerardia* sp. and *C. lauuense*. Although the correspondence of coral types to relief types was good, it is possible that relief serves as a proxy or modifier for near-bottom water flow conditions. Flow and particulate load were variables not considered in this study, but have been identified as important in other studies (Leverette and Metaxas, 2005). Future work should assess the effect of the relief variables on flow characteristics and suspended particles as they pertain to deep corals. Different taxa of corals may thrive in different flow conditions and bottom relief can influence flow directionality, intensity, and character (e.g., laminar, turbulent). Finally, this appraisal of corals and their habitat is subject to considerable uncertainty about coral life history. *Gerardia* sp. is a particularly good example in that it often settles on and grows over other corals that have colonized an area, suggesting that *Gerardia* sp. could be host-limited. The localized structural and environmental conditions found around *Gerardia* sp. may be more critical for the initial colonization of the host coral than to the *Gerardia* sp. itself. Future work is planned to address threshold flow values and particle suspension of the environment around deep corals. Increased insight into the habitat and environmental needs of deep-sea corals will improve our ability to locate these isolated patches for management and protection.

## Acknowledgments

Dive time for Pisces submersibles and the Rcv-150 was awarded by the National Undersea Research Program's Hawaii Undersea Research Laboratory. Thanks are due to the HURL submersible staff and the crew of the KOK for the outstanding support provided to this project. Field assistance was graciously provided by R. Moffitt, R. Humphreys, R. Grigg, F. Oishi, W. Ikehara, and B. Mundy. Helpful comments were provided by T. Hourigan, M. Randall, and two anonymous reviewers.

## Literature Cited

Baco, A. R. and T. M. Shank. 2005. Population genetic structure of the Hawaiian precious coral *Corallium Lauuense* (Octocorallia: Corallidae) using microsatellites. Pages 663–678 *in* A. Freiwald and J. M. Roberts, eds. Cold-water corals and ecosystems Springer-Verlag Berlin Heidelberg.

Bayer, F. M. 1956. Descriptions and redescriptions of the Hawaiian octocorals collected by the U.S. Fish Commission Steamer "Albatross" (2. Gorgonacea: Scleraxonia) Pac. Sci. 10: 67–95.

Gass, S. E. and J. H. Willison. 2005. An assessment of the distribution of deep-sea coral in Atlantic Canada by using both scientific and local forms of knowledge. Pages 223–245 *in* A. Freiwald and J. M. Roberts, eds. Cold-water corals and ecosystems. Springer-Verlag Berlin Heidelberg.

Genin, A., C. K. Paull, and W. P. Dillon. 1992. Anomalous abundances of deep-sea fauna on a rocky bottom exposed to strong currents. Deep-Sea Res. 39: 293–302.

Grigg, R. W. 1974. Distribution and abundance of precious corals in Hawaii. Pages 235–240 *in* Proc. Second Int. Symp. Coral Reefs. Brisbane.

_____. 1988. Recruitment limitation of a deep benthic hard-bottom octocoral population in the Hawaiian Islands. Mar. Ecol. Prog. Ser. 45: 121–126.

_____. 1993. Precious coral fisheries of Hawaii and the US Pacific Islands. Mar. Fish. Rev. 55: 50–60.

_____. 2002. Precious coral in Hawaii: Discovery of a new bed and revised management measures for existing beds. Mar. Fish. Rev. 64:13–20.

_____ and Bayer. 1976. Present knowledge of the systematics and zoogeography of the order Gorgonacea in Hawaii. Pac. Sci. 30: 167–175.

Leverette, T. L. and A. Metaxas. 2005. Predicting habitat for two species of deep-water coral on the Canadian Atlantic continental shelf and slope. Pages 467–479 *in* A. Freiwald and J. M. Roberts, eds. Cold-water corals and ecosystems. Springer-Verlag Berlin Heidelberg.

Moffitt, R. B. and F. A. Parrish. 1992. An assessment of the exploitable biomass of *Heterocarpus laevigatus* in the main Hawaiian Islands. Part 2: Observations from a submersible. Fish. Bull. 90: 476–482.

Oksanen, L. 2001. Logic of experiments in ecology: is pseudo replication a pseudoissue. Oikos 94: 27–38.

Parrish, F. A., K. Abernathy, G. J. Marshall, and B. M. Buhleier. 2002. Hawaiian Monk Seals (*Monachus schauinslandi*) foraging in deep-water coral beds. Mar. Mamm. Sci. 18: 244–258.

Roark, E. B., T. P. Guilderson, R. B. Dunbar, and B. L. Ingram. 2006. Radiocarbon based ages and growth rates: Hawaiian deep-sea corals. Mar. Ecol. Prog. Ser. 327: 1–14.

Watling, L. and P. J. Auster. 2005. Distribution of deep-water Alcyonacea off the northeast coast of the United States. Pages 279–296 *in* A. Freiwald and J. M. Roberts, eds. Cold-water corals and ecosystems. Springer-Verlag Berlin Heidelberg.

ADDRESS: *Pacific Islands Fisheries Science Center, NMFS, NOAA 2570 Dole Street, Honolulu, Hawaii 96822. E-mail: <Frank.Parrish@noaa.gov>.*

# Large assemblages of cold-water corals in Chile: a summary of recent findings and potential impacts

Verena Häussermann and Günter Försterra

## Abstract

Basic data on distribution, composition, and size of coral communities are extremely scarce for the Southeast Pacific. With only 23 known species of azooxanthellate Scleractinia, 12 antipatharians, 13 hydrocorals, and approximately 10–15 gorgonians, Chilean coral fauna have been considered to be rather poor. Contradictory to this assumption, recent studies record extensive, diverse, and hitherto unknown coral communities in shallow water of the Chilean fjord region. The distribution of stony corals was mapped along bathymetric and horizontal transects within the Comau Fjord (42°30´S), as well as along east-west and north-south transects in large parts of the Chilean fjord region (42°S–55°S). Habitat-forming hydrocoral (*Errina antarctica* Gray, 1872) and gorgonian communities were documented from shallow waters. In addition, we provide a list of coral species that were accidentally sampled as bycatch of fisheries on demersal species. Seven of the 23 Chilean scleractinians have been added through these recent findings; two of them have recently been described. Five of the antipatharians represent newly added records for Chile with one new species included. These numbers indicate that the diversity and abundance of cold-water corals in Chile might be much higher than previously assumed. Based on the findings, we discuss potential threats and protection measures for these newfound communities.

Benthic systems along the shelf and slope areas of the southeast Pacific, a major portion of which lie in Chilean waters, include many poorly studied marine habitats. Most publications dealing with the cold-water coral fauna of this region are historical (Philippi, 1866; Verrill, 1869; Pourtalès, 1874; Brook, 1889; Philippi, 1892, 1894; Carlgren, 1899; Kükenthal, 1924; Molander, 1929), and only two scientific publications in the second half of the 20th century deal with Chilean corals (Cairns, 1982, 1983). Material included in early studies came from vessel-based expeditions, which could not sample rocky substratum due to gear restrictions and occurred mainly in depths > 50 m. The results of these studies led to the impression of a rather depauperate coral fauna without important assemblages. Only Cairns (1982) and Stanley and Cairns (1988) suggest that large assemblages of *Desmophyllum dianthus* (Esper, 1794) could have the same character as deep-sea coral banks. Only recently has interest in Chilean cold-water fauna increased, mainly promoted by the discovery of many large and dense assemblages of cold-water coral communities in shallow water of the Chilean fjord region (Försterra and Häussermann, 2001, 2003, 2005; Pérez and Zamponi, 2004; Wehrmann et al., 2004; Cairns et al., 2005; Försterra et al., 2005; McCulloch et al., 2005; Häussermann and Försterra, 2007). Further evidence was found in the coral bycatch from demersal fisheries which indicate an unexpected deep-water coral diversity and abundance (Bravo et al., 2005).

George, R. Y. and S. D. Cairns, eds. 2007. Conservation and adaptive management of seamount and deep-sea coral ecosystems. Rosenstiel School of Marine and Atmospheric Science, University of Miami.

Here we present a list of Scleractinia, Antipatharia, Stylasteridae, and Gorgonacea from Chilean continental waters, based on specimens that have been sampled by SCUBA diving and recorded by biological observers on fishing boats. This list includes not only several recently added species, but also more new records for Chile. We also report results on the geographic and bathymetric distribution of scleractinian steep wall communities in the Chilean fjord region. We discuss necessary protection measures for these areas based on these results and potential threats.

## Material and Methods

Specimens, photos, and data on shallow-water corals were collected during SCUBA surveys at depths to 40 m. Along the exposed coast of northern and central Chile from 5°S to 41°S, approximately 30 sites were sampled down to 25 m in 1997–1998 with a distance of approximately 200 km between sites. In the southern Chilean fjord region (42°S–53°S), more than 140 sites were studied from 1997 to 2006, down to 40 m (Figs. 1, 2A). Twenty-one of these were situated within the Comau Fjord (42°30´S); for detailed information on these study sites see Försterra and Häussermann (2003) and Häussermann (2005). In the Comau Fjord, 15 additional vertical transects were studied with ROV down to 255 m in both November 2004 and February 2005. To detect shallow-water distribution patterns within a fjord, overlapping photographs were taken along vertical transects during SCUBA dives to 30 m. These photos included scale, depth, and substrate inclination for quantitative analyses of benthic coverage. The locations for the first set of SCUBA transects were randomly chosen along both shores of the Comau Fjord and at several sites along the southern shore of the Reñihué fjord to get data on the relative abundance and assemblages of scleractinians. Locations for the second set of SCUBA transects were all steep wall habitats which were more likely to have scleractinian communities. The latter criteria was also used to place the 15 ROV transects in the Comau Fjord. A 1-wk expedition was undertaken in March 2005 to survey the coral distributional patterns along an east-west gradient (continental fjords—channel region—exposed coast) at Chiloé Island and the Guaitecas archipelago (Försterra et al., 2006). During this expedition, qualitative studies of species composition in the benthic communities at five sites, consisting of both channels and more exposed areas, were carried out by means of SCUBA diving to depths of 35–40 m. Two 2-wk expeditions to the fjords and inner channels south of Golfo de Penas (48°–52°S) were undertaken to describe the distribution patterns of cold-water corals along the latitudinal extension of the Chilean fjord region. During these expeditions, SCUBA dives were performed at 14 and 33 sites, respectively (Försterra et al., 2006; Willenz et al., 2007). At all 47 sites, qualitative surveys on the species composition of the benthic communities were performed. Additionally at two of these sites, overlapping photographs were taken along vertical transects, including information on scale, depth, and inclination.

Most bycatch samples of deep-water corals (Gorgonacea, Antipatharia, Scleractinia, and Stylasteridae) off the Patagonian coast were fragments of colonies that were entangled in long-line gear used for fishing demersal species such as Patagonian toothfish (*Dissostichus eleginoides* Smitt, 1998) in depths up to 2500 m. These samples were saved either by biological observers or crew members on the fishing boats and were then archived at the Universidad Austral de Chile (A. Bravo). Samples and distributional data on some Antipatharians and Alcyonceans came from bottom trawls (550 m depth) for orange roughy (*Hoplostethus atlanticus* Collett, 1889) habitat assessment on the O'Higgins Seamount off the central Chilean coast (Y. Cañete, Universidad de Magallanes, pers. comm.). All specimens were either dried or preserved in 95% ethanol and were sent to taxonomic specialists for identification. Gorgonian and antipatharian samples that were preserved dry left the polyp tissue in more or less poor condition.

Figure 1. Distribution of Scleractinia in shallow water of the Chilean fjord region. Squares: study sites without corals, triangles: study sites where corals (*Desmophyllum dianthus, Caryophyllia huinayensis, Tethocyathus endesa*) were found.

## Results

SCUBA-diving surveys along the Chilean coast, ROV transects in the Comau fjord, and bycatch of demersal fisheries indicated that cold-water species typically found in deep water were found to extend into the shallow water of South Chilean fjord region where habitat-forming hydrocoral (*Errina antarctica*) and gorgonian (*Primnoella* spp., *Thouarella* spp., *Acanthogorgia* sp.) communities were documented (Figs. 2B–D; 3D–F).

**Scleractinia.**—The scleractinian corals *Caryophyllia huinayensis* and *Tethocyathus endesa* were reported from Concepción (~37°S) to the southern fjords (~51°S) (Table 1,

Figure 2. (A) Map of cold-water coral regions along continental Chile detected during recent studies. PP: Peruvian Province, MP: Magellanic Province. Roman numerals mark sites from fisheries bycatch: I: O'Higgins Seamount, II: off Valdivia, III: Golfo de Ancud, IV: Golfo de Penas, V: Straits of Magellan, VI: Islas Diego Ramirez. Arabic numerals mark dive sites. (B) *Errina antarctica* and *Adeonella patagonica* Hayward, 1988 in the Messier Channel, south Chile, 20 m. (C) *Errina antarctica* harvested by fishermen at the Madre de Dios Archipelago (10–20 m). (D) Meadow of *Thouarella* sp., channels south of the Peninsula Taitao, 25 m. (E) *Lillipathes* sp., Islas Diego Ramirez, 2000 m.

Table 1. Species list of cold-water corals (Anthozoa and Hydrozoa) included in this survey in Chile, including depths, and distribution. Legend: PP: Peruvian Province, MP: Magellanic Province, JF: Juán Fernández Islands; *: specimens entangled during long-line fishery

| Scleractinia (Anthozoa) | Depth (m) | Location |
|---|---|---|
| *Bathycyathus chilensis* Milne-Edwards and Haime, 1848 | 26–420 | JF, PP |
| *Caryophyllia huinayensis* Cairns et al., 2005 | 11–800 | PP, MP |
| *Tethocyathus endesa* Cairns et al., 2005 | 11–240 | PP, MP |
| *Desmophyllum dianthus* (Esper, 1794) | 4–2,460 | JF, MP |
| *Madrepora oculata* Linnaeus, 1758* | 600–2,500 | JF, PP |
| Antipatharia (Anthozoa) | | |
| *Plumapathes* aff. *fernandezi* (Pourtalès, 1874) | 40 | JF |
| *Bathypathes patula* Brook, 1889* | 600 | PP |
| *Antipathes speciosa* (Brook, 1889)* | 600 | PP |
| *Cladopathes* sp., cf. *C. plumosa* Brook, 1889* | 1,300 | MP |
| *Lillipathes* sp.* | 2,000 | MP |
| *Chrysopathes* sp. | ~550 | PP |
| *Chrysopathes* aff. *formosa* Opresko, 2003 | ~550 | PP |
| *Leiopathes* sp. | ~550 | PP |
| Alcyonacea (Anthozoa) | | |
| *Acanella* sp. | ~550 | PP |
| *Keratoisis* sp. 1* | 1,800 | MP |
| *Keratoisis* sp. 2* | 1,800 | MP |
| *Isididae* sp. 1* | 1,200 | MP |
| *Isididae* sp. 2* | 1,300 | PP |
| *Leptogorgia platyclados* (Philippi, 1866) | 5–30 | PP |
| *Primnoella chilensis* (Philippi, 1894) | 10–40 (80?) | MP |
| *Thouarella* sp.* Kükenthal, 1912 | 130 | PP |
| *Thouarella koellikeri* Wright and Studer, 1889 | 15–45 | MP |
| *Thouarella* sp. 1 | 15–45 | MP |
| *Thouarella* sp. 2 | 15–45 | MP |
| Gorgonacea sp. 1 | 250 | PP |
| *Callogorgia* sp.* | 600 | PP |
| *Plumarella* sp.* | 300–600 | PP |
| Gorgonacea sp. 2* | 1,900 | MP |
| *Acanthogorgia* sp. cf. *A. laxa* Wright and Studer, 1889 | 15–40 | MP |
| *Acanthogorgia* sp. | 15–40 | MP |
| Plexauridae sp. | 30–40 | MP |
| Paragorgiidae sp. | ~550 | PP |
| *Sibogagorgia* sp. | ~550 | PP |
| Stylasteridae (Hydrozoa) | | |
| *Stylaster densicaulis* Moseley, 1879* | 30–1,244 | MP |
| *Errina antarctica* Gray, 1872 | 15–771 | MP |
| *Errina* sp.* | 80 | MP |
| Stylasteridae sp.* | 80 | MP |

Fig. 3D) at depths from ~800 m and 250 m, respectively (off Concepción), up to 16 m in the fjords. They are a regular component of the shallow-water coral communities in Chilean fjords and channels, but due to their small size do not play a structure-forming role.

*Desmophyllum dianthus* (form *ingens*) was reported from the Seno de Reloncaví at the northern end of the Chilean fjord region (~41°35′S) to the Straits of Magellan (~54°S), at depths between 2460 m (Cairns, 1982) and 8 m at Chiloé Island (Försterra and Häussermann, 2003). Large accumulations covering more than 1000 m² were found below 20 m on steep, overhanging walls with substrate inclinations exceeding 80°. Massive banks with densities exceeding 1500 specimens m⁻², many in the form of pseudo-colonies consisting of up to more than five polyp generations and lengths in excess of 40 cm, were found in the Comau Fjord (Fig. 3D), Reñihué Fjord, and reported from the Reloncaví Fjord. These findings suggest that *D. dianthus* is a framework species for a diverse benthic community. Uranium-thorium age analysis of specimens from dense aggregations in shallow water indicated 60–70 yrs of age for mid-sized (25 cm) individuals (McCulloch et al., 2005).

**Bathymetric distribution of *Desmophyllum dianthus* in the Comau Fjord.**—ROV transects revealed that the shallow occurrence of the framework-forming species, *D. dianthus* is merely the upper portion of a continuous population inhabiting steep walls of fjords and channels down to a depth of at least 255 m (greatest depth examined by ROV). The occurrence appears more related to substrate slope (> 80°) than to depth. The upper limit for the presence of corals coincides with the average extension of the low salinity layer (LSL, 2‰–10‰) that is characteristic for the inner fjords. The upper limit for dense assemblages coincides with the absolute maximum extension of the LSL (below 20 m). Although extension and density of assemblages as well as shape of the pseudo-colonies of *D. dianthus* appeared to be comparable in all depths, the associated benthic communities changed strongly along a bathymetrical gradient (own data from ongoing inventory projects). While the upper end of the coral population, reaching well into the euphotic zone, has diverse associated fauna and flora (see Cairns et al., 2005), the associated benthic communities in the aphotic deeper parts (> 50 m) seem to be less diverse in macrobenthic species. Several of the species associated with the *Desmophyllum* banks belong to genera such as the bivalve *Acesta* sp., the sponge *Geodia* sp., and the fish *Sebastes* sp. which are typical members of deep-water coral communities. Some of the associated species can reach high abundances, for instance the bivalve *Acesta* aff. *patagonica* (Dall, 1902).

Neither the qualitative benthic analyses of the SCUBA and ROV transects nor the quantitative analyses of vertical transects along the fjord showed a significant gradient in the distribution of scleractinian corals within the fjord. Corals were present in relatively equal densities whenever the substrate slope exceeded 80°, independent of the distance to the head of the fjord.

**Distribution of shallow-water *Desmophyllum dianthus* from inner fjords to the exposed coast.**—Dive surveys along the channels and exposed parts of the Guaitecas Islands and at Islets and channels in southeast Chiloé Island showed a decline from the inner fjords to the exposed coast in abundance, density, and average size of both individuals and pseudo-colonies. Similar to the inner fjords, presence of *D. dianthus* was restricted to steep, overhanging walls exceeding 80°. In the fjords, corals inhabit almost all areas of suitable substrate (defined by slope), whereas in the channel, corals have not colonized many areas with apparently appropriate slope characteristics.

Pseudo-colonies were rare and never consisted of more than three generations. When a LSL was present, the upper limit of coral occurrence coincided with its maximum bathymetric extension but never was shallower than 8 m (Cailin Island, SE of Chiloé Island).

At all five study sites in the Guaitecas Islands (March 2005), only dead specimens of *D. dianthus* were found. Large accumulations and small clusters were affected in the same way. From the state of erosion and the epibiontic overgrowth, it is inferred that the die-off must have taken place simultaneously at least some 3–5 yrs prior to the study (before 2002). From the size of the largest corallites, we calculated a maximum population age of at least 60 yrs. No signs of recovery or new recruitment could be seen, nor could a cause of this large scale die-off be detected. *Tethocyathus endesa* and *C. huinayensis* were present in slightly less dense populations than in the inner fjords, and seemed not to have been affected by a die-off.

**Latitudinal distribution of scleractinian assemblages in the Chilean fjord region.**—Surveys at Puyuhuapi and Aysén Fjord, and the fjord/channel system south of the Golfo de Penas showed a significant north to south decline in abundance, lateral extension, and density of shallow-water scleractinian assemblages of the three above-mentioned species. Furthermore, a decline in individual size of corallites of *D. dianthus* was also found along the north to south gradient. Although all three coral species were present south of the Golfo de Penas, coral patches rarely exceeded 10 m$^2$ in size and sizes of corallites of *D. dianthus* never exceeded 10 cm in length. South of the Golfo de Penas, corals were exclusively found under overhanging portions of rock walls or bigger boulders. Similar to sites farther north, the presence of corals was restricted to the high salinity water below the pycnocline. In the fjords south of the Golfo de Penas this was generally below 18 m.

**Gorgonacea.**—At most study sites in the fjords and channels of Chile, unbranched primnoids, especially *Primnoella chilensis*, form an important faunal component on moderately steep rocky slopes between 15 m and 50 m depths, mainly in the northern Patagonian fjord region (Fig. 3D). The upper limit of these populations coincides with either the average maximum extension of the LSL (if present) or 11 m depth. The deepest specimens were observed around 100 m. At some sites, *P. chilensis* dominates the substratum for large areas (> 1000 m$^2$) with densities exceeding 40 colonies m$^{-2}$. The dead branches of gorgonian colonies are a popular substratum for many epibenthic organisms.

In the channels south of the Golfo de Penas, extensive fields of large branched primnoids and acanthogorgiids, especially *Thouarella* spp. and *Acanthogorgia* spp., and soft corals were found in depth below 20 m (Figs. 2D, 3E). The soft corals (mainly *Alcyonium* spp.) either form rather small, three-dimensional colonies or cover primary and biogenic substratum in epizoic colonies. *Acanthogorgia* colonies that exceed 1 m fan diameter formed popular substratum for a variety of epizoic organisms (Fig. 3E). The lower bathymetric limit of these communities was below SCUBA diving depths and therefore could not be censused during this survey.

During this SCUBA study, fields of *Leptogorgia platyclados* were found at several sites along the exposed coast off Concepción (37°S) on rocky substratum beginning in 12 m depth (Fig. 3F). At one site in the Coliumo Bay, a dense population of *L. platycla-*

Figure 3. Chilean cold-water anthozoans. (A–C) Specimens from long-line bycatch. (A) *Madre-pora oculata* off Valdivia, 2000 m. (B) Isididae from Golfo de Penas, 1200 m. (C) *Callogorgia* sp. with ophiuroid. (D) Dense coral bank of *Desmophyllum dianthus* in Comau Fjord, 25 m; with sea whips *Primnoella* sp. on upper part of rock. (E) Field of large gorgonians of the genus *Acan-thogorgia* sp., Canal Messier, 30 m. (F) *Leptogorgia platyclados* in Bahia de Coliumo, 12 m.

*dos* was found in a cave at 3 m depth. Due to the weak slope, lower limits of the bathymetric distribution were out of reach (below 20 m) and therefore could not be censused.

**Stylasteridae.**—Stylasteridae of the species *E. antarctica* were a common component of shallow-water benthic communities in channels that are subject to stronger currents at depths below 10 m south of the Golfo de Penas (Fig. 2B). At some well flushed sites, density and size of colonies (diameter > 40 cm) reach levels that make them an important structure-forming element of the benthic communities. During the 2006 expedition, reef-like assemblages of hydrocorals were discovered which are described in a separate paper (Häussermann and Försterra, 2007). Dead portions of the colonies were densely covered by epibiontic organisms. The lower bathymetric distribution limit of these *E. antarctica* populations was below diving depths (40 m) and therefore were not censused. Within the distribution range, two extremes of colony morphologies were identified. Colonies found on vertical walls were mainly two-dimensional fan-shaped with flattened tips (Fig. 2B), while samples from U-formed channels in the Madre de Dios Archipelago show three-dimensional colony structures with rounded tips (Fig. 2C). These growth forms seem to be extremes of a continuum of colony morphologies rather than two distinct morphotypes.

## Discussion

**Distribution of cold-water corals.**—Findings of the present study indicate that the shallow-water existence of scleractinian cold-water corals, formally considered to be deep-water corals, is a phenomenon that is typical for fjords and channels of the Chilean fjord region. Scleractinian corals are also known from shallow water of New Zealand and Norwegian fjords. This shallow-water occurrence is part of the deep-water emergence trend that can be observed within several taxa in this region. Nevertheless, massive banks of scleractinians, representing a structure-forming framework organism for an entire benthic community, were only found in the three northernmost fjords (Reloncaví, Comau, and Renihué). Neither in the channels that lie southeastwards of these continental fjords nor in the fjords farther south do scleractinian assemblages reach comparable extent, structure, and densities. The most obvious distinguishing characteristic that these three fjords share is their comparably isolated location at the northern end of the Chilean fjord region and direct connection with the Golfo de Ancud. Which exact factors cause the extraordinary growth of corals in these three fjords in contrast to channels or fjords farther south remains to be studied.

The shallow-water populations of *D. dianthus* are restricted to near vertical and overhanging substratum, which might indicate sensitivity to sedimentation (Försterra and Häussermann, 2003). Most fjords south of the Golfo de Penas are under strong influence from glacial melt water. This glacier run-off carries large amounts of fine glacial sediment which produces a strong gradient from the head of a fjord to its mouth. Consequently, the stress for benthic organisms due to sediment increases towards the head of these fjords (Försterra et al., 2006). This situation is very well reflected by the distribution patterns of corals that we found in the fjords south of the Golfo de Penas where presence, size, and density of scleractinian assemblages were negatively correlated with the apparent sedimentation rate. This suggests that the continental fjords south of the Golfo de Penas are a less suitable habitat for Scleractinia than fjords north of the Peninsula Taitao.

The presence of stylasterid and some branched gorgonian communities in shallow-water of channels south of the Golfo de Penas, and their absence in comparable habitats north of the Peninsula Taitao supports the hypothesis of a zoogeographic barrier at 47°S–48°S (Pickard, 1973; Lancellotti and Vásquez, 1999; Häussermann and Försterra, 2005). The Peninsula Taitao and the Golfo de Penas might represent an obstacle for larval transport of species adapted to conditions of the protected, freshwater-influenced inner fjords and channels. Larval transport past these obstacles may be further impeded through the currents. The West Wind Drift hits the continent at approximately the latitude of Pensinsula Taitao, and then splits into the strong northward flowing Humboldt Current and the weaker southward flowing Cape Horn Current (see Häussermann and Försterra, 2005). This oceanographic situation suggests that latitudinal larval transport past the splitting zone would require movement against one of these currents, especially for those brooding species (such as *Stylaster* and some gorgonians) that release short-lived larvae with low dispersal potential.

## Potential threats to Chilean cold-water corals

**Bottom trawling.**—Bottom trawling is a destructive fishing technique with much associated bycatch. Since this method can heavily damage and destroy ancient coral reefs (Fossa et al., 2002; Hall-Spencer et al., 2002), its use has been very controversially disputed over the last decade. As a consequence, bottom trawling has been banned in several areas of the worlds' oceans where corals occur (e.g., in the North Atlantic to protect *Lophelia* reefs). In Chile, bottom trawling mainly occurs between 5 nmi and the Exclusive Economic Zone around the Juan Fernandez Islands down to > 2000 m depths. The targets species are crustaceans and fishes, e.g., hake species, orange roughy, and prawn (see www.subpesca.cl). These organisms often live and/or aggregate for spawning at sites where deep-water corals occur. Paragorgiids, isidids, and antipatharians have been sampled from the O'Higgins Seamount (550 m) off central Chile during a survey of the orange roughy population. In November 2004, the Chilean government agreed to support a temporary suspension of bottom trawling outside the EEZ proposed by the Deep-Sea Conservation Coalition.

**Longline fishery.**—The longline method is a common and widespread fishing technique for demersal species like Patagonian toothfish (*D. eleginoides*) and ling cod (*Genypterus blacodes* (Forster in Bloch and Schneider, 1801) in Chilean waters. Depending on the precise techniques and the depth, longline fisheries do not necessarily bring a lot of bycatch. However, Table 1 shows that a considerable number of specimens of structure-forming corals like *Madrepora oculata*, gorgonians, and antipatharians are entangled in bottom gear and landed despite the low probability of corals being brought up this way from great depth, landed, and then sampled by a biological observer. The large sizes of bycatch specimens and fragments of Isididae and Scleractinia such as *M. oculata* and the high diversity of other gorgonians in relation to the comparably low absolute number of samples brought to the museum provide evidence that there are extensive old populations of deep-sea corals in Chilean waters with higher diversity than previously expected.

**Aquaculture.**—The fast development of an aquaculture industry in Chilean Patagonia with the absence of sufficient biological and ecosystemic knowledge carries a lot of risk for vulnerable and unique communities. Fish and mussel farms are known to produce sediments and point source nutrient pollution that affect nearby benthic systems (Macleod et al., 2004). In semi-closed environments like fjords, fish farms may cause eutrophication (Aure and Stigebrandt, 1990) and oxygen depletion near the bottom and in deeper water. These impacts might be serious threats to coral communities in Chilean fjords. In the last few years, all three fjords in Chile where massive coral banks are found were densely covered with aquaculture concessions, which most likely will or already have changed the special ecological conditions in these fjords. The consequence may be the disappearance of these unique communities before they are even really studied.

**Local market for display items.**—In the Madre de Dios Archipaelgo, local fishermen are known to harvest large quantities of the stylasterid *E. antarctica* for selling as decorative objects. From New Zealand fjords, the gross growth rate of similar stylasterid corals was estimated 1–7 mm per year (Miller, 1998; Stratford et al., 2001; Miller et al., 2004). If growth rates of Chilean stylasterids are comparable, recovery of harvested populations may take decades to hundreds of years if it occurs at all. Although harvesting of shallow-water pseudo-colonies of *D. dianthus* still satisfies a comparably small market, expanding knowledge of their existence might raise demand, and destruction through harvesting may become a more important issue. To date, fishery agencies have expressed interest in banning harvesting of corals species. Nevertheless, control of such laws might be difficult as most corals are sold on a national market.

**Protection of cold-water coral habitats.**—In other parts of the world, marine protected areas have been and are being established to protect the unique and often very old ecosystems of cold-water corals, e.g., in the high seas to protect *Lophelia* reefs in the North Atlantic Ocean or in nearshore areas to protect hydrocorals in New Zealand fjords (Miller et al., 2004). Considering the broad distribution of cold-water corals along the Chilean coast, a network of near-shore and high seas marine protected areas in combination with the control and/or ban of destructive fishing and aquaculture activities may be the most effective measures for protecting cold-water coral communities in Chile. Priority should be given to areas where aquaculture development or bottom trawling has reached critical levels and where hotspots in coral diversity and abundance can be found. Bycatch monitoring, studies on the massive *D. dianthus* assemblages, and mapping efforts of the Chilean fjord region should have research priority to gather data that are needed to assess and minimize dangers for these unique and important ecosystems.

## Acknowledgments

We are very grateful to A. Bravo, Universidad Austral de Chile, Y. Cañete, Universidad de Magallanes, J. Sellanes, Universidad de Coquimbo, and L. Prado, Pontificia Universidad Católica de Santiago for providing specimens and data from their collections. Thanks to S. Cairns and D. Opresko, Smithsonian Institution, J. Sanchez, Universidad de los Andes, and O. Breedy, Universidad de Costa Rica for the identification of the scleractinians, hydrocorals, antipatharians, and gorgonians. The expeditions on which these studies were based were supported by the PADI Foundation, Centro Aclimatización Zoológica, the Huinay Foundation, and by a Prosul grant from the Brazilian government to E. Hajdu, National Museum of Rio de Janeiro. Many thanks to CONAF (Corporación Nacional Forestal de Chile) and NAVIMAG for their logistic support, and the crew of the YEPAYEK for the great help. We also thank George Institute for Biodiversity which gave support for the participation of G. Försterra at the 3rd International Symposium on Deep-Sea

Corals (ISDSC3). Many thanks to the reviewers for the helpful comments. This survey has also been supported by project DID/UACH S-200518. This is publication number 16 of the Huinay Scientific Field Station.

## Literature Cited

Aure, J. and A. Stigebrandt. 1990. Quantitative estimates of the eutrophication effects of fish farming on fjords. Aquaculture 90: 135–156.

Bravo, A., G. Försterra, and V. Häussermann. 2005. Fishing in troubled waters – evidence for higher diversity and high abundance of cold water corals along the Chilean coast. Page 234 *in* R. Brock and R. Y. George, eds. Deep-sea corals science and management. 3rd Int. Symp. Deep-Sea Corals, Miami, FL.

Brook, G. 1889. Report on the Antipatharia. Reports of the scientific results of the voyage of the CHALLENGER. Zoology 32: 1–222.

Cairns, S. D. 1982. Antarctic and Subantarctic Scleractinia. Antarct. Res. Ser. 34: 1–74.

_____. 1983. Antarctic and Subantarctic Stylasterina (Coelenterata: Hydrozoa). Antarct. Res. Ser. 38: 61–164.

_____, V. Häussermann, and G. Försterra. 2005. A review of the Scleractinia (Cnidaria: Anthozoa) of Chile, with the description of two new species. Zootaxa 1018: 14–46.

Carlgren, O. 1899. Zoantharien. Hamb. Magelh. Sammelr. 4: 1–48.

Försterra, G. and V. Häussermann. 2001. Large assemblages of azooxanthellate Scleractinia (Cnidaria: Anthozoa) in shallow waters of South Chilean fjords. Page 155 *in* 4th Ann. Meet. Gesellschaft für Biologische Systematik, GfBS, Oldenburg, Germany.

_____ and _____. 2003. First report on large scleractinian (Cnidaria: Anthozoa) accumulations in cold-temperate shallow water of south Chilean fjords. Zool. Verh. (Leiden) 345: 117–128.

_____ and _____. 2005. In sight, still out of mind! - Coral banks in Chilean fjords: characteristics, distribution, threats. Page 125 *in* R. Brock and R. Y. George, eds. Deep-sea corals science and management. 3rd Int. Symp. Deep-Sea Corals, Miami, FL.

_____, _____, and G. Foley, Jr. 2006. Adding pieces to a complex puzzle - discovering the benthic life in the channels and fjords of Chilean Patagonia. Glob. Mar. Env. 3: 18–21.

_____, L. Beuck, V. Häussermann, and A. Freiwald. 2005. Shallow water *Desmophyllum dianthus* (Scleractinia) from Chile: characteristics of the biocenoses, the bioeroding community, heterotrophic interactions and (palaeo)-bathymetrical implications. Pages 937–977 *in* A. Freiwald and J. M. Roberts, eds. Cold-water corals and ecosystems. Springer-Verlag, Berlin, Heidelberg.

Fossa, J. H., P. B. Mortensen, and D. M. Furevik. 2002. The deep-water coral *Lophelia pertusa* in Norwegian waters: distribution and fishery impacts. Hydrobiology 471: 1–12.

Hall-Spencer, J., V. Allain, and J. H. Fossa. 2002. Trawling damage to Northeast Atlantic ancient coral reefs. Proc. R. Soc. Lond., Ser. B: Biol. Sci. 269: 507–511.

Häussermann, V. 2005. The sea anemone genus *Actinostola* Verrill, 1883: variability and utility of traditional taxonomic features, and a re-description of *Actinostola chilensis* McMurrich, 1904. Polar Biol. 28: 338–350.

_____ and G. Försterra. 2005. Distribution patterns of Chilean shallow-water sea anemones (Cnidaria: Anthozoa: Actiniaria, Corallimorpharia); with a discussion of the taxonomic and zoogeographic relationships between the actinofauna of the South East Pacific, the South West Atlantic and Antarctica. Pages 91–102 *in* W. E. Arntz, G. A. Lovrich, and S. Thatje, eds. The Magellan-Antarctic connection: links and frontiers at high southern latitudes. Sci. Mar. 69 (Suppl. 2). 218 p.

_____ and _____. 2007. Extraordinary abundance of hydrocorals (Cnidaria, Hydrozoa, Stylasteridae) in shallow water of the Patagonian fjord region. Polar Biol. 30: 487–492.

Kükenthal, W. 1924. Gorgonaria. Pages 1–478 *in* F. E. Schultze and W. Kükenthal, eds. Das Tierreich. Walter de Gruyter and Co, Berlin and Leipzig.

Lancellotti, D. A. and J. A. Vásquez. 1999. Biogeographical patterns of benthic macroinvertebrates in the Southeastern Pacific littoral. J. Biogeogr. 26: 1001–1006.

Macleod, C. K., C. M. Crawford, and N. A. Moltschaniwskyj. 2004. Assessment of long term change in sediment condition after organic enrichment: defining recovery. Mar. Pollut. Bull. 49: 79–88.

McCulloch, M., P. Montagna, G. Försterra, G. Mortimer, V. Häussermann, and C. Mazzoli. 2005. Uranium-series dating and growth rates of the cool-water coral *Desmophyllum dianthus* from the Chilean fjords. Page 191 *in* R. Brock and R. Y. George, eds. Deep-sea corals science and management. 3rd Int. Symp. Deep-Sea Corals, Miami, FL.

Miller, K. 1998. Population structure and growth rates of red coral in Doubtful Sound, Fiordland. Contract Report WLG 98/14 for Department of Conservation, Invercargill, NIWA, Wellington, New Zealand.

_____, C. N. Mundy, and L. W. Chadderton. 2004. Ecological and genetic evidence of the vulnerability of shallow water populations of the stylasterid hydrocoral *Errina novaezelandiae* in New Zealand's fiords. Aquat. Cons.: Mar. Freshwat. Ecosyst. 14: 75–94.

Molander, A. R. 1929. Die Octactiniarien. P. A. Norstedt & Söner, Stockholm. 86 p.

Pérez, C. D. and M. O. Zamponi. 2004. New records of octocorals (Cnidaria, Anthozoa) from the south western Atlantic Ocean, with zoogeographic considerations. Zootaxa 630: 1–12.

Philippi, R. A. 1866. Kurze Beschreibungen einiger chilenischer Zoophyten. Arch. Naturgesch. 32: 118–120.

_____. 1892. Los zoofitos chilenos del Museo Nacional. Anales Mus. Nac. Chile, primera sección (Zool.) 5: 1–11.

_____. 1894. *Callirhabdos*, ein neuer Genus der gorgonienartigen Pflanzenthiere? Arch. Naturgesch. 60: 211–213.

Pickard, G. L. 1973. Water structure in Chilean fjords. Pages 95–104 *in* R. Fraser, ed. Oceanography of the South Pacific. New Zealand National Commission for UNESCO, Wellington.

Pourtalès, L. F. d. 1874. Crinoids and corals. Zoological results of the Hassler Expedition, III. Catalogue Mus. Comp. Zool. Harvard 8: 25–50.

Stratford, P., B. G. Stewart, and A. Chong. 2001. In situ growth rate measurements on the red hydrocoral, *Errina novaezelandiae*, in Doubtful Sound. N. Z. J. Mar. Freshwat. Res. 35: 659–660.

Stanley, G. D. and S. D. Cairns. 1988. Contructional azooxanthellate coral communities, an overview with implications for the fossil record. Palaios 3: 233–242.

Verrill, A. E. 1869. No. 6. - Review of the corals and polyps of the west coast of America. Transac. Connect. Acad. Arts Sci. 1: 377–567.

Wehrmann, A., A. Schmidt, and G. Försterra. 2004. The end of a tropical story: carbonate production by coral assemblages and related benthic communities in the shallow waters of South Chilean fjords. Schriftenr. Dt. Geol. Ges. - Kurzfassungen und Exkursionsführer 33: 174. Sediment 2004, Hannover.

Willenz, P., V. Häussermann, G. Försterra, M. Schrödl, R. Melzer, L. Atwood, and C. Jorda. 2007. Finding more pieces of the Chilean puzzle. Glob. Mar. Env. 5: 26–29.

ADDRESSES: (V.H.) *Fundación Huinay (Huinay Scientific Field Station), Casilla 462, Puerto Montt, Chile and Universidad Austral de Chile, Departamento de Biología Marina, Campus Isla Teja, Ave. Inés de Haverbeck, Casilla 456, Valdivia, Chile.* (G.F.) *Fundación Huinay (Huinay Scientific Field Station), Casilla 462, Puerto Montt, Chile and Ludwig-Maximilians-Universität Munich, Department Biologie II, Großhadernerstr. 2, 82152 Planegg-Martinsried, Germany.* CORRESPONDING AUTHOR: (V.H.) *E-mail: <vreni_haeussermann@yahoo.de>.*

# Note

# Azooxanthellate hard corals (Scleractinia) from India

Krishnamoorthy Venkataraman

India has a coastline of 7516 km, of which the mainland accounts for 5422 km. The Lakshadweep coast extends up to 132 km and Andaman and Nicobar Islands have a coastline of 1962 km. The coastal zone of India is endowed with a very wide range of coastal ecosystems such as estuaries, lagoons, mangroves, backwaters, salt marshes, rocky coasts, sandy stretches, and coral reefs which are characterized by unique biotic and abiotic properties and processes (Venkataraman, 2003). The mainland coast of India has two widely separated areas containing shallow-water coral reefs: the Gulf of Kachchh (=Kutch) at northwest, which has some of the most northerly reefs in the world and Palk Bay and Gulf of Mannar in the southeast.

In addition to these reefs, Andaman and Nicobar Islands have fringing reefs around many islands, and a long barrier reef (329 km) on the west coast of these islands. These reefs are poorly known scientifically but are the most diverse and in the best condition in Indian waters (Venkataraman et al., 2004). The Lakshadweep Islands also have extensive atoll reefs but these are also inadequately known (Venkataraman et al., 2003).

Knowledge of the occurrence and distribution of deep-water coral reefs from India is very poor, being largely based on few surveys in limited geographic areas. Wood-Mason and Alcock (1891a,b) reported deep-water corals of Indian Ocean collected during the expeditions RIMS Investigator I and II from the Indian Ocean. Gardiner (1904) examined over 2000 specimens collected off South Africa and reported 15 species. Van der Horst (1921) reported eight species of dendrophylliids. Gardiner and Waugh (1938, 1939) published results of the John Murray expedition (H.M.S. Mabihiss stations 102–133) discussing 28 species of deep-water corals. Other records that include useful information on deep-water corals from India and the Indian Ocean were those of Alcock (1894, 1898, 1902), Bourne (1905), Gardiner (1929), and Wells (1956). Some of the more recent studies on Indian deep-water corals are those of Fricke and Schuhmacher (1983); Pillai and Scheer (1976); Scheer and Pillai (1974, 1983); Sheppard and Sheppard (1991); Zibrowius (1980), and Zibrowius and Gili (1990).

The seas and the deep shelf regions adjacent to India have had few surveys and limited reports on the occurrence of deep-water corals except the studies of Pillai and Scheer (1976) from Nicobar, Pillai (1967a,b; 1983, 1986, 1988) from the Gulf of Kachchh, Gulf of Mannar, Lakshadweep, and Andaman and Nicobar Islands, and Venkataraman et al., (2003) from India. Here I use the samples collected off the Chennai coast (from 10–30 m depth) by R/V Sagara Sampatha (trawl) in 2003 to report the occurrence of eight species of deep-water Scleractinia and discuss these in the context of earlier records.

George, R. Y. and S. D. Cairns, eds. 2007. Conservation and adaptive management of seamount and deep-sea coral ecosystems. Rosenstiel School of Marine and Atmospheric Science, University of Miami.

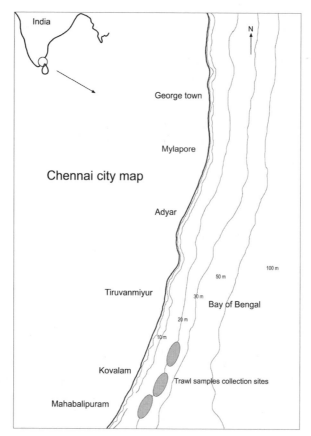

Figure 1. Map showing the collection sites off Chennai Coast.

## Material and Methods

Trawl samples were collected during 2003 at 10–30 m depth along the Chennai coast between Madras Harbour and Mahabalipuram (Fig. 1). Azooxanthellate corals collected and reported in the present study are from (1) Kovalam, 12°52´N, 80°18´E, 19 m, 9 January, 2003 (28 specimens); (2) Mahabalipuram, 12°49´N, 80°19´E, 20 m, on 9 February, 2003 20 m (32 specimens); (3) Mahabalipuram, 12°50´N, 80°18´E, 26 m (7 specimens), and (4) Chidiyatappu, Port Blair, South Andaman 25 m, on December 1999 (one colony of *Tubastraea* sp.) (collected by hand by SCUBA divers). The identified corals were washed in freshwater, sun-dried, and labeled for identification. The publications of Pillai (1967a,b), Cairns (1989, 1998), and Cairns and Keller (1993) were used for identification of the collected samples. The identified samples were deposited at the Zoological Survey of India, Chennai.

## Results and Discussion

The existence of azooxanthellate scleractinian corals has been known for nearly two centuries. However, despite some early attempts to study the deep-water coral reefs or banks in Indian Ocean, the wide distribution of these habitats and their contribution to biodiversity in the deep ocean remain largely unknown. From 1888–1892, Lt. Col. A. W. Alcock (Investigator II and Siboga expeditions) published studies on the deep-water corals of the Indian Ocean. These publications formed the basis of the deep-water coral knowledge from the Indian Ocean and the seas around India. Recently, a compilation of deep-water corals from the world (Cairns, 1999; Cairns et al., 1999) indicates that out of 1314 extant species of scleractinians, 669 are deep water, of which 227 species comprising 12 families and 71 genera (30% of world deep-water corals) are reported from Indian Ocean by several authors.

Within the family Caryophylliidae, 86 species have been reported from the Indian Ocean (Cairns, 1999), of which seven genera and 16 species are reported in the present study. Species such as *Paracyathus indicus indicus*, *Paracyathus profundus*, and *Paracyathus stokesii* have been reported so far only from Indian waters. *Heterocyathus sulcatus*, *Heterocyathus alternatus*, *Deltocyathus rotulus*, and *Polycyathus fuscomarginatus* are reported from the Chennai coast in the present study. *Heterocyathus aequicostatus* earlier reported by Pillai, 1983 is also recorded in the Chennai Coast (Kovalam and Mahalipuram). Out of the 46 species of the family Dendrophylliidae reported from Indian Ocean (Cairns, 1999), six genera and 17 species are reported in the present study (Table 1). Species such as *Balanophyllia scabra*, *Dendrophyllia indica*, and *Dendrophyllia minuscula* have so far been reported only off India. *Eguchipsammia gaditana* and *Tubastraea coccinea* are reported from Gulf of Mannar and *Enallopsammia rostrata* (Pourtalès, 1878) reported from Nicobar are found to be common in other oceans of the world also. *Heteropsammia cochlea* and *Tubastraea aurea* are reported from Chennai in the present study and *Tubastraea micrantha* is reported only from Nicobar. Of the 35 species of the family Flabellidae reported from the Indian Ocean (Cairns, 1999), six species in four genera are reported in the present study (Table 1). *Flabellum deludens* is reported from Chennai coast (Kovalam). In the present study, one species of each of the families Fungiacyathidae and Oculinidae is reported. *Madrepora oculata* reported from the Lashadweep Sea by Alcock (1898) has also been reported world over (Cairns, 1999). Of the nine species reported from the family Rhizangiidae from Indian Ocean, only two species were found during the present study (Table 1). Only 43 species in six families and 21 genera have been reported from the shallow coasts of India including those from the present study. Of these six families, most species were from the families Caryophylliidae and Dendrophylliidae.

**Human threats to corals.**—Little over five million people live within 50 km of the Chennai coast. The Chennai coast is thus a place of high human activity and the coastal and marine ecosystems are being highly disturbed and threatened. Between Ennore estuary in the northern sector and the fish landing center at Royapuram, many industries bordering the coastline discharge their effluents both in the form of chemical byproducts, fly-ash, and heated coolant waters. The Chennai Harbour situated south of Royapuram, is one of the most active ports of the sub-continent. To the south of Chennai Harbour, the shore is devoid of any major structures up to Adyar estuary wherein the mouth of Cooum River is also situated. Neither the Cooum River nor the Adyar estuary discharge their contents into the sea year-round. Nevertheless, their effect is felt in the sea immediately after heavy rainfall during the late southwest monsoon period and northeast monsoon period.

The major human impact on azooxanthellate scleractinian coral reefs is due to fishing activities throughout the continental shelf and deep-sea regions of Chennai coast as well as other coasts of India. The depletion of many traditional shallow-water fisheries has led to the exploitation of fish species that inhabit deeper waters. Since modern deep-

Table 1. Azooxanthellate scleractinians reported from four major coral reefs and other regions in India (BOB = Bay of Bengal, IO = Indian Ocean, GOM = Gulf of Mannar, AN = Andaman and Nicobar, GOK = Gulf of Kutch, LOK = Lakshadweep) (* eight reports from the present study)

| Species name | GOM | AN | GOK | LOK | Other regions of India |
|---|---|---|---|---|---|
| **Caryophylliidae** | | | | | |
| *Caryophyllia grandis* Gardiner and Waugh, 1938 | – | AN | – | LOK | – |
| *Caryophyllia arcuata* Milne-Edwards and Haime, 1848 | – | AN | – | LOK | – |
| *Caryophyllia grayi* Milne-Edwards and Haime, 1848 | – | AN | – | – | – |
| *Deltocyathus andamanicus* Alcock, 1898 | – | AN | – | – | – |
| *\*Deltocyathus rotulus* (Alcock, 1898) | – | – | – | – | Chennai |
| *\*Heterocyathus aequicostatus* Milne-Edwards and Haime, 1848 | GOM | AN | – | – | Chennai Coast |
| *\*Heterocyathus alternatus* Verrill, 1865 | – | – | – | – | Chennai Coast |
| *\*Heterocyathus sulcatus* (Verrill, 1866) | – | – | – | – | Chennai Coast |
| *Paracyathus indicus indicus* Duncan, 1889 | – | AN | – | – | – |
| *Paracyathus profundus* Duncan, 1889 | GOM | – | – | – | – |
| *Paracyathus stokesii* Milne-Edwards and Haime, 1848 | – | – | GOK | – | – |
| *Polycyathus andamanensis* Alcock, 1893 | GOM | AN | – | – | – |
| *\*Polycyathus fuscomarginatus* (Klunzinger, 1879) | – | – | – | – | Chennai Coast |
| *Polycyathus verrilli* Duncan, 1889 | GOM | AN | GOK | – | – |
| *Solenosmilia variabilis* Duncan, 1873 | – | – | – | LOK | – |
| *Stephanocyathus nobilis* (Moseley, 1873) | – | – | – | LOK | – |
| **Dendrophylliidae** | | | | | |
| *Balanophyllia affinis* (Semper, 1872) | GOM | – | – | – | – |
| *Balanophyllia imperialis* Kent, 1871 | – | AN | – | – | – |
| *Balanophyllia ponderosa* van der Horst, 1926 | – | – | – | – | IO |
| *Balanophyllia scabra* Alcock, 1893 | – | AN | – | – | – |
| *Balanophyllia stimpsonii* (Verrill, 1865) | GOM | AN | – | – | – |
| *Dendrophyllia arbuscula* van der Horst, 1922 | – | AN | – | – | – |
| *?Dendrophyllia coarctata* (Duncan, 1873) | GOM | – | – | – | – |
| *Dendrophyllia indica* Pillai, 1967 | GOM | – | – | – | – |
| *Dendrophyllia minuscula* (Bourne, 1905) | – | – | GOK | – | – |
| *Eguchipsammia gaditana* (Duncan, 1873) | GOM | – | – | – | – |
| *Enallpsammia rostrata* (Portualès, 1878) | – | AN | – | – | – |
| *Endopsammia philippensis* Milne-Edwards and Haime, 1848 | GOM | – | – | – | – |
| *\*Heteropsammia cochlea* (Spengler, 1781) | – | – | – | – | Chennai Coast |
| *Heteropsammia michelini* Milne-Edwards and Haime, 1848 | GOM | AN | – | – | – |
| *\*Tubastraea aurea* (Quoy and Gaimard, 1833) | GOM | AN | GOK | – | Chennai Coast |
| *Tubastraea coccinea* Lesson, 1829 | GOM | – | – | – | – |
| *Tubastraea diaphana* Dana, 1846 | GOM | – | – | – | – |
| *Tubastraea micrantha* (Ehrenberg, 1834) | – | AN | – | – | – |
| **Flabellidae** | | | | | |
| *\*Flabellum deludens* Marenzeller, 1904 | – | – | – | – | Chennai Coast |
| *Flabellum pavoninum* Lesson, 1831 | – | – | – | LOK | – |
| *Placotrochus laevis* Milne-Edwards and Haime, 1848 | GOM | AN | – | – | – |
| *Rhizotrochus typus* Milne-Edwards and Haime, 1848 | – | – | – | – | BOB |
| *Truncatoflabellum paripavonium* (Alcock, 1894) | – | – | – | LOK | – |
| *Truncatoflabellum stokesii* (Milne-Edwards and Haime, 1848) | – | AN | – | – | – |
| **Fungiacyathidae** | | | | | |
| *Fungiacyathus stephanus* (Alcock, 1893) | – | – | – | LOK | – |
| **Oculinidae** | | | | | |
| *Madrepora oculata* Linnaeus, 1758 | – | – | – | LOK | – |
| **Rhizangiidae** | | | | | |
| *Cladangia exusta* Lütken, 1873 | GOM | – | – | – | – |
| *Culicia rubeola* Quoy and Gaimard, 1833 | GOM | – | – | – | – |
| Total | 17 | 18 | 4 | 8 | 8+2 |

sea trawlers are designed to fish in different seascapes, they are able to trawl over coral habitats. Trawl fishing destroys the corals, reducing or eliminating the reef habitat. One of the major bycatches of deep-sea fishing vessels is corals, although over the past few years, the bycatch of corals has reduced, reflecting their depletion. So far, very few sites have been protected on the seas of India and in Indian Ocean. Exploitation of fisheries associated with deep-sea habitats in the Indian Ocean is continuing in an uncontrolled and unmanaged trend. The government of India, scientists, and even the fishing industry itself are unaware of the environmental damage being caused by these activities. It is probable that many deep-water coral habitats and associated animals are being trawled before they are fully studied and their species diversity assessed. Human activity related to the gathering of biological or mineral resources in the Indian Ocean by India and other nations has impacts on the environment which are poorly understood or managed. Hence it is very important to protect the deep-sea corals found in the seas around India and their associated fauna for the future.

## Acknowledgments

I thank the Chairman, National Biodiversity Authority, and the Director, Zoological Survey of India for providing necessary facility to carryout the work. My sincere thanks are due to George Institute for Biodiversity and Sustainability Science Committee for providing me the travel grant to attend the 3rd ISDSC. Special thanks are due to R. Y. George, Co-Organizer, 3rd ISDEC 2005, and S. Cairns, Smithsonian Institution, NMNH, Washington for their encouragement.

## Literature Cited

Alcock, A. 1893. On some newly-recorded corals from Indian Seas. J. Asiat. Soc. Bengal (Nat. Hist.) 2: 138–149.

_____. 1894. Natural history notes from HM Indian Marine Survey Streamer 'INVESTIGATOR' Series 2, number 15: on some new and rare corals from the deep waters of India. J. Asiat. Soc. Bengal (Nat. Hist.) 2: 186–188.

_____. 1898. An account of the deep-sea Madreporaria collected by the Royal Indian Marine Survey Ship 'INVESTIGATOR'. Calcutta: Trustees of the Indian Museum. 29 p. + 3 plates.

_____. 1902. Report on the deep-sea Madreporaria of the Siboga Expedition. Siboga-Expeditie 16: 1–52.

Bourne, G. C. 1905. Report on the solitary corals collected by Professor Herdman, at Ceylon in 1902. Ceylon Pearl Oyster Fisheries (Suppl. Rep. 29): 187–242.

Cairns, S. D. 1989. Discriminant analysis of Indo-West Pacific Flabellum. Mem. Assoc. Australas. Paleontols. 8: 61–68.

_____. 1998. Azooxanthellate Scleractinia (Cnidaria: Anthozoa) of Western Australia. Rec. of West. Austr. Mus. 18: 361–417.

_____. 1999. Species richness of Recent Scleractinia. Atoll Res. Bull. 459: 1–46.

_____ and N. B. Keller. 1993. New taxa and distributional records of azooxanthellate Scleractinia from the tropical south-west Indian Ocean, with comments on their zoogeography and ecology. Ann. S. Afr. Mus. 103: 213–292.

_____, B. W. Hoeksema, and J. van der Land 1999. List of extant stony corals. Atoll Res. Bull. 459: 13–45.

Fricke, H. W. and H. Schuhmacher. 1983. The depth limits of Red Sea corals: an ecophysiological problem (a deep diving survey by submersible). Mar. Ecol. 4: 163–194.

Gardiner, J. S. 1904. The turbinolid corals of South Africa, with notes on their anatomy and variation. Mar. Invest. South Africa 3: 93–129.

_____. 1929. Corals of the genus *Flabellum* from the Indian Ocean. Rec. Indian Mus. 31: 301–310.

_____ and P. Waugh. 1938. The flabellid and turbinolid corals. Sci. Rep. John Murray Exped. 1933-34, Brit. Mus. (Nat. Hist.) 5: 167–202.

_____ and _____. 1939. Madreporaria excluding Flabellidae and Turbinolidae, Sci. Rep. John Murray Exped, 1933–1934, Brit. Mus. (Nat. Hist.) 6: 225–242.

Pillai, C. S. G. 1967a. Studies on Indian Corals — 3. Report on a new species of *Dendrophyllia* (Scleractinia, Dendrophyllidae) from Gulf of Mannar. J. Mar. Biol. Assoc. 9: 407–409.

_____. 1967b. Studies on Indian Corals — 4. Redescription of *Cladangia exusta* Lutken (Scleractinia, Rhizangiidae) J. Mar. Biol. Assoc. India 9: 410–411.

_____. 1983. Structure and generic diversity of recent Scleractinia of India. J. Mar. Biol. Assoc. India. 25: 78–90.

_____. 1986. Recent corals from the south-east coast of India. Pages 107–198 *in* P. S. B. R. James, ed. Recent Advances in Marine Biology. Today and Tomorrow's Printers and Publishers, New Delhi.

_____. 1988. Scleractinian corals from the Gulf of Kutch. J. Mar. Biol. Assoc. India. 30: 54–74.

_____ and G. Scheer. 1976. Report on the stony corals from the Maldive Archipelago. Zoologica, Stuttgart 126: 1–83.

Scheer, G. and C. S. G. Pillai. 1974. Report on the Scleractinia of the Nicobar Islands. Zoologica, Stuttgart 122: 1–75.

_____ and _____. 1983. Report on the stony corals from the Red Sea. Zoologica, Stuttgart 45: 1–198.

Sheppard, C. R. C. and A. L. S. Sheppard. 1991. Corals and coral communities of Arabia. Fauna of Saudi Arabia 12: 1–170.

Van der Horst, C. J. 1921. The Madreporaria of the Siboga Expedition, Part 2: Madreporaria Fungida. Siboga-Expeditie, 16b: 46 p.

Venkataraman, K. 2003. Natural Aquatic Ecosystems of India, National Biodiversity Strategy Action Plan, Thematic Biodiversity Strategy and Action Plan. Zool. Surv. India 1–272.

_____, R. Jeyabaskaran, K. P. Raghuram, and J. R. B Alfred. 2004. Bibliography and checklists of coral and coral reef associated organisms of India, Rec. Zool. Surv. India, Occ. Paper. No 226: 1–468.

_____, C. H. Satyanarayana, J. R. B. Alfred, and J. Wolstenholme. 2003. Handbook on Hard Corals of India. Publ. Director, Zool. Surv. India, Kolkata. 1–266.

Wells, J. W. 1956. Scleractinia. Pages 328–440 *in* R. C. Moore, ed. Treatise on invertebrate paleontology. Part F. Coelenterata, Geol. Soc. Am. and Univ. Kansas Press. Lawrence.

Wood-Mason, J. and A. Alcock. 1891a. Natural history notes from H.M. Indian Marine Survey streamer 'Investigator', Series 2, Number 1: on the results of the deep-sea dredging during season 1890–1891. Ann. Mag. Nat. Hist. (6) 8: 427–452.

_____ and _____. 1891b. Natural history notes from H.M. Indian Marine Survey streamer 'Investigator', Series 2, number 1: On the results of the last season's deep-sea dredging. Ann. Mag. Nat. Hist. (6) 7: 1–19.

Zibrowius, H. 1980. Les Scletactiniaries de la Mediterranee et de l'Atlandique Nord-Oriental. Mem. Inst. Oceanogr. Monaco 11 (1429): 284 p.

_____ and J. M. Gili. 1990. Deep-water Scleractinia from Namibia, South Africa and Walvis Ridge, southern Atlantic. Scientia Mar. 54: 19–46.

Address: (K.V.) *Member Secretary, National Biodiversity Authority, #475, 9th South Cross Street, Kapaleeswar Nagar, Neelankarai, Chennai 600 041, India. E-mail: <nba_india@vsnl.net>.*

# Taxonomy and zoogeography of cold-water corals in coastal British Columbia

G. S. Jamieson, N. Pellegrin, and S. Jesson

**Abstract**

The current state of knowledge of cold-water corals in British Columbia is summarized. Pacific Canada has a more diverse coral community than does Atlantic Canada, as is the case for most taxonomic groups. A list of Pacific Canada's known coral species and potential species based on records from adjacent jurisdictions is presented, along with maps derived from existing records showing all currently known locations of corals in British Columbia. To date, five orders, 24 families, and 59 species of corals are documented from British Columbian waters, but an additional three families and 36 species may also occur in British Columbia, as these species have been documented from adjacent areas, i.e., southeast Alaska, Gulf of Alaska seamounts, and Washington/ Oregon.

Cold-water corals occur throughout the world's oceans, from the Antarctic to the Arctic and within the Mediterranean Sea. Diversity is greater in the northern hemisphere than in the southern, and the age of the fauna is a key factor in determining its latitudinal distribution (Keller and Pasternak, 2001). Depth, currents, productivity, and sedimentation all influence large-scale distributional patterns (Roberts et al., 2006), and the depth distribution of a species may vary across an ocean basin, e.g., eastern vs western Atlantic (Cairns and Chapman, 2001).

Typically, these corals are found in relatively shallow continental shelf and slope waters (Roberts et al., 2006), 50–1000 m depth, including shelf-edge canyons, deep channels between fishing banks (MacIsaac et al., 2001) and on fjord walls. However, cold-water corals, mostly gorgonians and hydrocorals, in the eastern North Pacific are most abundant in a much narrower depth zone. The most extensive coral surveys in the eastern North Pacific have been conducted in the Aleutian Islands. Limited submersible data to 367 m there found that corals were most abundant between 100–200 m, with mean coral abundance (3.85 colonies m$^{-2}$ over 25 transects) far exceeding that reported for other high-latitude ecosystems (Stone, 2006).

We use the term "coral" for members of the class Anthozoa, subclasses Octocorallia (soft corals, black corals, and sea fans) and Hexacorallia (stony and cup corals); and class Hydrozoa, order Filifera (fire and lace corals). Although we include the octocoral orders Alcyonacea, Antipatharia, and Pennatulacea as corals, only the hexacoral order Scleractinia is included, because other hexacoral orders (e.g., Actiniaria, Zoanthidea) are not skeletonized. Our octocoral classification follows Williams and Cairns (last revised January 2006), which uses a system of higher taxa after Bayer (1981) for Helioporacea and Alcyonacea, and Kükenthal (1915) and Williams (1995) for Pennatulacea. Because it is quite recent, this classification does not always agree with current Integrated Taxonomic Information System (ITIS) data on the web; e.g., species in the order Gorgonacea

George, R. Y. and S. D. Cairns, eds. 2007. Conservation and adaptive management of seamount and deep-sea coral ecosystems. Rosenstiel School of Marine and Atmospheric Science, University of Miami.

have been included in the order Alcyonacea by Williams and Cairns (2006), though the order is still shown as separate in ITIS. Octocorals, scleractinians and hydrozoan corals can be either zooxanthellate or azooxanthellate, while antipatharians are strictly azooxanthellate. Cold-water corals that exist below the photic zone do not form symbiotic associations and are thus azooxanthellate.

The majority of cold-water corals are solitary as well as azooxanthellate. However, a few colonial forms are hermatypic, e.g., *Lophelia pertusa* (Linnaeus, 1758) and *Dendrophyllia* sp., that is, their skeletons accumulate as biogenic build-ups. Deep- and cold-water coral accumulations have been called reefs, banks, mounds, and bioherms. The term reef, while widely used, traditionally implies a navigational hazard. Tropical coral reefs are biogenic structures built by both reef-building scleractinians and coralline algae, each having a key "cementing" function in reef development. In contrast, cold-, and usually deep-water scleractinian reefs (e.g., *Lophelia* reefs) are built mainly by the corals themselves without any input from algae. However, shallow cold-water coral reefs have been poorly studied, so this generalization may not be true for all cold-water corals. Bioherm is a more general term for any biogenic build-up of skeletal remains on the site where the organisms lived. They may consist of coral, sponge, or other skeletonised invertebrate fragments that have been built on and added to by successive generations. We also distinguish between true scleractinian coral reefs and octocoral "forests", i.e., concentrations of octocorals, which seem to be more common in British Columbian waters than are true scleractinian reefs.

Cold-water corals can form an extensive structural habitat, but their spatial occurrence, biology, and ecological significance are relatively poorly known due to their predominance in deep waters. However, Roberts at al. (2006) suggest that cold-water corals are arguably the most three-dimensionally complex habitats in the deep ocean, noting that over 1300 species have been found living on *L. pertusa* reefs in the NE Atlantic. While their role in the marine benthic ecosystem has yet to be fully defined, cold-water corals are often found in association with numerous other species (Cimberg et al., 1981; Koslow et al., 2001; Witherell and Coon, 2001; Etnoyer and Morgan, 2003; Gass, 2003; Hyland et al., 2005; and Stone, 2006), leading to the precautionary concern that coral concentrations in particular may have an important ecosystem role and thus should be particularly conserved.

Cold-water coral structures range from small, solitary individuals to massive reef habitats, often in relatively barren surroundings. Habitat-forming cold-water corals include octocorals, hexacorals (hermatypic scleractinian corals), and hydrocorals (Roberts et al., 2006). Live and dead portions of a coral's matrix or lattice framework can create substratum and shelter for other corals, sponges, brachiopods, bivalves, crustaceans, bryozoans, crinoids and tunicates (Koslow et al., 2001; Hall-Spencer et al., 2002). The complex branching morphology of many cold-water corals creates structures of sufficient size to provide substrate or refuge for other species (Etnoyer and Morgan, 2003; Roberts et al., 2006; Stone, 2006).

The majority of cold-water corals exhibit preference for rocky substrate or hard surfaces with moderate to strong currents, although pennatulaceans generally prefer unconsolidated substrates. Preferred coral substrate thus ranges from fine, well-sorted sand, gravel areas, and shell deposits, to slump deposits with rock outcrops, boulders, crevices, rock pinnacles, over hangs, living habitat, sheer cliffs, and iceberg furrows (MacIsaac et al., 2001).

In Canada, cold-water corals have been reported from both Atlantic (Breeze et al., 1997) and Pacific waters (McAllister and Alfonso, 2001; Lamb and Handby, 2005). In Nova Scotia, distributions are limited by both geography and bathymetry (Breeze and Davis, 1998). Off Nova Scotia, highest coral diversity has so far been observed at the "Gully" and the "Stone Fence" (Breeze et al., 1997; MacIsaac et al., 2001). Reyes-Bonilla (2002) suggested that there were about 79 species of azooxanthellate corals along the western coast of the Americas, representing ten families, with 29 species from six families having an average depth distribution greater than 200 m. Gass (2003) reported that the northeastern Pacific coral fauna was dominated by large gorgonian octocorals. In the Northeast Pacific, the most extensive regional surveys have been conducted in Alaskan waters (Heifetz, 2002; Etnoyer and Morgan, 2003; Stone, 2006).

Increased awareness of concentrations of cold-water corals and their high vulnerability to damage from human activities such as benthic fishing gear, coupled with new legislation (e.g., Canada's Oceans Act) requiring ecosystem-based approaches to management, are providing a new impetus for describing coral occurrence and distribution off British Columbia. Better conservation of biogenic marine habitats is high on the agenda of environmental non-governmental organisations (e.g., Ardron, 2005), including the minimization of impacts from fishing gear on benthic habitats (MacIssac et al., 2001).

Little published data exist on cold-water corals in British Columbia. Levings and McDaniel (1974) noted corals were one of many benthic organisms on an underwater cable in the Strait of Georgia, Austin (1985) provided an extensive list of marine invertebrates (60 corals) for the north-eastern Pacific, and McAllister and Alfonso (2001) listed 21 species in a preliminary assessment of the cold-water corals of British Columbia. Most recently, Canessa et al. (2003) reported corals in the ecosystem overview report of Bowie Seamount, Ardron et al. (2007) summarized and mapped coral occurrences in bycatch analyses of groundfish trawl data, and Conway et al. (2005) reported corals from their ROV surveys in the 100–300 m depth range.

Seamounts, i.e., extinct underwater volcanoes rising over 1000 m from the seafloor, provide opportunity for patchy, wide-scale occurrence of benthic species in a primarily homogeneous environment. Canada's Pacific Bowie Seamount has been recently surveyed, with research efforts targeted on a limited number of species (primarily rockfish and sablefish), but alcyonacean and scleractinian corals were observed there (Canessa et al., 2003). In 2002, seven seamounts (Patton, Murray, Chirikof, Marchand, Campbell, Scott, and Warwick) from the Cobb Hotspot in the Gulf of Alaska were explored using submersibles and multibeam bathymetric surveys (Etnoyer and Morgan, 2003).

Conway et al. (2007) documented the remains of a large dead coral reef in the Strait of Georgia at 255 m, identified as *L. pertusa* (S. Cairns, Smithsonian National Museum of Natural History, pers. comm.) (Jamieson et al., 2006). While this particular reef did not contain live corals, its presence in British Columbian waters strongly suggests that other as-of-yet undiscovered *Lophelia* reefs likely exist in British Columbian waters. Austin (1985) reported this species, initially misidentified as *Solenosmilia variabilis* Duncan, 1873, from Alberni Inlet, Vancouver Island. Rogers (1999) and Hyland et al. (2005) listed other *Lophelia* records for the North Pacific, including Cobb Seamount and the American side of Juan de Fuca Canyon, but there are no other records from Canadian waters.

To date, no summary exists of available data on cold-water corals of British Columbia, Canada, similar to that for Nova Scotia (Breeze et al., 1997). Here, we list both observed corals from Pacific Canadian waters and potential corals based on records from adjacent areas, and identify and map locations of corals from existing records and anecdotal information. Most data presented here are from deeper waters and specifically from either commercial fishing data or research surveys. Scuba diving observations are inadequately represented. Modern sport divers can dive within the depth ranges of some cold-water coral species in coastal regions, and valuable information is likely available from interested citizens.

## Field taxonomic identification challenges

Generally, coral genera and species are distinguished by skeletal morphology. However, many features, such as septal arrangements in scleractinians and sclerite form and distribution in octocorals require tissue preparation and a microscope, an approach that takes time and cannot be implemented at sea by observers reporting commercial fishing bycatches. Observers, if present, can retain some reference specimens for later identification, but they, like fishers, have time and space constraints. Thus, precise taxonomic identification in the field is unlikely to occur except for the most common species. To date, observers in British Columbia have not had convenient field identification guides for even common species, so observer-reported data have not been usable taxonomically. Efforts are now underway to produce such coral identification information. The following references have been used to determine taxa groupings: Orders Alcyonacea and Pennatulacea (Williams and Cairns, 2002), Order Antipatharia (Opresko, 2005, ITIS Taxonomic Report, November 2005), Order Scleractinia (Cairns, 1994), and order Filifera (ITIS Taxonomic Report, November 2005). Cairns et al. (2002) was also utilized.

## Methods

Data were obtained from surveys, by-catch information, and personal and museum collections. For reporting consistency, corals were mapped only to orders due to identification inconsistencies. Mapped data points included areas where identified corals were collected during research surveys and/or fishing activities. As mentioned above, published literature on cold-water corals in British Columbia is limited. Bibliographic databases included those of Fisheries and Oceans (called WAVES), museum libraries, and scientific journals. Literature requests from cold-water coral researchers were also made, and on-line resources were utilized when available from reliable sources.

Cold-water coral records from museums (Table 1) were included in the maps. These data, along with species listed in the published literature or from private collections where specimens were identified by recognized experts, were the only data used to determine taxa present at the species level, as taxonomic identification in other databases were not at the species level. Records unidentifiable to species, either from British Columbian waters or as potential species from adjacent areas, were included in the list as "sp." only if no other species of that genus was listed in any of the shaded columns in Table 2B; if other species were present, these records, identified in Table 2B, were assumed to be from existing species. The exceptions were for unidentified species records if either they appeared different from known species by the person reporting them (typically listed as sp. A, B, etc.) or they were geographically distant from other known species records from that genus. These latter records, also identified (Table 2B), were assumed here to represent other species, but this will only be confirmed when proper identifications are completed.

The Fisheries and Oceans Canada (DFO) trawl observer database is a major source of British Columbian fisheries data that includes by-catch records. Beginning in 1954 and prior to the establishment of complete observer coverage in 1996, fishers logged only two occurrences of coral.

Table 1. Number of cold-water coral records mapped for British Columbia, their source and years. DFO = Fisheries and Oceans Canada, Pacific Region.

| Data source | Data years | No. Records mapped |
|---|---|---|
| Royal British Columbia Museum | 1965–2001 | 19 |
| Canadian Museum of Nature | 1900–1991 | 7 |
| National Museum of Natural History–Smithsonian Institution | 1888–2001 | 9 |
| Centre for Marine Biodiversity and Parks Canada–Gwaii Haanas | 1888–2000 | 262 |
| GFBio Database–DFO Groundfish research data | 1966–2002 | 152 |
| Tanner Crab Database–DFO | 1999–2003 | 29 |
| Shrimp Database–DFO | 1966–2003 | 114 |
| International Pacific Halibut Commission–Stock Assessment Data | 1995–2003 | 15 |
| DFO PacHarv Observer Trawl Database | 1996–2003 | 860 |

Recording of data has since greatly improved, although by-catch species identification remains an issue. All records of corals in DFO's "PacHarvTrawl" (groundfish trawl) and "PacHarvSable" (sablefish trap) databases represent species (or groups of organisms) observed during fishing events, combined here as PacHarv, between the first quarter of 1996 and the fourth quarter of 2003. International Pacific Halibut Commission (IPHC) annual stock assessment surveys provided an additional 15 records (Tracee Geernaert, IPHC, pers. comm.). Spatial data from the available fishery databases contained "rolled-up" records representing information from three or more vessels within a defined spatial area (to protect the privacy of an individual's records). To optimize the number of records within the smallest data area, records were rolled-up into 16-km² bins by year. The centers of each spatially referenced bin (latitude and longitude) containing coral records identifiable to order were plotted and represented as point records.

Only a few DFO scientific research surveys contain records of cold-water corals. Groundfish, shrimp, and tanner crab surveys prior to 2004 contained 295 records of observed corals as incidental catch. The latitude and longitude of these locations were mapped and represent the midpoint of each set line. To date, there have been no directed research initiatives on cold-water corals in British Columbia.

The limited literature on British Columbian cold-water corals includes a taxonomic list by Austin (1985) and a report by McAllister and Alfonso (2001). Since cold-water marine environments are generally similar in adjacent waters, species presence and distribution can, to some extent, be inferred from literature on corals from nearby northeast Pacific regions, notably Washington and southeastern Alaska, and to some extent from global observations. Maps were produced to complement the earlier maps produced by McAllister and Alfonso (2001) and more recently produced maps by Ardron et al. (2007) for coral bycatch in British Columbia. All available coral data within the Canadian Exclusive Economic Zone (EEZ) were mapped. This encompasses all of British Columbia's territorial waters. Maps were produced in ArcGIS (9.0) with base map bathymetric intervals of 500 m.

## Results and Discussion

Cold-water corals found to date in British Columbian waters include representatives of five orders, 24 families, and 59 species (Table 2). Inclusion of potential species from southeastern Alaska, Gulf of Alaska seamounts, and Washington/Oregon brings the total to a likely 27 families and 95 species. Additional likely families are Scleroptilidae, Chrysogorgiidae, and Coralliidae.

To date, 55 species of corals have been reported from south-eastern Alaskan waters and Gulf of Alaska seamounts (Table 2; B. Stone, NOAA, pers. comm.) and 30 species from Washington/Oregon (C. E. Whitmire and M. E. Clarke, NOAA, pers. comm.). Only two taxa are known from both Southeast Alaska/Gulf of Alaska seamounts and

Table 2. Geographical distributions of (A) family and (B) species coral records in north-eastern Pacific waters. Families and species known to occur in British Columbia are in the darkly shaded column, and those likely to occur in British Columbia include those reported from Washington and Oregon, southeastern Alaska and the Alaska seamounts (lightly shaded columns). BS = Bering Sea, AI = Aleutian Islands, WG = western Gulf of Alaska, EG = eastern Gulf of Alaska, SM = seamounts, BC = British Columbia, WO = Washington and Oregon, Loc = total estimated number to date of local (EG to WO) species, Reg = total estimated number to date of regional species. ● = present. Alaskan records from R. Stone, NOAA, pers. comm., adapted by him from Heifetz et al. (in press); Washington and Oregon records from C. E. Whitmire and M. E. Clarke, NOAA, pers. comm. and Rogers (1999); British Columbian records include those from Kosloff (1974), Austin (1985), Sloan et al. (2001), and Etnoyer and Morgan (2003). ◆ = unidentified species records assumed to be a known listed species from the same general area. ♣ = unidentified species records assumed to be a non-listed species, because either they appeared to be different from known listed species by the person reporting them or they were geographically distant from other known listed species records from that genus.

| A. Taxa | BS | AI | WG | EG | SM | BC | WO | Loc | Reg |
|---|---|---|---|---|---|---|---|---|---|
| Order Alcyonacea (gorgonians and true soft corals) | 8 | 46 | 14 | 14 | 7 | 21 | 13 | 37 | 66 |
| Acanthogorgiidae | | 4 | 1 | 2 | | 2 | 2 | 5 | 7 |
| Alcyoniidae | 1 | 4 | 1 | 1 | | 4 | 2 | 4 | 7 |
| Chrysogorgiidae | | 2 | | | | | 1 | 1 | 2 |
| Clavulariidae | 1 | 2 | 1 | 1 | | 1 | | 1 | 2 |
| Coralliidae | | | | | 1 | | | 1 | 1 |
| Isididae | | 2 | 3 | 3 | 4 | 3 | 2 | 6 | 6 |
| Nephtheidae | 1 | 3 | 1 | 1 | | 2 | 1 | 2 | 3 |
| Paragorgiidae | | 1 | 1 | 1 | 1 | 3 | 1 | 3 | 4 |
| Paramuriceidae | | 1 | | | | | | | 1 |
| Plexauridae | | 8 | 1 | 1 | 1 | 5 | 2 | 7 | 12 |
| Primnoidae | 5 | 19 | 5 | 4 | | 1 | 2 | 7 | 21 |
| Order Pennatulacea (sea pens) | 4 | 3 | 4 | 7 | | 17 | 10 | 23 | 24 |
| Anthoptilidae | 2 | | | | | 2 | 1 | 2 | 3 |
| Funiculinidae | | | | | | 1 | 1 | 1 | 1 |
| Halipteridae | 1 | 1 | 1 | 2 | | 1 | | 3 | 3 |
| Kophobelemnidae | | | | | | 2 | 2 | 3 | 3 |
| Pennatulidae | | 1 | 1 | 2 | | 2 | 1 | 2 | 2 |
| Protoptilidae | | | 1 | 1 | | 2 | 1 | 3 | 3 |
| Scleroptilidae | | | | | | | 1 | 1 | 1 |
| Stachyptilidae | | | | | | 1 | | 1 | 1 |
| Umbellulidae | 1 | 1 | 1 | 1 | | 1 | 1 | 1 | 1 |
| Virgulariidae | | | | 1 | | 5 | 3 | 6 | 6 |
| Order Antipatharia (black corals) | | 2 | 1 | 9 | 2 | 6 | 3 | 13 | 14 |
| Antipathidae | | 1 | | 1 | | 1 | 2 | 2 | 2 |
| Cladopathidae | | | 1 | 2 | 1 | 1 | | 3 | 4 |
| Schizopathidae | | 1 | | 5 | 1 | 4 | 1 | 8 | 8 |
| Order Scleractinia (stony and cup corals) | 2 | 9 | 6 | 6 | 2 | 9 | 1 | 11 | 14 |
| Caryophylliidae | | 4 | 3 | 4 | 1 | 7 | 1 | 8 | 9 |
| Dendrophylliidae | | 1 | 1 | 1 | | 1 | | 1 | 1 |
| Flabellidae | 1 | 2 | 1 | | | | | | 2 |
| Fungiacyathidae | 1 | 2 | 1 | 1 | 1 | 1 | | 2 | 2 |
| Order Filifera (hydrocorals) | 3 | 25 | 2 | 8 | | 6 | 3 | 11 | 31 |
| Stylasteridae | 3 | 25 | 2 | 8 | | 6 | 3 | 11 | 31 |
| Total | 17 | 85 | 27 | 44 | 11 | 59 | 30 | 95 | 149 |

Table 2. Continued.

| B. Taxa | BS | AI | WG | EG | SM | BC | WO |
|---|---|---|---|---|---|---|---|
| Order Alcyonacea (gorgonians and true soft corals) | | | | | | | |
| Acanthogorgiidae | | | | | | | |
| cf. *Acanthogorgia* | | ● | | | | | |
| *Calcigorgia beringi* | | ● | | ● | | | |
| *Calcigorgia spiculifera* | | ● | ● | ● | | ● | ● |
| *Calcigorgia* sp. A | | ● | | | | | |
| *Calcigorgia* sp. | | | | | | | ● |
| *Calcigorgia kinoshitae* | | | | | | ● | |
| Alcyoniidae | | | | | | | |
| *Alcyonium* sp. | | | | | | ● | |
| *Anthomastus* cf. *grandiflora* | | | | | | ● | |
| *Anthomastus japonicus* | ● | ● | | | | | |
| *Anthomastus* cf. *japonicus* | | ● | | | | | |
| *Anthomastus ritteri* | | ● | ● | ● | | ● | ● |
| *Anthomastus* sp. A (red) | | ● | | | | | |
| *Anthomastus* sp. B (gray) | | | | | | ● | ● |
| Chrysogorgiidae | | | | | | | |
| cf. *Chrysogorgia* | | ● | | | | | ● |
| *Radicipes* sp. A | | ● | | | | | |
| Clavulariidae | | | | | | | |
| *Clavularia moresbii* | | ● | ● | ● | | ◆ | |
| *Sarcodictyon incrustans* | ● | ● | | | | | |
| Coralliidae | | | | | | | |
| *Corallium* sp. A | | | | | ● | | |
| Isididae | | | | | | | |
| *Acanella* sp. A | | | | | ● | | |
| *Isidella paucispinosa* | | ● | ● | ● | | | |
| *Isidella* sp. A | | | | | ● | | |
| *Isidella* sp. | | | | | | ● | ● |
| *Keratoisis profunda* | | ● | ● | ● | ◆ | ◆ | ◆ |
| *Lepidisis* sp. | | | ◆ | ◆ | ◆ | ● | |
| Nephtheidae | | | | | | | |
| cf. *Eunephthya*[?] (*Gersemia*) | | ● | | | | | |
| *Eunephthea rubiformis* | ● | ● | ● | ● | | ● | |
| *Eunephthea* sp. | | ● | | | | ◆ | ◆ |
| Paragorgiidae | | | | | | | |
| *Paragorgia arborea* (=*P. pacifica*) | | ● | ● | ● | ◆ | ● | ◆ |
| *Paragorgia yutlinux* | | | | | | ● | |
| *Paragorgia stephencairnsi* | | | | | | ● | |
| Paramuriceidae | | | | | | | |
| *Paramuricea* sp. A | | ● | | | | | |
| Plexauridae | | | | | | | |
| *Alaskagorgia aleutiana* | | ● | | | | | |
| *Euplexaura marki* | | ● | ● | ● | | | ● |
| *Muriceides cylindrica* | | ● | | | | | |
| *Muriceides* cf. *cylindrica* | | ● | | | | | |
| *Muriceides nigra* | | ● | | | | | |

Table 2. Continued.

| B. Taxa | BS | AI | WG | EG | SM | BC | WO |
|---|---|---|---|---|---|---|---|
| *Swiftia (Psammogorgia) spauldingi* | | | | | | ● | |
| *Swiftia beringi* | | ● | | | | | |
| *Swiftia kofoidi* | | | | | | | ● |
| *Swiftia pacifica* | | ● | | | | ● | |
| *Swiftia simplex* | | ● | | | ● | ● | |
| *Swiftia torreyi* | | | | | | ● | |
| Primnoidae | | | | | | | |
| *Amphilaphis* sp. A | | ♣ | | | | | |
| *Amphilaphis* sp. B | | ♣ | | | | | |
| *Amphilaphis* sp. C | | ♣ | | | | | |
| *Arthrogorgia kinoshitai* | ● | ● | ● | ● | | | |
| *Arthrogorgia otsukai* | ● | ● | | | | | |
| *Arthrogorgia utinomii* | | ● | | | | | |
| *Callogorgia* sp. | | | | | | | ● |
| *Fanellia compressa* | ● | ● | | | | | |
| *Fanellia fraseri* | | ● | ● | | | | |
| *Narella* sp. A | | ● | | | | | |
| *Parastenella* sp. A | | ♣ | | | | | |
| *Parastenella* cf. *doederleini* | | | | | | | ● |
| *Plumarella flabellata* | | ● | | | | | |
| *Plumarella longispina* | ● | ● | ● | ● | | | |
| *Plumarella spicata* | | ● | | | | | |
| *Plumarella spinosa* | | ● | | | | | |
| *Primnoa pacifica* | ● | ● | ● | ● | | | |
| *Primnoa pacifica* var. *willeyi* | | | | | | ● | |
| *Primnoa wingi* | | ● | ● | ● | | | |
| *Thouarella hilgendorfi* | | ● | | | | | |
| *Thouarella striata* | | ● | | | | | |
| *Thouarella superba* | | ● | | | | | |
| *Thouarella* sp. | | | | | | | ♣ |
| Order Pennatulacea (sea pens) | | | | | | | |
| Anthoptilidae | | | | | | | |
| *Anthoptilum grandiflorum* | ● | | | | | ● | ● |
| *Anthoptilum murrayi* | ● | | | | | | |
| *Anthoptilum* cf *murrayi* | | | | | | ● | |
| Funiculinidae | | | | | | | |
| *Funiculina parkeri* | | | | | | ● | ● |
| Halipteridae | | | | | | | |
| *Halipteris californica* | | | | ● | | | |
| *Halipteris* cf. *californica* | | | | | | ● | |
| *Halipteris willemoesi* | ● | ● | ● | ● | | | |
| Kophobelemnidae | | | | | | | |
| *Kophobelemnon hispidum* | | | | | | ● | |
| *Kophobelemnon affine* | | | | | | ● | ● |
| *Kophobelemnon biflorum* | | | | | | | ● |

Table 2. Continued.

| B. Taxa | BS | AI | WG | EG | SM | BC | WO |
|---|---|---|---|---|---|---|---|
| Pennatulidae | | | | | | | |
| *Pennatula phosphorea* | | | | • | | • | |
| *Ptilosarcus gurneyi* | | • | • | • | | • | • |
| Protoptilidae | | | | | | | |
| *Distichoptilum rigidum* | | | | | | | • |
| *Distichoptilum* cf *rigidum* | | | | | | • | |
| *Protoptilum* sp. A | | | • | • | | • | |
| Scleroptilidae | | | | | | | |
| *Scleroptilum* sp. | | | | | | | • |
| Stachyptilidae | | | | | | | |
| *Stachyptilum superbum* | | | | | | • | |
| Umbellulidae | | | | | | | |
| *Umbellula lindahli* | • | • | • | • | | • | ♦ |
| Virgulariidae | | | | | | | |
| *Acanthoptilum gracile* | | | | | | • | |
| *Balticina californica* | | | | | | • | |
| *Balticina septentrionalis* | | | | | | • | |
| *Stylaulta elongate* | | | | | | • | |
| *Stylatula gracile* | | | | | | | • |
| *Virgularia cystiferum* | | | | ♦ | | • | ♦ |
| Order Antipatharia (black corals) | | | | | | | |
| Antipathidae | | | | | | | |
| *Antipathes* sp. | | | | | | • | • |
| *Parantipathes* sp. | | • | | • | | | • |
| Cladopathidae | | | | | | | |
| *Chrysopathes formosa* | | | | • | | | |
| *Chrysopathes speciosa* | | | | • | | • | |
| *Heliopathes pacifica* | | | | | • | | |
| *Trissopathes pseudotristicha* | | | • | | | | |
| Schizopathidae | | | | | | | |
| *Bathypathes alternata* | | | | • | | | |
| *Bathypathes patula* | | | | • | | • | ♦ |
| *Dendrobathypathes boutillieri* | | | | | | • | |
| *Dendrobathypathes* n. sp. A | | • | | • | | | |
| *Lillipathes lilliei* | | | | • | | | |
| *Lillipathes wingi* | | | | | | • | |
| *Lillipathes* n. sp. A | | | | • | | | |
| *Umbellapathes helioanthes* | | | | | • | ♦ | |
| Order Scleractinia (stony and cup corals) | | | | | | | |
| Caryophylliidae | | | | | | | |
| *Caryophyllia alaskensis* | | • | • | • | | • | |
| *Caryophyllia arnoldi* | | • | • | • | | • | |
| *Crispatotrochus foxi* | | • | • | • | | | |
| *Labyrinthocyathus quaylei* | | | | | | • | |
| *Desmophyllum dianthus* (=*D. cristagalli*) | | | | | | • | |
| *Leptopenus discus* | | • | | | | | |
| *Lophelia pertusa* | | | | | • | • | • |

Table 2. Continued.

| B. Taxa | BS | AI | WG | EG | SM | BC | WO |
|---|---|---|---|---|---|---|---|
| *Paracyathus caltha* | | | | | | ● | |
| *Paracyathus stearnsi* | | | | ● | | ● | |
| Dendrophylliidae | | | | | | | |
| *Balanophyllia elegans* | | ● | ● | ● | | ● | |
| Flabellidae | | | | | | | |
| *Flabellum* sp. | | ● | | | | | |
| *Javania borealis* | ● | ● | ● | | | | |
| *Javania cailleti* | ● | ● | ● | ● | | ● | |
| Fungiacyathidae | | | | | | | |
| *Fungiacyathus marenzelleri* | | ● | | | ♦ | | |
| Order Filifera (hydrocorals) | | | | | | | |
| Stylasteridae | | | | | | | |
| *Crypthelia trophostega* | ● | ● | | | | | |
| *Cyclohelia lamellata* | ● | ● | | | | | |
| *Cyclohelia* sp. A | | ♣ | | | | | |
| *Distichopora borealis* | | ● | | | | | |
| *Distichopora* sp. A | | ♣ | | | | | |
| *Errinopora nanneca* | | ● | | | | | |
| *Errinopora pourtalesii* | | ● | ● | ● | | ● | |
| *Errinopora stylifera* | | ● | | | | | |
| *Errinopora zarhyncha* | | ● | | | | | |
| *Errinopora* sp. A | | ♣ | | | | | |
| cf. *Stenohelia* sp. [?] | | ● | | | | | |
| *Stylantheca papillosa* | | ● | | | | | |
| *Stylantheca porphyra* | | | | ● | | ● | |
| *Stylantheca pterograpta* | | ● | | ● | | ● | ● |
| *Stylaster alaskana* | | ● | | | | | |
| *Stylaster brochi* | | ● | | | | | |
| *Stylaster californicus* | | | | | | | ● |
| *Stylaster campylecus campylecus* | | ● | ● | ● | | | |
| *Stylaster campylecus parageus* | | | | ● | | | |
| *Stylaster campylecus trachystomus* | | ● | | | | | |
| *Stylaster campylecus tylotus* | | ● | | | | | |
| *Stylaster cancellatus* | | ● | | ● | | | |
| *Stylaster elassotomus* | | ● | | | | | |
| *Stylaster moseleyanus* | | ● | | | | | |
| *Stylaster norvegicus* [?] | | | | | | ● | |
| *Stylaster polyorchis* | | ● | | | | | |
| *Stylaster porphyra* | | | | | | ● | |
| *Stylaster stejnegeri* | ● | ● | | | | | |
| *Stylaster venustus* | | | | ● | | ● | ● |
| *Stylaster verrillii* | | ● | | ● | | | |
| *Stylaster* sp. A | | ♣ | | | | | |

continental US waters (but not British Columbia), which constitutes a more conservative number of potential species for British Columbia than those known only from either Southeast Alaska/Gulf of Alaska seamounts or Washington/Oregon. In total, 149 taxa are now reported in the northeast Pacific north of California. The disparities in regional numbers of taxa likely largely reflects differing levels of survey effort conducted in the different jurisdictions, with significant changes in numbers of taxa present south of Alaska likely to occur when future survey effort is directed towards corals. Apart from Alaska, most coral records to date have been obtained incidentally during the course of surveys targeted on other taxa, which were mostly commercial species.

Among the 149 regional taxa listed here, 66 (44%) were Alcyonaceans (gorgonians or true soft corals), 24 (16%) were Pennatulaceans (sea pens and sea whips), 14 (9%) each were Antipatharians (black corals) and Scleractinians (stony and cup corals), and 31 (21%) were filiferans (hydrocorals). These relative proportions may largely reflect proportions from the locations, depth ranges (50–250 m), and gear types (e.g., groundfish trawls, traps) where commercial fishing has occurred to date. For example, taxa such as the scleractinians are likely less efficiently captured by commercial gears than the generally larger octocorals. In British Columbia, proportions of known occurrences of species to date of the different orders above were 36, 29, 15, 10, and 10%, respectively. These proportions are lower in the alcyonaceans and higher in the pennatalaceans than regionally, which probably reflects the greater reporting emphasis in British Columbia from groundfish trawl bycatch records, and the focusing by fishers on relatively smooth substrate areas that have been fished for decades. Accurate commercial bycatch reporting has mostly been since 1996, and in many areas, large structural species such as gorgonians that might have occurred in previously undisturbed habitat would likely have been destroyed by trawl activity decades ago.

Within coral orders in British Columbia, in the Alcyonacea, the families Plexauridae and Alcyoniidae had the most genera, but no family had more than five species listed. In the Pennatulacea, the Virgularidae had five species listed, and four other families each had two species. The Caryophylliidae dominated the Scleractinia at seven species, the Achizopathidae dominated the Antipatharia with four species, and all six filiferan species were in the family Stylasteridae.

In total, 1826 British Columbia cold-water coral records were mapped. Table 1 presents their data source, years of data collection, and the number of records with geospatial information. These maps identify areas where corals have been encountered and expand on those of McAllister and Alfonso (2001). Figure 1 illustrates the distribution of all cold-water coral documented in British Columbian waters. The concentration of filiferans around the Queen Charlotte Islands in northern British Columbia (Fig. 1E) likely reflects the extensive marine biodiversity surveys conducted recently off Gwai Haanas National Park Reserve, i.e., sampling effort distribution, while the spatial concentration of pennatalaceans just east of the Queen Charlotte Islands (Fig. 1C) can be explained by the extensive occurrence of soft substrate (sand and gravel) in this relatively heavily fished area, i.e., substrate type. The clustering of many coral records along the continental shelf break and slope reflects the presumed relative abundance of suitable coral habitat at the greater depths that occurs there, i.e., both the presence of more exposed rock and stronger currents that occur at depth in the canyons that transverse this region. These bathymetric and oceanographic features create habitat particularly suitable for many corals, i.e., hard substrates devoid of sediment and readily available suspended food particles. Finally, because effort distribution to date has been biased and methods

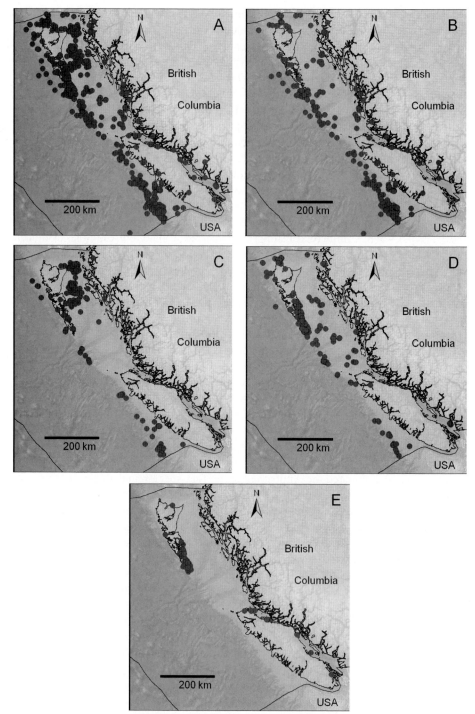

Figure 1. Locations of corals reported from British Columbia, Canada. (A) All, (B) Alcyonacea, (C) Pennatulacea, (D) Scleractinia, (E) Filifera.

of data collection have been limited, documented existing coral order distributions may only partially reflect actual spatial distributions, which will no doubt become more apparent as additional data are acquired from areas not extensively fished to date.

The distributional scale of much existing data is currently too large to define discrete locations of corals (e.g., groundfish trawl tows average ~10 km; tow midpoints were mapped, as it is not known where along the tow paths corals landed were actually caught by the gear), although Ardron (2005) and Ardron et al. (2007) suggest where coral concentrations are based on groundfish trawl by-catch data. Surveys utilizing video equipment are needed for ground-truthing and identifying coral species and other vulnerable benthic organisms at an acceptable resolution. Such surveys would also allow coral abundance assessments in areas with relatively low fishing disturbance or in areas not well suited for current fishing methods (e.g., rough ground) due to potential gear damage or loss.

Given the relative lack of dedicated surveys for corals in the north-eastern Pacific and the relatively large number of species that exist in adjacent areas to British Columbia but which are as yet undocumented in British Columbia, it is expected that many new coral species will be found in Pacific Canada. Cold-water corals are now recognised as important and worth conserving by the general public, and future research and management efforts to survey their presence and abundance and to minimize human impacts on them are being developed.

## Acknowledgments

We thank the museums listed for making a list of their British Columbia specimens available to us. We also thank J. Boutillier for reviewing this manuscript, and adding his new records of British Columbian corals; B. Stone and C. Whitmire for allowing us to reference their most recent lists of Alaskan and Washington/Oregon coral species, respectively; and anonymous reviewers for constructive comments.

## Literature Cited

Austin, W. C. 1985. An annotated checklist of marine invertebrates in the cold temperate northeast Pacific. Khoyatan Marine Laboratory 1: 218 p.

Ardron, J. 2005. Protecting British Columbia's corals and sponges. Living Oceans Society, British Columbia, Canada. 22 p. http://www.livingoceans.org/PDFs/Protecting%20BC%27s%20 Corals%20and%20Sponges.pdf

_____, G. S. Jamieson, and D. Hangaard. 2007. Spatial identification of closures to reduce the bycatch of corals and sponges in the groundfish trawl fishery, British Columbia, Canada. Pages 157–167 in R. Y. George and S. D. Cairns, eds. Conservation and adaptive management of seamount and deep-sea coral ecosystems. Rosenstiel School of Marine and Atmospheric Science, University of Miami. Miami. 324 p.

Bayer, F. M. 1981. Key to the genera of Octocorallia exclusive of the Pennatulacea (Coelenterata: Anthozoa), with descriptions of new taxa. Proc. Biol. Soc. Wash. 94: 902–947.

Breeze, H. and D. S. Davis. 1998. Section 6.5.2. Deep sea corals. Pages 113–120 in W. G. Harrison and D. G. Fenton, eds. The gully: a scientific review of its environment and ecosystem. CSAS Res. Doc. 1998/83.

_____, D. S. Davis, M. Butler, and V. Kostylev. 1997. Distribution and status of deep sea corals off Nova Scotia. Marine Issues Committee Special Publication Number 1. Ecology Action Centre: 58 p.

Cairns, S. D. 1994. Scleractinia of the temperate North Pacific. Smithson. Contrib. Zool. 557: 150 p.

_____ and R. E. Chapman. 2001. Biogeographic affinities of the North Atlantic deep-water Scleractinia. Pages. 30–57 *in* J. H. Willison, J. Hall, S. E. Gass, E. L. R. Kenchington, M. Butler, and P. Doherty, eds. Proc. First Int. Symp. on Deep-Sea Corals, Halifax, Nova Scotia. Ecology Action Centre and Nova Scotia Museum.

_____, D. R. Calder, A. Brinckmann-Voss, C. B. Castro, D. G. Fautin, P. R. Pugh, C. E. Mills, W. C. Jaap, M. N. Arai, S. H. D. Haddock, and D. M. Opresko. 2002. Common and scientific names of aquatic invertebrates from the United States and Canada: Cnidaria and Ctenophora. 2nd Edition. AFS Spec. Publ. 28: 115 p.

Canessa, R. R., K. W. Conley, and B. D. Smiley. 2003. Bowie Seamount pilot marine protected area: An ecosystem overview report. Canadian Tech. Rep. Fish. Aquat. Sci. 2461: 85 p.

Cimberg, R. L., T. Gerrodette, and K. Muzik. 1981. Habitat requirements and expected distribution of Alaska coral. Final Report, Research Unit No. 601, U.S. Office of Marine Pollution Assessment, Alaska Office. 54 p.

Conway, K. W., J. V. Barrie, W. C. Austin, P. R. Ruff, and M. Krautter. 2005. Deep-water sponge and coral habitats in the coastal waters of British Columbia, Canada: multibeam and ROV survey results. Abstract: Third Int. Symp. on Deep-sea Corals, Miami, Nov 28–Dec 2, 2005: 32 p.

_____, J. V. Barrie, P. R. Hill, W. C. Austin, and K. Picard. 2007. Mapping sensitive habitats in the Strait of Georgia, coastal British Columbia: deep-water sponge and coral reefs. Geological Survey of Canada, Current res. 2007-A2, 6 p.

Etnoyer, P. and L. E. Morgan. 2003. Occurrences of habitat-forming deep water corals in the Northeast Pacific Ocean. Final Report to NOAA Office of Habitat Protection, Washington D.C. 32 p.

Gass, S. E. 2003. Conservation of deep-sea corals in Atlantic Canada. World Wildlife Fund Canada, Toronto, Canada. 60 p. Available from: www.wwf.ca/NewsAndFacts/Resources. asp?type=resources.

Hall-Spencer, J., V. Allain, and J. H. Fossa. 2002. Trawling damage to Northeast Atlantic ancient coral reefs. Proc. R. Soc. Lond. 269: 507–511.

Heifetz, J. 2002. Coral in Alaska: distribution, abundance and species associations. Hydrobiologia 471: 19–28.

Hyland, J., C. Cooksey, E. Bowlby, and M. S. Brancato. 2005. A Pilot survey of deepwater coral/ sponge assemblages and their susceptibility to fishing/harvest impacts at the Olympic Coast National Marine Sanctuary (OCNMS). NOAA Cruise report: NOAA Ship McArthur II Cruise AR-04-04: Leg 2. 18 p.

ITIS Taxonomic Report, November 2005. National Benthic Inventory. January, 2006. Available from: http://www.nbi.noaa.gov/(iowwou55nny44d55ldczufe2)/itis.aspx?tsn=51940.

Jamieson, G. S., N. Pellegrin, and S. Jessen. 2006. Taxonomy and zoogeography of cold-water corals in explored areas of coastal British Columbia. Centre for Science Advice, Pacific Region, Fisheries and Oceans Canada, Res. Doc 2006/062: 49 p.

Keller, N. B. and F. A. Pasternak. 2001. Coral polyps (Scleractinia, Alcyonacea, Gorgonacea and Pennatulacea) and their role in the formation of the landscape of the Reykjanes Ridge rift zone. Oceanology 41: 531–539.

Kosloff, E. N. 1974. Keys to the marine invertebrates of Puget Sound, the San Juan Archipelago and adjacent regions. U. Washington Press. 226 p.

Koslow, J. A., K. Gowlett-Holmes, J. K. Lowry, T. O'Hara, G. C. B. Poore, and A. Williams. 2001. Seamount benthic macrofauna off southern Tasmania: community structure and impacts of trawling. Mar. Ecol. Prog. Ser. 213: 111–125.

Kükenthal, W. 1915. Pennatularia. Das Tierreich 43: i-xv + 132 pp. Berlin, Verlag von R. Friedlander und Sohn.

Lamb, A. and B. Hanby. 2005. Marine life of the Pacific northwest: a photographic encyclopedia of saltwater invertebrates, seaweeds and selected fishes. Harbour Pub. Co, Madeira Park. 398 p.

Levings, C. D. and N. G. McDaniel. 1974. A unique collection of baseline biological data: Benthic Invertebrates from and under-water cable across the Strait of Georgia. Fisheries Research Board of Canada; Technical Report No. 441. 10 p.

MacIsaac, K., C. Bourbonnais, E. Kenchington, D. Gordon, Jr., and S. Gass. 2001. Observations on the occurrences and habitat preference of corals in Atlantic Canada. Pages 58–75 *in* J. H. Willison, J. Hall, S. E. Gass. E. L. R. Kenchington, M. Butler, and P. Doherty, eds. Proc. First Int. Symp. on Deep-Sea Corals, Halifax, Nova Scotia. Ecology Action Centre and Nova Scotia Museum.

McAllister, D. E. and N. Alfonso. 2001. The distribution and conservation of deep-water corals on Canada's west coast. Pages 126–144 *in* J. H. Willison, J. Hall, S. E. Gass. E. L. R. Kenchington, M. Butler, and P. Doherty, eds. Proc. First Int. Symp. on Deep-Sea Corals, Halifax, Nova Scotia. Ecology Action Centre and Nova Scotia Museum.

Opresko, D. M. 2005. New genera and species of antipatharian corals (Cnidaria: Anthozoa) from the North Pacific. Zool. Meded. Leiden 79: 129–165.

Reyes-Bonilla, H. 2002. Checklist of valid names and synonyms of stony corals (Anthozoa: Scleractinia) from the eastern Pacific. J. Nat. Hist. 36: 1–13.

Roberts, J. M., A. J. Wheeler, and A. Freiwald. 2006. Reefs of the Deep: the biology and geology of cold-water coral ecosystems. Science 312: 543–547.

Rogers, A. D. 1999. The biology of *Lophelia pertusa* (Linnaeus, 1758) and other deep-water reef-forming corals and impacts from human activities. Int. Rev. Hydrobiol. 84: 315–406.

Sloan, N. A., P. M. Bartier, and W. C. Austin. 2001. Living marine legacy of Gwaii Haanas. II: Marine invertebrate baseline to 2000 and invertebrate-related management issues. Report 035. Parks Canada-Technical Reports in Ecosystem Science. # 035. 331 p.

Stone, R. P. 2006. Coral habitat in the Aleutian Islands of Alaska: depth distribution, fine-scale species associations, and fisheries interactions. Coral Reefs 25: 229–238.

Williams, G. C. 1995. Living genera of sea pens (Coelenterata: Octocorallia: Pennatulacea): illustrated key and synopsis. Zool. J. Linn. Soc. 113: 93–140.

_____ and S. D. Cairns. 2006. Systematic list of valid octocoral genera (last revised January 2006). http://www.calacademy.org/research/izg/OCTOCLASS.htm#penna.

Witherell, D. and C. Coon. 2001. Protecting gorgonian corals off Alaska from fishing impacts. Pages 117–125 *in* J. H. Willison, J. Hall, S. E. Gass. E. L. R. Kenchington, M. Butler, and P. Doherty, eds. Proc. First Int. Symp. on Deep-Sea Corals, Halifax, Nova Scotia. Ecology Action Centre and Nova Scotia Museum.

ADDRESSES: (G.S.J.) *Fisheries and Oceans Canada, Pacific Biological Station, Nanaimo, BC, V9T 6N7, Canada.* (N.P., S.J.) *Canadian Parks and Wilderness Society, 410 - 698 Seymour Street, Vancouver, BC, V6B 3K6, Canada.* CORRESPONDING AUTHOR: (G.S.J.) *E-mail: <jamiesong@pac.dfo-mpo.gc.ca>.*

# Bathymetric distribution and biodiversity of cold-water octocorals (Coelenterata: Octocorallia) in Sagami Bay and adjacent waters of Japan

Asako K. Matsumoto, Fumihito Iwase,
Yukimitsu Imahara, and Hiroshi Namikawa

**Abstract**

in total, 260 octocoral species, including 144 gorgonians, 80 alcyonaceans, and 36 pennatulaceans have been recorded in the literature at Sagami Bay, Japan. Fifteen of the octocoral species were newly recorded at littoral to bathyal depths (13 gorgonians, one alcyonacean, and one pennatulacean). Thirty families of octocorals were recorded between the depth of 100–200 m (deep littoral) and 23% of all recorded species and 37% of newly recorded species were obtained at this depth. A total of 114 species of octocoral is endemic to Japan and adjacent waters, while 74 of these octocorals are only found in or around Sagami Bay. These results reconfirmed the high biodiversity of octocoral fauna in Sagami Bay, and suggest there are two main explanations for these high levels of biodiversity in this bay and in this area. One explanation appears to be that the recorded species represent a faunal boundary which is formed by converging elements of temperate, sub-tropical, and sub-arctic regions. Another explanation may be that the 100–200 m depth range may be a benthic/pelagic ecotone, a transitional zone between two ecosystems, which results in a high degree of biodiversity due to benthic-pelagic coupling.

Sagami Bay (approximately 33–35°N and 138–140°E), located near Tokyo, has played very important roles historically, commercially, and culturally for Japan (Fig. 1A). The sea bottom in the area is composed of the North American Plate and Philippine Sea Plate and has very complicated geographical features. Research on octocoral (Coelenterata: Octocorallia) fauna in Sagami Bay began in 1868 with a report on the gorgonian coral *Acanella gregori* by Gray collected from Enoshima by B. Gregory, the surgeon onboard the H.M.S. Rattler (Imahara, 2006). Since then, several studies on octocoral biodiversity in Sagami Bay have been conducted, including many expeditions such as those of the Challenger, Tiefsee, Albatross, and Sixten Bock, the research by Döderlain, Doflein, Mortensen, Japanese emperor Hirohito (Isono, 1988; Fujita and Namikawa, 2003; Takeda et al., 2006), reported by Holm (1894), May (1899), Moroff (1902), Kishinouye (1903), Kükenthal (1906, 1909), Kinoshita (1908, 1909, 1913), Kükenthal and Gorzawsky (1908b), Balss (1910), Kükenthal and Broch (1911), Nutting (1912), Okubo (1929), Thomson and Rennet (1929), Kumano (1937), Utinomi (1952, 1955, 1962), Suzuki (1971), and Utinomi and Imahara (1976) (see additional references in Imahara, 1996, 2006; Iwase and Matsumoto, 2006).

George, R. Y. and S. D. Cairns, eds. 2007. Conservation and adaptive management of seamount and deep-sea coral ecosystems. Rosenstiel School of Marine and Atmospheric Science, University of Miami.

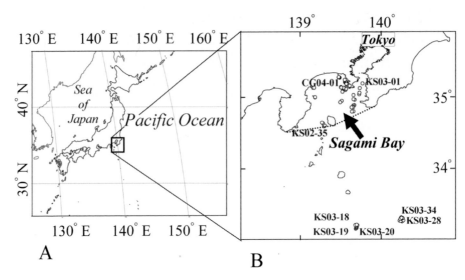

Figure 1. Location of Sagami Bay and distribution of octocorals. (A) Locations studied, Sagami Bay, Japan, northwest Pacific. (B) Stations at Sagami Bay and adjacent waters from the data of 2001–2005 survey presented in Table 1.

Through these surveys and research expeditions, the known number of species of octocorals from Sagami Bay and adjacent waters is 260 (Imahara, 1996), demonstrating a high degree of octocoral biodiversity. However, it has been more than 100 yrs since the 19th century studies on the fauna of the bay, suggesting the need for a revision of the taxonomical catalogue of fauna. A 5-yr comprehensive biodiversity study entitled "Study on Environmental Changes in Sagami Sea and Adjacent Coastal Area with Serial Comparison of Fauna and Flora" was conducted from 2001 to 2005 by the National Science Museum of Tokyo (NSMT, Japan). A cold-water octocoral biodiversity study was included in this large project. Here we report on the known and new occurrences of cold-water octocorals from Sagami Bay.

## Methods

The areas studied include Sagami Bay, Suruga Bay, Tokyo Bay, Izu Archipelago, and offshore areas near Omae-Saki, the Izu peninsula, and Chiba (Fig. 1). Samples were collected using crab-baskets (EB), gill-net fishing by the fishing boats of Nagai Fishery Port in Kanagawa Prefecture in 2001 and 2002; dredging by the research vessels RINKAI-MARU (RK) of the Misaki Marine Biological Station of the University of Tokyo in 2002, T/V SHINYO-MARU (KS) of the Tokyo University of Marine Science and Technology in 2002 and 2003, and R/V TACHIBANA-MARU (TC) of the Manazuru Marine Biological Station of Yokohama National University in 2004, and by scuba diving (SC) off the Manazuru coast of Kanagawa Prefecture in 2004 (Table 1).

We compared the Sagami Bay octocoral species biodiversity measured during 2001–2005 with previous literature records including the literature listed in Table 2, the work of Matsumoto (1997, 2004, 2005, 2006), and results of surveys by R/V TANSEI-MARU (1986–2005 especially KT98-14, KT02-03, KT04-06, KT05-30) and R/V YAYOI (1998–2005) of the Ocean Research Institute of the University of Tokyo, and the commercial fishing boats DAI-8-TACHI-MARU and KIRYO-MARU. We also compared species biodiversity with Indo-Pacific, NE Pacific, and Atlantic species using data compiled by Broch (1935), Imahara (1996), Grasshoff (2000), Grasshoff and Bargibant (2001), Krieger (2001), MacIsaac et al. (2001), McAllister and Alfonso (2001), Gass and Willison

Table 1. Station List for Octocoral collection. Station name is based on "vessel or locality–Year–Station No". CG: CHOUGORO–MARU, DA: DAI–2–AOKI–MARU, DI: Dai–3–Ido = INKYO–MARU, DN: TAN–MARU, DY: DAIYAMA–MARU, EB: Clab–basket, IM = station no. of Imahara, 2006, KO: KOSHIN–MARU, KS: SHIN–YO–MARU, KY: KIYOMATSU–MARU, MR: MARUSE–MARU, MT: MARUTATSU–MARU, NB: NOBORU–MARU, RK: RINKAI–MARU, SC: SCUBA diving, SG: Sagami–bay, TC: TACHIBANA–MARU, YH: YOHEI–MARU. For more details of the station (Date, locality name etc., see Imahara, 2006 and Iwase and Matsumoto, 2006).

| Station | Longitude | Latitude | Depth | Rem. |
|---|---|---|---|---|
| MR01-01 | 139°34.530'E–139°34.625'E | 35°11.516'N–35°11.537'N | 49.3–48.7 m | IM06 |
| MR02-01 | 139°34.81'E–139°34.60'E | 35°11.49'N–35°11.48'N | 46.9–53.2 m | IM07 |
| DI01-01 | 139°40'37.1"E | 35°03'52.4"N | 520 m | |
| DI01-02 | 139°44.436'E–139°43.502'E | 35°03.556'N–35°03.821'N | ca 500 m | |
| DI02-01 | 139°44.61'E | 35°04.68'N | ca 250 m | |
| DI03-01 | – | – | ca 250–300 m | |
| DI03-02 | – | – | ca 450–490 m | |
| MT02-01 | 139°33'54"E–139°34'03"E | 35°12'13"N–35°12'04"N | – | IM15 |
| MT02-02 | 139°34'07"E–139°34'17"E | 35°12'12"N–35°12'10"N | 45m | |
| MT02-03 | – | – | 45 m | IM14 |
| DY01-02 | 139°34'57.8"E | 35°12'11.6"N | 15 m | |
| DY01-03 | 139°34'51.8"E | 35°12'11.6"N | 23 m | |
| DY01-04 | 139°34'38.2"E | 35°12'19.2"N | 28 m | |
| DY01 (unknown) | – | – | 20 m | SG |
| KY01-04 | 139°30.9'E–139°30.9'E | 35°08.8'N–35°08.7'N | 730–725 m | IM03 |
| NB02-05 | 139°34.72'E–139°34.96'E | 35°11.31'N–35°11.30'N | 63–68 m | |
| NB02 (unknown) | – | – | – | SG |
| KG01-02 | – | – | – | IM04 |
| EB01-01 | 139°44.877'E | 35°08.065'N | 300 m | IM05 |
| EB02-01 | – | – | 40–60 m | IM16 |
| KO02-01 | – | – | 10–20 m | SG? |
| DN02-08 | 139°33.76'E–139°33.95'E | 35°13.96'N–35°13.89'N | 66.0–68.4 m | IM17 |
| DA02-02 | 139°34'55"E–139°35'12"E | 35°11'02"N–35°11'01"N | 60–67 m | |
| YH02-01 | 139°29.9'E–139°30.0'E | 35°15.5'N–35°15.5'N | – | IM08 |
| KS02-07 | 139°32'E–139°32'E | 35°05.4'N–35°05.3'N | 410–699 m | |
| KS02-19 | 139°11.8'E–139°11.8'E | 35°01.3'N–35°01.4'N | 197–173 m | IM29 |
| KS02-25 | 139°11.9'E–139°11.8'E | 35°02.0'N–35°01.1'N | 150–135 m | IM30 |
| KS02-29 | 139°18.6'E–139°18.4'E | 34°40.2'N–34°40.4'N | 307–289 m | IM31 |
| KS02-30 | 139°19.3'E–139°19.0'E | 34°40.7'N–34°40.8'N | 252–228 m | IM32 |
| KS02-34 | 139°17.7'E–139°17.7'E | 34°42.5'N–34°42.5'N | 151–154 m | IM33 |
| KS02-35 | 139°16.9'E–139°16.9'E | 34°43.3'N–34°43.3'N | 171–181 m | IM34 |
| KS02-41 | 139°40.1'E–139°40.3'E | 34°51.3'N–34°51.1'N | 172–135 m | IM35 |
| KS02-42 | 139°38.9'E–139°39.3'E | 34°51.4'N–34°50.8'N | 452–381 m | IM36 |
| KS02 (unknown) | – | – | 100–1000 m | SG |
| KS03-01 | 139°41.1'E–139°41.1'E | 34°59.6'N–34°59.7'N | 81–78 m | IM37 |
| KS03-03 | 139°40.2'E–139°40.3'E | 34°60.0'N–35°0.0'N | 97–108 m | |
| KS03-05 | 139°40.2'E–139°40.5'E | 35°00.3'N–35°00.2'N | 300–274 m | |
| KS03-07 | 139°40.9'E–139°40.9'E | 34°55.7'N–34°55.7'N | 108–104 m | IM38 |
| KS03-09 | 139°40.2'E–139°40.5'E | 34°55.5'N–34°55.4'N | 375–275 m | IM39 |
| KS03-10 | 139°39.7'E–139°39.9'E | 34°54.8'N–34°54.8'N | 348–312 m | IM40 |
| KS03-11 | 139°39.9'E–139°39.3'E | 34°54.2'N–34°54.3'N | 315–365 m | |
| KS03-12 | 139°38.8'E–139°38.9'E | 34°54.1'N–34°54.4'N | 460–534 m | |
| KS03-17 | 139°41.9'E–139°41.6'E | 33°26.0'N–33°26.1'N | 160–190 m | IM41 |
| KS03-18 | 139°42.3'E–139°42.0'E | 33°26.3'N–33°26.5'N | 157–172 m | IM42 |
| KS03-19 | 139°42.7'E–139°42.4'E | 33°26.8'N–33°27.0'N | 170–176 m | IM43 |
| KS03-20 | 139°42.6'E–139°42.4'E | 33°27.3'N–33°27.7'N | 200–211 m | IM44 |
| KS03-21 | 139°42.6'E–139°42.4'E | 33°28.0'N–33°28.5'N | 340–446 m | IM45 |
| KS03-22 | 139°41.6'E–139°41.4'E | 33°28.6'N–33°28.9'N | 445–547 m | IM46 |

Table 1. Continued.

| Station | Longitude | Latitude | Depth | Rem. | Station | Longitude | Latitude | Depth | Rem. |
|---|---|---|---|---|---|---|---|---|---|
| YH02-02 | 139°29.5'E–139°29.6'E | 35°15.7'N–35°15.6'N | 60–70 m | IM09 | KS03-23 | 139°43.2'E–139°43.1'E | 33°28.5'N–33°29.3'N | 462–571 m | IM47 |
| YH02-03 | 139°29.1'E–139°29.2'E | 35°15.8'N–35°15.7'N | 60–70 m | IM10 | KS03-25 | 139°42.7'E–139°42.6'E | 33°28.8'N–33°29.5'N | 512–600 m | IM48 |
| YH02-04 | 139°29.2'E–139°29.3'E | 35°16.0'N–35°15.9'N | – | IM11 | KS03-26 | 140°14.6'E–140°14.6'E | 33°31.7'N–33°31.5'N | 158–158 m |  |
| RK02-01 | 139°34.5'E–139°34.5'E | 35°07.9'N–35°07.7'N | 94–95 m | IM12 | KS03-27 | 140°16.5'E–140°16.5'E | 33°31.2'N–33°31.2'N | 161–163 m | IM49 |
| RK02-02 | 139°33.5'E–139°33.4'E | 35°07.9'N–35°07.7'N | 104–117 m | IM13 | KS03-28 | 140°16.6'E–140°16.7'E | 33°31.2'N–33°31.4'N | 168–166 m | IM50 |
| RK02-03 | 139°32.92'E–139–32.76'E | 35–08.09'N–35–07.57'N | 240–418 m |  | KS03-29 | 140°15.5'E–140°15.4'E | 33°31.5'N–33°31.3'N | 161–160 m | IM51 |
| RK02-04 | 139°32.9'E–139–32.6'E | 35–08.0'N–35–07.5'N | 282–453 m | IM18 | KS03-30 | 140°17.5'E–140°17.5'E | 33°31.9'N–33°31.6'N | 345–313 m | IM52 |
| RK02-05 | 139°34.6'E–139–34.0'E | 35–06.7'N–35–06.8'N | 336–447 m | IM19 | KS03-31 | 140°16.0'E–140°15.9'E | 33°32.9'N–33°32.8'N | 170–171 m | IM53 |
| RK02-06 | 139°34.8'E–139–34.7'E | 35–07.6'N–35–07.8'N | 91 m | IM20 | KS03-32 | 140°15.2'E–140°15.0'E | 33°32.9'N–33°32.8'N | 166–160 m | IM54 |
| RK02-07 | 139°34.28'E–139°34.05'E | 35°07.55'N–35°07.40'N | 99–108 m |  | KS03-33 | 140°17.0'E–140°16.8'E | 33°33.5'N–33°33.2'N | 187–177 m | IM55 |
| RK02-08 | 139°34.7'E–139°34.6'E | 35°08.7'N–35°08.5'N | 86–89 m | IM21 | KS03-34 | 140°15.9'E–140°16.0'E | 33°34.1'N–33°34.4'N | 179–182 m | IM56 |
| RK02-09 | 139°37.4'E–139°37.4'E | 35°09.7'N–35°09.7'N | 8.8–8.8 m | IM22 | KS03-35 | 140°16.1'E–140°16.2'E | 33°33.7'N–33°33.4'N | 180–173 m |  |
| RK02-10 | 139°34.6'E–139°34.4'E | 35°07.6'N–35°07.9'N | 93.3–94.7 m | IM23 | KS03-36 | 139°31.5'E–139°31.2'E | 34°58.8'N–34°59.2'N | 900–950 m |  |
| RK02-11 | 139°33.6'E–139°33.2'E | 35°07.4'N–35°07.8'N | 121–156 m | IM24 | KS03-37 | 139°30.4'E–139°30.0'E | 34°57.8'N–34°58.3'N | 1195–1,200 m | IM57 |
| RK02-12 | 139°34.7'E–139°34.8'E | 35°07.4'N–35°07.7'N | 92–90 m | IM25 | CG04-01 | 139°47.4'E–139°46.4'E | 35°10.9'N–35°11.8'N | 32–55 m | IM58 |
| SG (unknown) | – | – | – | SG | TC04-01 | 139°10.7'E–139°10.8'E | 35°08.4'N–35°08.4'N | 67–68 m |  |
| KS02-01 | 139°32'E–139°32'E | 35°08.06'N–35°08.05'N | 185–216 m | IM26 | TC04-02 | 139°10.4'E–139°10.3'E | 35°08.4'N–35°08.3'N | 52–56 m | IM59 |
| KS02-02 | 139°32.9'E–139°32.7'E | 35°08.3'N–35°08.3'N | 177–148 m | IM27 | TC04-03 | 139°11.1'E–139°11.1'E | 35°08.3'N–35°08.4'N | 115–120 m | IM60 |
| KS02-03 | 139°31.4'E–139°30.9'E | 35°08.5'N–35°08.4'N | 549–683 m | IM28 | SC-01 | – | – | 5–15 m | IM61 |

(2005), Mortensen and Buhl-Mortensen (2005), Ofwegen (2005), Sanchez (2005), Watling and Auster (2005), and Etnoyer (Aquanautix, USA, pers. comm.).

The average annual temperature of Sagami Bay is 19.02 °C at 0 m, 14.67 °C at 100 m, 10.37 °C at 250 m, 6.42 °C at 500 m, and 3.23 °C at 1000 m (data from the Japan Oceanographic Data Centre). Bathymetric data includes data from upper littoral (0–40 m), deep littoral (40–200 m), upper bathyal (200–500 m), deep bathyal (500–3000 m), and abyssal (3000–6000 m) zones (Table 2).

The classification of octocorals used here is basically that of Bayer (1951, 1956a, 1981, 1982), Bayer and Grasshoff (1994), Williams (1995, 1999), Grasshoff (1999, 2000), Fabricius and Alderslade (2001), Cairns and Bayer (2005) and Ofwegen (2005). Gorgonacea and Alcyonacea have now been combined into the order Alcyonacea, however, since it has been suggested that abandoning the name Gorgonacea would be impractical and confusing (Grasshoff and Bargibant, 2001), here we continue to use the term "gorgonian corals" for this group of the Order Alcyonacea.

For old specimens or records, we conducted a literature survey based only on the data in Imahara (1996). However, we have attempted to compare our specimens with old or type materials, especially for the collection of Utinomi in the Seto Marine Biological Laboratory (SMBL, Japan) and for the collections of family Primnoidae which are deposited at several museums listed in the Acknowledgments. All new specimens collected as part of the larger NSMT project have been deposited at NSMT, however, they have not yet been catalogued.

## Results

**Species diversity.**—In Japanese territorial seas, approximately 620 octocoral species, including 260 gorgonians, 291 alcyonaceans, and 72 pennatulaceans, of the worldwide total of 2000 known octocorals have been reported (Bayer, 1981; Imahara, 1996). However, the worldwide total is now suggested to be closer to 2900 octocorals (S. Cairns, National Museum of Natural History (USNM), pers. comm.). In the literature, 260/620 species of octocorals have been recorded in Sagami Bay and adjacent waters, while 404 specimens comprising at least 88 species were collected during the present study (Table 2).

There are a total of 144 species and 13 families of gorgonian corals reported in the literature that are found in Sagami Bay. Twenty-five species already recorded in the literature, 13 newly recorded species, and 19 unidentified species were collected in the present study. The newly recorded species are *Arthrogorgia otsukai*, *Bebryce boninensis*, *Chrysogorgia* cf. *aurea*, *Echinomuricea peterseni*, *Ellisella laevis*, *Euplexaura nuttingi*, *Euplexaura robusta*, *Narella leilae*, *Paracis spinifera*, *Placogorgia dendritica*, *Plumarella* cf. *gracilis*, *Subergorgia thomsoni*, and *Villogorgia japonica*. The unidentified species contained probably two new records of species and two new records of the gorgonian genus, *Astrogorgia* Verrill, 1868 and *Guaiagorgia* Grasshoff and Alderslade, 1997. *Dicholaphis deliacta* Kinoshita, 1907 is the second record of this species after its first description. The most widespread cold-water octocoral species in the world belong to the genera *Primnoa* and *Paragorgia*. Species of both genera have been previously observed in Sagami Bay as older specimens, however, they were not collected during our study. We were able to find these specimens in an additional, separate survey of this area (Matsumoto et al., unpubl. data). In all, we succeeded at collecting at least 42 gorgonian species in the current survey.

There are 80 species and 10 families of alcyonacean corals reported in the literature to be in Sagami Bay. We identified 30 previously recorded species and one newly recorded species, and observed more than 30 unidentified species. *Clavularia ornata* was found for the first time in Sagami Bay area and Japanese waters. We also collected three new unidentified species belonging to the alcyonacean genera, *Bellonella* Gray, 1862, *Par-*

Table 2. Bathymetrical distribution and literature records of octocoral species observed in Sagami Bay and adjacent waters. Taxon, depth range, literature record (R), this study record (TS), new record at Sagami-Bay (NRS), new record in Japanese waters (NRJ), number of recording station (S), (E) Endemic, (IP) Indo-Pac., (NP) NE-Pac., (A) Atlantic, and references are listed. Data are from Imahara (1996, 2006), Iwase and Matsumoto (2006), this project, and the literature. Upper littoral: 0–40 m, deep littoral: 40–200 m, upper bathyal: 200–500 m, deep bathyal 500–3,000 m, abyssal: > 3,000 m. Long names have been abbreviated as follows: Wright and Studer (Wr. and St.), Kükenthal and Gorzawsky (Kük. and Gorz.), Quoy and Gaimard (Q. and G.), Thomson and Mackinnon (Thom. and Mack.), Thomson and Henderson (Th. and Hend.), Tixier- Durivault and Prevorsek (T. -D. and Prev.).

| Taxon | depth range | R | TS | NRS | NRJ | S | E | IP | NP | A |
|---|---|---|---|---|---|---|---|---|---|---|
| Octocorallia | | | | | | | | | | |
| Alcyonarian Corals | | | | | | | | | | |
| Suborder Stolonifera | | | | | | | | | | |
| Family Cornulariidae | | | | | | | | | | |
| *Cornularia aggregata* Utinomi, 1955 | upper littral | + | – | – | – | – | + | – | – | – |
| *Cornularia komaii* Utinomi, 1950 | – | – | + | – | – | – | + | – | – | – |
| *Cornularia sagamiensis* Utinomi, 1955 | deep littoral | + | – | – | – | – | + | – | – | – |
| Family Clavulariidae | | | | | | | | | | |
| *Clavularia eburnea* Kükenthal, 1906 | ?deep bathyal | + | – | – | – | – | + | – | – | – |
| *Clavularia mikado* Utinomi, 1962 | u.lit.–u. bath. | + | + | – | – | 9 | + | – | – | – |
| *Clavularia multispiculosa* Utinomi, 1955 | upper bathyal | + | + | – | – | 1 | + | – | – | – |
| *Clavularia ornata* (Thomson and Dean, 1931) | upper bathyal | – | + | + | + | 1 | + | – | – | – |
| *Clavularia spongicola* Utinomi, 1955 | littoral | + | – | – | – | – | + | – | – | – |
| *Clavularia* spp. | d.lit.–u. bath. | – | + | – | – | 3 | – | – | – | – |
| *Sarcodictyon gotoi* (Okubo, 1929) | – | – | + | – | – | – | – | + | – | – |
| Family Telestidae | | | | | | | | | | |
| *Telesto sagamina* Kinoshita, 1909 | deep littoral | + | + | – | – | 1 | + | – | – | – |
| *Telesto trichostemma* (Dana, 1846) | – | – | + | – | – | – | – | – | + | – |
| *Telesto tubulosa* Kinoshita, 1909 | – | + | + | – | – | 2 | + | – | – | – |
| Family Coelogorgiidae | | | | | | | | | | |
| *Paratelesto rosea* (Kinoshita, 1909) | – | – | + | – | – | – | + | – | – | – |
| Family Pseudocladochonidae | | | | | | | | | | |
| *Pseudocladochonus hicksoni* Versluys, 1907 | – | – | + | – | – | – | – | + | – | – |
| Suborder Alcyoniina | | | | | | | | | | |
| Family Alcyoniidae | | | | | | | | | | |
| *Alcyonium acaule* Marion, 1878 | – | – | + | – | – | – | – | – | – | + |
| *Anthomastus granulosa* Kükenthal, 1910 | deep littoral | + | + | – | – | 4 | – | + | – | – |
| *Anthomastus muscarioides* Kükenthal, 1910 | – | – | + | – | – | – | + | – | – | – |
| *Anthomastus* sp. | deep littoral | – | + | – | – | 2 | – | – | – | – |
| *Bathyalcyon robustum* Versluys, 1906 | deep bathyal | + | – | – | – | – | – | + | – | – |
| *Bellonella* sp. | upper bathyal | – | + | – | – | 1 | – | – | – | – |
| *Cladiella digitulata* (Klunzinger, 1877) | – | – | + | – | – | – | – | + | – | – |
| *Cladiella arborea* (Utinomi, 1954) | upper littoral | + | – | – | – | – | + | – | – | – |
| *Eleutherobia albiflora* (Utinomi, 1957) | deep littoral | + | – | – | – | – | + | – | – | – |
| *Eleutherobia dofleini* (Kükenthal, 1906) | deep littoral | + | – | – | – | – | – | + | – | – |
| *Eleutherobia flava* (Nutting, 1912) | deep littoral | + | + | – | – | 4 | – | + | – | – |
| *Eleutherobia grandiflora* (Kükenthal, 1906) | ? Deep sea | + | – | – | – | – | + | – | – | – |
| *Eleutherobia rigida* Pütter, 1900 | deep littoral | + | + | – | – | 2 | – | + | – | – |
| *Eleutherobia* rubra Brundin, 1896 | upper littoral | + | + | – | – | 3 | – | + | – | – |
| *Eleutherobia* cf. *sumbawaensis* Verseveldt and Bayer, 1988 | deep littoral | + | + | – | – | 4 | – | + | – | – |
| *Eleutherobia unicolor* (Kükenthal, 1906) | upper littoral | + | – | – | – | – | – | – | – | – |
| *Eleutherobia* spp. | deep littoral | – | + | – | – | 10 | – | – | – | – |
| *Minabea robusta* Utinomi and Imahara, 1976 | upper littoral | + | – | – | – | – | + | – | – | – |
| *Minabea* sp. | deep littoral | – | + | – | – | 1 | – | – | – | – |
| *Protodendron* sp. | deep bathyal | – | + | + | – | 1 | – | – | – | – |

Table 2. Continued.

| Taxon | depth range | R | TS | NRS | NRS | S | E | IP | NP | A |
|---|---|---|---|---|---|---|---|---|---|---|
| **Family Nephtheidae** | | | | | | | | | | |
| *Chromonephthea hirotai* (Utinomi, 1951) | – | + | – | – | – | – | – | + | – | – |
| *Coronephthya macrospiculata* (Thom. and Mack.1910) | deep littoral | + | + | – | – | 12 | – | + | – | – |
| *Coronephthya* sp. | | – | + | – | – | 1 | – | – | – | – |
| *Dendronephthya (D.) aurea* Utinomi, 1952 | – | + | – | – | – | – | – | + | – | – |
| *Dendronephthya (D.) carnea* (Wr. and St., 1889) | – | + | – | – | – | – | – | + | – | – |
| *Dendronephthya (D.) celosia* (Lesson, 1931) | uppper littoral | + | – | – | – | – | – | + | – | – |
| *Dendronephthya (D.) doederleini* (Kükenthal, 1905) | – | + | – | – | – | – | – | + | – | – |
| *Dendronephthya (D.) gigantea* (Verrill, 1864) | littoral | + | + | – | – | 1 | – | + | – | – |
| *Dendronephthya (D.) nipponica* Utinomi, 1952 | upper littoral | + | + | – | – | 1 | + | – | – | – |
| *Dendronephthya (D.)* spp. | | – | + | – | – | 2 | – | – | – | – |
| *Dendronephthya (Morchellana) acaulis* Kükenthal, 1906 | d.lit.–d.bath. | + | + | – | – | 1 | – | + | – | – |
| *Dendronephthya (M.) castanea* Utinomi, 1952 | upper littoral | + | – | – | | – | – | + | – | – |
| *Dendronephthya (M.) densa* Kükenthal, 1906 | deep littoral | + | – | – | – | – | + | – | – | – |
| *Dendronephthya (M.) flabellifera* (Studer, 1888) | – | + | – | – | – | – | – | + | – | – |
| *Dendronephthya (M.) habereri* Kükenthal, 1905 | upper littoral | + | + | – | – | 1 | – | + | – | – |
| *Dendronephthya (M.) maxima* Kükenthal, 1905 | deep littoral | + | – | – | – | – | + | – | – | – |
| *Dendronephthya (M.) pectinata* (Holm, 1895) | littoral | + | + | – | – | 4 | – | – | – | – |
| *Dendronephthya (M.) pumilio* (Studer, 1888) | littoral | + | + | – | – | 4 | – | + | – | – |
| *Dendronephthya (M.) querciformis* Kükenthal, 1906 | littoral | + | – | – | – | – | + | – | – | – |
| *Dendronephthya (M.) sinensis* (Pütter, 1900) | deep littoral | + | – | – | – | – | – | + | – | – |
| *Dendronephthya (M.) stolonifera* (May, 1899) | littoral | + | + | – | – | 1 | – | + | – | – |
| *Dendronephthya (M.)* spp. | u.lit.–u.bath. | – | + | – | – | 7 | – | – | – | – |
| *Dendronephthya (Roxasia) decussatospinosa* Utinomi, 1952 | – | + | – | – | – | – | + | – | – | – |
| *Dendronephthya (R.) filigrana* Kükenthal, 1906 | deep littoral | + | + | – | – | 1 | – | + | – | – |
| *Dendronephthya (R.) golgotha* Utinomi, 1952 | upper littoral | + | + | – | – | 1 | – | + | – | – |
| *Dendronephthya (R.) gracillima* Kükenthal, 1905 | deep littoral | + | – | – | – | – | – | + | – | – |
| *Dendronephthya (R.) japonica* Kükenthal, 1905 | deep littoral | + | – | – | – | – | – | + | – | – |
| *Dendronephthya (R.) jucunda* T.–D. and Prev., 1960 | deep bathyal | + | – | – | – | – | + | – | – | – |
| *Dendronephthya (R.) mollis* (Holm, 1895) | deep littoral | + | + | – | – | 2 | – | + | – | – |
| *Dendronephthya (R.) punctata* Kükenthal, 1906 | d. lit.–d. bath. | + | + | – | – | 5 | – | + | – | – |
| *Dendronephthya (R.) putteri* (Kükenthal, 1905) | d. lit.–d. bath. | + | + | – | – | 5 | – | + | – | – |
| *Dendronephthya (R.) rigida* (Studer, 1888) | deep littoral | + | + | – | – | 1 | – | + | – | – |
| *Dendronephthya (R.) suensoni* (Holm, 1895) | upper littoral | + | – | – | – | – | – | + | – | – |
| *Dendronephthya (R.) surugaensis* Imahara, 1977 | u.lit.–d. bath. | + | + | – | – | 1 | + | – | – | – |
| *Dendronephthya (R.) tenera* (Holm, 1895) | upper littoral | + | – | – | – | – | + | – | – | – |
| *Dendronephthya (R.)* spp. | u.lit.–d. bath. | – | + | – | – | 6 | – | – | – | – |
| *Daniela koreni* von Koch, 1891 | littoral | + | + | – | – | 1 | – | + | – | + |
| *Duva bicolor* (Utinomi, 1951) | upper littoral | + | + | – | – | 2 | + | – | – | – |
| *Gersemia rubiformis* (Ehrenberg, 1834) | lit.–bath. | + | – | – | – | – | – | – | + | + |
| *Lamnalia laevis* Thomson and Dean, 1931 | – | + | – | – | – | – | – | + | – | – |
| *Paraspongodes spiculosa* (Kükenthal, 1906) | – | + | – | – | – | – | – | + | – | – |
| *Scleronephthya gracillima* (Kükenthal, 1906) | u.lit.–d. bath. | + | + | – | – | 5 | – | + | – | – |
| *Scleronephthya* spp. | d. lit.–u. bath. | – | + | – | – | 6 | – | – | – | – |
| *Stereacanthia spiculosa* (Kükenthal, 1906) | deep littoral | + | – | – | – | – | – | + | – | – |
| *Stereacanthia* spp. | deep littoral | – | + | – | – | 4 | – | – | – | – |
| *Stereonephthya japonica* Utinomi, 1954 | upper littoral | + | – | – | – | – | + | – | – | – |
| *Stereonephthya* spp. | d.lit.–d. bath | – | + | – | – | 10 | – | – | – | – |
| *Umbellulifera graeffei* Kükenthal, 1906 | – | + | – | – | – | – | – | + | – | – |
| *Umbellulifera striata* Thomson and Henderson, 1905 | – | + | – | – | – | – | – | + | – | – |

Table 2. Continued.

| Taxon | depth range | R | TS | NRS | NRS | S | E | IP | NP | A |
|---|---|---|---|---|---|---|---|---|---|---|
| *Umbellulifera* sp. | deep littoral | – | + | – | – | 1 | – | – | – | – |
| Family Nidaliidae | | | | | | | | | | |
| *Nidalia macrospina* (Kükenthal, 1906) | deep littoral | + | + | – | – | 4 | – | + | – | – |
| *Nidalia* spp. | deep littoral | – | + | – | – | 4 | – | – | – | – |
| *Chironephthya crassa* (Wr. and St., 1889) | deep bathyal | + | – | – | – | – | + | – | – | – |
| *Chironephthya dipsacea* (Wr. and St., 1889) | u.lit.–d. bath. | + | – | – | – | – | – | + | – | – |
| *Chironephthya dofleini* Kükenthal, 1906 | u.lit.–d. bath. | + | + | – | – | 3 | – | + | – | – |
| *Chironephthya scoparia* (Wr. and St., 1889) | d.lit.–d. bath. | + | + | – | – | 4 | + | – | – | – |
| *Chironephthya* spp. | deep littoral | – | + | – | – | 13 | – | – | – | – |
| *Siphonogorgia godeffroyi* Kölliker, 1874 | deep bathyal | + | – | – | – | – | – | + | – | – |
| *Siphonogorgia splendens* Kükenthal, 1906 | deep littoral | + | – | – | – | – | – | + | – | – |
| *Siphonogorgia* spp. | deep littoral | – | + | – | – | 14 | – | – | – | – |
| Family Xeniidae | | | | | | | | | | |
| *Anthelia japonica* Kükenthal, 1906 | deep littoral | + | – | – | – | – | + | – | – | – |
| *Cespitularia multipinnata* (Q. and G., 1833) | upper littoral | + | – | – | – | – | – | + | – | – |
| Family Paralcyoniidae | | | | | | | | | | |
| *Paralcyonium* sp. | d. lit.–u. bath. | – | + | + | – | 4 | – | – | – | – |
| *Carotalcyon sagamianum* Utinomi, 1952 | upper bathyal | + | – | – | – | – | + | – | – | – |
| *Carotalcyon* sp. | deep littoral | – | + | – | – | 2 | – | – | – | – |
| Gorgonian Corals | | | | | | | | | | |
| Suborder Scleraxonia | | | | | | | | | | |
| Family Subergorgiidae | | | | | | | | | | |
| *Anella reticulata* (Ellis and Solander, 1786) | deep bathyal | + | – | – | – | – | – | + | – | – |
| *Subergorgia koellikeri* Wr. and St., 1889 | deep bathyal | + | – | – | – | – | – | + | – | – |
| *Subergorgia thomsoni* Nutting, 1911 | deep littoral | – | + | + | – | 1 | – | + | – | – |
| *Subergorgia verriculata* (Esper, 1794) | – | | + | – | – | – | – | + | – | – |
| Family Paragorgiidae | | | | | | | | | | |
| *Paragorgia regalis* Nutting, 1912 | deep bathyal | + | – | – | – | – | – | + | – | – |
| *Paragorgia tenuis* Kinoshita | – | | + | – | – | – | + | – | – | – |
| Family Coralliidae | | | | | | | | | | |
| *Corallium boshuense* Kishinouye, 1903 | – | | + | – | – | – | – | + | – | – |
| *Corallium pusillum* Kishinouye, 1903 | – | | + | – | – | – | – | + | – | – |
| *Corallium sulcatum* Kishinouye, 1903 | deep littoral | + | – | – | – | – | – | + | – | – |
| *Pleurocoralloides confusum* (Moroff, 1902) | – | | + | – | – | – | – | + | – | – |
| *Pleurocoralloides formosum* Moroff, 1902 | – | | + | – | – | – | – | + | – | – |
| Family Melithaeidae | | | | | | | | | | |
| *Acabaria abyssicola* Kükenthal, 1911 | bathyal | + | – | – | – | – | – | + | – | – |
| *Acabaria corymbosa* Kükenthal, 1908 | – | | + | – | – | – | – | + | – | – |
| *Acabaria habereri* Kükenthal, 1908 | deep bathyal | + | – | – | – | – | – | + | – | – |
| *Acabaria japonica* (Verrill, 1865) | – | | + | – | – | – | – | + | – | – |
| *Acabaria modesta* Kükenthal, 1908 | deep littoral | + | – | – | – | – | + | – | – | – |
| *Acabaria tenuis* Kükenthal, 1908 | lit.–bath. | + | – | – | – | – | – | + | – | – |
| *Acabaria undulata* Kükenthal, 1908 | deep bathyal | + | – | – | – | – | – | + | – | – |
| *Acabaria* spp. | deep littoral | – | + | – | – | 11 | – | – | – | – |
| *Melithaea arborea* (Kükenthal, 1908) | – | | + | – | – | – | – | + | – | – |
| *Melithaea densa* (Kükenthal, 1908) | – | | + | – | – | – | – | + | – | – |
| *Melithaea flabellifera* (Kükenthal, 1908) | upper littoral | + | + | – | – | 1 | – | + | – | – |
| *Melithaea flabellifera var. cylindrata* (Kükenthal, 1908) | lit.–bath. | + | – | – | – | – | – | + | – | – |
| *Melithaea flabellifera var. reticulata* (Kükenthal, 1908) | lit.–bath. | + | – | – | – | – | + | – | – | – |
| *Melithaea nodosa* (Wr. and St., 1889) | deep bathyal | + | – | – | – | – | – | + | – | – |

Table 2. Continued.

| Taxon | depth range | R | TS | NRS | NRS | S | E | IP | NP | A |
|---|---|---|---|---|---|---|---|---|---|---|
| *Melithaea* spp. | – | – | + | – | – | 1 | – | – | – | – |
| *Mopsella dichotoma* (Pallas, 1766) | deep littoral | + | – | – | – | – | – | + | – | – |
| *Mopsella* sp. | deep littoral | – | + | – | – | 2 | – | – | – | – |
| Family Parisididae | | | | | | | | | | |
| *Parisis fruticosa* Verrill, 1864 | deep littoral | + | – | – | – | – | – | + | – | – |
| *Parisis minor* Wr. and St., 1889 | d.lit.–d. bath. | + | + | – | – | 13 | – | + | – | – |
| *Parisis* sp. | upper bathyal | – | + | – | – | 1 | – | – | – | – |
| Suborder Holaxonia | | | | | | | | | | |
| Family Keroeididae | | | | | | | | | | |
| *Keroeides koreni* Wr. and St., 1889 | deep littoral | + | + | – | – | 5 | – | + | – | – |
| *Keroeides* sp. | deep littoral | – | + | + | – | 2 | – | – | – | – |
| Family Acanthogorgiidae | | | | | | | | | | |
| *Acalycigorgia densiflora* Kük. and Gorz., 1908 | upper littoral | + | + | – | – | 3 | + | – | – | – |
| *Acalycigorgia grandiflora* Kük. and Gorz., 1908 | upper littoral | + | – | – | – | – | – | + | – | – |
| *Acalycigorgia inermis* (Hedlund, 1890) | – | + | – | – | – | – | – | + | – | – |
| *Acalycigorgia irregularis* Kük. and Gorz., 1908 | littoral | + | + | – | – | 3 | – | + | – | – |
| *Acalycigorgia radians* Kük. and Gorz., 1908 | upper littoral | + | – | – | – | – | – | + | – | – |
| *Acanthogorgia angustiflora* Kük. and Gorz., 1908 | deep bathyal | + | – | – | – | – | + | – | – | – |
| *Acalycigorgia dofleini* Kük. and Gorz., 1908 | deep littoral | + | – | – | – | – | – | + | – | – |
| *Acalycigorgia gracillima* Kükenthal, 1909 | deep bathyal | + | – | – | – | – | – | + | – | – |
| *Acalycigorgia gracillima* var. *lata* Kükenthal, 1909 | – | + | – | – | – | – | + | – | – | – |
| *Acalycigorgia japonica* Kük. and Gorz., 1908 | upper bathyal | + | + | – | – | 1 | – | + | – | – |
| *Acalycigorgia multispina* Kük. and Gorz., 1908 | deep littoral | + | – | – | – | – | – | + | – | – |
| *Acalycigorgia multispina* var. *iridescens* Kükenthal, 1908 | deep littoral | + | – | – | – | – | + | – | – | – |
| *Acalycigorgia paradoxa* Nutting, 1912 | deep littoral | + | – | – | – | – | + | – | – | – |
| *Acalycigorgia procera* (Moroff, 1902) | – | + | – | – | – | – | + | – | – | – |
| *Acalycigorgia spissa* Kükenthal, 1909 | – | + | – | – | – | – | – | + | – | – |
| *Acalycigorgia vegae* Aurivillius, 1931 | deep littoral | + | – | – | – | – | – | + | – | – |
| *Acalycigorgia* spp. | deep littoral | – | + | – | – | 14 | – | – | – | – |
| *Anthogorgia bocki* Aurivillius, 1931 | littoral | + | + | – | – | 16 | – | + | – | – |
| *Anthogorgia japonica* Studer, 1889 | – | + | – | – | – | – | – | – | – | – |
| *Anthogorgia* sp. | deep littoral | – | + | – | – | 1 | – | – | – | – |
| Family Plexauridae | | | | | | | | | | |
| *Anthoplexaura dimorpha* Kükenthal, 1908 | upper littoral | + | – | – | – | – | + | – | – | – |
| *Astrogorgia* sp. | upper littoral | – | + | + | – | 1 | – | – | – | – |
| *Bebryce boninensis* Aurivillius, 1931 | deep littoral | – | + | + | – | 4 | + | – | – | – |
| *Bebryce* sp. | deep littoral | – | + | + | – | 2 | – | – | – | – |
| *Calicogorgia granulosa* Kük. and Gorz., 1908 | deep bathyal | + | – | – | – | – | – | + | – | – |
| *Echinogorgia armata* (Kükenthal, 1909) | lit.–bath. | + | – | – | – | – | + | – | – | – |
| *Echinogorgia asper* (Moroff, 1902) | – | + | – | – | – | – | + | – | – | – |
| *Echinogorgia* sp. | d.lit.–u. bath. | – | + | – | – | 2 | – | – | – | – |
| *Echinomuricea peterseni* Hedlund, 1890 | upper littoral | – | + | + | – | 1 | – | + | – | – |
| *Echinomuricea spinifera* Nutting, 1910 | littoral | + | + | – | – | 2 | – | + | – | – |
| *Euplexaura abietina* Kükenthal, 1908 | – | + | – | – | – | – | – | + | – | – |
| *Euplexaura anastomosans* Brundin, 1896 | deep littoral | + | – | – | – | – | + | – | – | – |
| *Euplexaura attenuata* (Nutting, 1910) | – | + | – | – | – | – | – | + | – | – |
| *Euplexaura crassa* Kükenthal, 1908 | deep littoral | + | + | – | – | 1 | + | – | – | – |
| *Euplexaura erecta* Kükenthal, 1908 | deep littoral | + | + | – | – | 5 | – | + | – | – |
| *Euplexaura nuttingi* Kükenthal, 1919 | deep littoral | – | + | + | + | 7 | – | + | – | – |
| *Euplexaura pinnata* Wr. and St., 1889 | deep littoral | + | – | – | – | – | + | – | – | – |

Table 2. Continued.

| Taxon | depth range | R | TS | NRS | NRS | S | E | IP | NP | A |
|---|---|---|---|---|---|---|---|---|---|---|
| *Euplexaura robusta* Kükenthal, 1908 | deep littoral | – | + | + | – | 1 | – | + | – | – |
| *Euplexaura sparsiflora* Kükenthal, 1908 | deep littoral | + | – | – | – | – | + | – | – | – |
| *Euplexaura* sp. | deep littoral | – | + | – | – | 2 | – | – | – | – |
| *Menella rigida* (Kükenthal, 1908) | – | + | – | – | – | – | – | + | – | – |
| *Menella indica* (Gray, 1870) | deep littoral | + | – | – | – | – | – | + | – | – |
| *Menella spinifera* (Kükenthal, 1911) | upper littoral | + | – | – | – | – | – | + | – | – |
| *Muriceides cylindrica* Nutting, 1912 | deep littoral | + | – | – | – | – | – | – | – | – |
| *Muricella abnormalis* Nutting, 1912 | deep littoral | + | – | – | – | – | – | + | – | – |
| *Muricella brunnea* Kükenthal, 1924 | deep littoral | + | – | – | – | – | + | – | – | – |
| *Muricella complanata* Wright and Studer, 1889 | deep bathyal | + | – | – | – | – | – | + | – | – |
| *Muricella nitida* (Verrill, 1868) | deep bathyal | + | – | – | – | – | – | + | – | – |
| *Muricella perramosa* Ridley, 1882 | deep bathyal | + | – | – | – | – | – | + | – | – |
| *Muricella reticulata* (Nutting, 1912) | deep bathyal | + | – | – | – | – | – | – | – | – |
| *Paracis ijimai* (Kinoshita, 1909) | d. lit.–d. bath. | + | + | – | – | 10 | – | + | – | – |
| *Paracis pustulata* (Wr. and St., 1889) | deep bathyal | + | – | – | – | – | – | + | – | – |
| *Paracis spinifera* (Nutting, 1912) | deep littoral | – | + | + | – | 1 | + | – | – | – |
| *Placogorgia dendritica* Nutting, 1910 | bathyal | – | + | + | + | 1 | – | + | – | – |
| *Placogorgia japonica* Nutting, 1912 | deep littoral | + | – | – | – | – | + | – | – | – |
| *Placogorgia placoderma* (Nutting, 1910) | deep littoral | + | – | – | – | – | – | + | – | – |
| *Thesea mitsukurii* (Kinoshita, 1909) | lit.–bath. | + | – | – | – | – | – | + | – | – |
| *Villogorgia arbuscula* (Gray, 1889) | deep bathyal | + | – | – | – | – | + | – | – | – |
| *Villogorgia japonica* Aurivillius, 1931 | deep littoral | – | + | + | – | 3 | + | – | – | – |
| *Villogorgia* spp. | d. lit.–d. bath. | – | + | – | – | 5 | – | – | – | – |
| Family Gorgoniidae | | | | | | | | | | |
| *Eunicella hendersoni* (Kükenthal, 1908) | upper bathyal | + | – | – | – | – | + | – | – | – |
| *Eunicella pendula* Kükenthal, 1908 | deep littoral | + | + | – | – | 4 | + | – | – | – |
| *Eunicella* sp. | deep bathyal | – | + | – | – | 1 | – | – | – | – |
| *Guaiagorgia* sp. | deep littoral | – | + | + | – | 1 | – | – | – | – |
| Suborder Calcaxonia | | | | | | | | | | |
| Family Ellisellidae | | | | | | | | | | |
| *Ellisella andamanensis* (Simpson, 1910) | – | + | – | – | – | – | – | + | – | – |
| *Ellisella gracilis* (Wr. and St., 1889) | deep littoral | + | – | – | – | – | – | + | – | – |
| *Ellisella laevis* (Verrill, 1865) | deep littoral | – | + | + | + | 2 | – | + | – | – |
| *Ellisella plexauroides* Toeplitz in Kük., 1919 | – | + | – | – | – | – | – | + | – | – |
| *Ellisella rubra* (Wr. and St., 1889) | lit.–bath. | + | – | – | – | – | – | + | – | – |
| *Ellisella* spp. | d.lit.–d.bath. | – | + | – | – | 6 | – | – | – | – |
| *Junceella juncea* (Pallas, 1766) | lit.–bath. | + | – | – | – | – | – | + | – | – |
| *Junceella racemosa* (Wr. and St., 1889) | – | + | – | – | – | – | – | + | – | – |
| *Junceella sanguinolenta* (Gray, 1859) | deep bathyal | + | – | – | – | – | – | + | – | – |
| *Junceella umbraculum* (Ellis and Solander, 1786) | – | + | – | – | – | – | – | + | – | – |
| Family Chrysogorgiidae | | | | | | | | | | |
| *Chrysogorgia agassizi* (Verrill, 1883) | deep bathyal | + | – | – | – | – | – | – | – | + |
| *Chrysogorgia cf. aurea* Kinoshita, 1913 | deep littoral | – | + | + | – | 1 | + | – | – | – |
| *Chrysogorgia cavea* Kinoshita, 1913 | deep bathyal | + | – | – | – | – | + | – | – | – |
| *Chrysogorgia comans* Kinoshita, 1913 | – | + | – | – | – | – | + | – | – | – |
| *Chrysogorgia debilis* Kükenthal, 1908 | – | + | – | – | – | – | + | – | – | – |
| *Chrysogorgia dichotoma* Thomson and Henderson, 1906 | – | + | – | – | – | – | – | + | – | – |
| *Chrysogorgia dispersa* Kükenthal,1908 | deep bathyal | + | – | – | – | – | + | – | – | – |
| *Chrysogorgia excavata* Kükenthal, 1908 | bathyal | + | – | – | – | – | + | – | – | – |
| *Chrysogorgia flexilis* (Wr. and St., 1889) | – | + | – | – | – | – | – | + | – | + |

Table 2. Continued.

| Taxon | depth range | R | TS | NRS | NRS | S | E | IP | NP | A |
|---|---|---|---|---|---|---|---|---|---|---|
| *Chrysogorgia geniculata* Wr. and St., 1889 | – | + | – | – | – | – | – | + | – | – |
| *Chrysogorgia japonica* (Wr. and St.,1889) | deep bathyal | + | – | – | – | – | – | + | – | – |
| *Chrysogorgia lata* (Versluys, 1902) | bathyal | + | – | – | – | – | – | + | – | – |
| *Chrysogorgia minuta* (Kinoshita,1913) | – | + | – | – | – | – | + | – | – | – |
| *Chrysogorgia okinosensis* Kinoshita, 1913 | upper bathyal | + | – | – | – | – | + | – | – | – |
| *Chrysogorgia papillosa* (Kinoshita,1913) | deep bathyal | + | – | – | – | – | + | – | – | – |
| *Chrysogorgia pellucida* Kükenthal, 1908 | deep littoral | + | – | – | – | – | + | – | – | – |
| *Chrysogorgia pyramidalis* (Kükenthal,1908) | d.lith.–d.bath. | + | – | – | – | – | + | – | – | – |
| *Chrysogorgia rotunda* Kinoshita, 1913 | deep bathyal | + | – | – | – | – | + | – | – | – |
| *Chrysogorgia sphaerica* (Aurivillius, 1931) | d.lith.–d.bath. | + | – | – | – | – | + | – | – | – |
| *Chrysogorgia versluys* Kinoshita, 1913 | deep bathyal | + | – | – | – | – | + | – | – | – |
| *Chrysogorgia* sp. | deep littoral | – | + | – | – | 2 | – | – | – | – |
| *Radicipes pleurocristatus* Stearns, 1883 | – | + | – | – | – | – | – | + | – | – |
| *Radicipes verrilli* (Wright,1885) | deep bathyal | + | – | – | – | – | – | + | – | – |
| Family Primnoidae | | | | | | | | | | |
| *Arthrogorgia ijimai* (Kinoshita, 1907) | bathyal | + | + | – | – | 2 | + | – | – | – |
| *Arthrogorgia otsukai* Bayer, 1952 | bathyal | – | + | + | – | 1 | – | – | + | – |
| *Callogorgia elegans* (Gray,1870) | – | + | – | – | – | – | – | + | – | – |
| *Callogorgia flabellum* (Ehrenberg, 1834) | deep bathyal | + | + | – | – | 1 | – | + | – | – |
| *Callogorgia flabellum var. grandis* Kük. and Gorz., 1908 | deep bathyal | + | – | – | – | – | + | – | – | – |
| *Callogorgia ramosa* Kük. and Gorz., 1907 | deep bathyal | + | – | – | – | – | + | – | – | – |
| *Calyptrophora japonica* Gray, 1866 | bathyal | + | + | – | – | 3 | – | + | + | – |
| *Calyptrophora* sp. | deep bathyal | – | + | – | – | 1 | – | – | – | – |
| *Dicholaphis delicata* Kinoshita, 1907 | bathyal | + | + | – | – | 1 | + | – | – | – |
| *Fanellia tuberculata* (Versluys, 1906) | bathyal | + | + | – | – | 3 | – | + | + | – |
| *Narella clavata var. japonensis* Aurivillius, 1931 | deep bathyal | + | – | – | – | – | + | – | – | – |
| *Narella leilae* Bayer, 1951 | upper bathyal | – | + | + | + | 1 | – | + | – | – |
| *Narella megalepis* (Kinoshita, 1908) | – | + | – | – | – | – | – | + | – | – |
| *Paracalyptrophora kerberti* (Versluys, 1906) | bathyal | + | + | – | – | 2 | – | – | + | – |
| *Parastenella doederleini* (Wr. and St., 1889) | deep bath.–abyss. | + | – | – | – | – | – | + | + | – |
| *Plumarella acuminata* Kinoshita, 1908 | d.lit.–bath. | + | + | – | – | 1 | + | – | – | – |
| *Plumarella adhaerans* Nutting, 1912 | deep littoral | + | – | – | – | – | + | – | – | – |
| *Plumarella alba* Kinoshita, 1908 | deep bathyal | + | + | – | – | 1 | + | – | – | – |
| *Plumarella carinata* Kinoshita, 1908 | – | + | – | – | – | – | + | – | – | – |
| *Plumarella cristata* Kük. and Gorz., 1908 | lit.–bath. | + | + | – | – | 5 | + | – | – | – |
| *Plumarella dofleini* Kük. and Gorz., 1908 | lit.–bath. | + | + | – | – | 11 | + | – | – | – |
| *Plumarella flabellata* Versluys, 1906 | – | + | – | – | – | – | + | – | – | – |
| *Plumarella cf. gracilis* Kinoshita, 1908 | upper bathyal | – | + | + | – | 1 | + | – | – | – |
| *Plumarella lata* Kük. and Gorz., 1908 | upper bathyal | + | – | – | – | – | + | – | – | – |
| *Plumarella longispina* Kinoshita, 1908 | d.lit.–d. bath. | + | + | – | – | 6 | – | + | + | – |
| *Plumarella rigida* Kük. and Gorz., 1908 | lit.–d.bath. | + | – | – | – | – | + | – | – | – |
| *Plumarella serta* Kük. and Gorz., 1908 | d.lit.–u.bath. | + | – | – | – | – | + | – | – | – |
| *Plumarella serta var. squamosa* Kük. and Gorz., 1908 | deep bathyal | + | – | – | – | – | + | – | – | – |
| *Plumarella spinosa* Kinoshita, 1907 | d.lit.–u.bath. | + | + | – | – | 4 | + | – | – | – |
| *Plumarella* spp. | d.lit.–d. bath. | – | + | – | – | 7 | – | – | – | – |
| *Primnoa pacifica* Kinoshita, 1907 | deep bathyal | + | – | – | – | – | – | – | + | – |
| *Thouarella alternata* Nutting, 1912 | deep bathyal | + | – | – | – | – | + | – | – | – |
| *Thouarella carinata* (Kükenthal, 1908) | deep bathyal | + | – | – | – | – | + | – | – | – |
| *Thouarella hilgendorfi* (Studer, 1878) | deep bathyal | + | – | – | – | – | – | + | – | + |
| *Thouarella hilgendorfi var. plumatilis* Aurivillius, 1931 | d.lit.–d. bath. | + | + | – | – | 1 | + | – | – | – |

Table 2. Continued.

| Taxon | depth range | R | TS | NRS | NRS | S | E | IP | NP | A |
|---|---|---|---|---|---|---|---|---|---|---|
| *Thouarella laxa* Versluys, 1906 | bathyal | + | – | – | – | – | – | + | – | – |
| *Thouarella parva* (Kinoshita, 1907) | – | + | – | – | – | – | + | – | – | – |
| *Thouarella recta* Nutting, 1912 | deep bathyal | + | – | – | – | – | + | – | – | – |
| *Thouarella typica* Kinoshita, 1907 | deep littoral | + | – | – | – | – | – | + | – | – |
| *Thouarella* sp. | d.lit.–d. bath. | – | + | – | – | 1 | – | – | – | – |
| Family Isididae | | | | | | | | | | |
| *Acanella japonica* Kükenthal, 1919 | deep bathyal | + | – | – | – | – | + | – | – | – |
| *Keratoisis japonica* (Studer, 1878) | deep bathyal | + | – | – | – | – | + | – | – | – |
| *Keratoisis paucispinosa* (Wr. and St.,1889) | deep bathyal | + | – | – | – | – | – | + | + | – |
| *Keratoisis profunda* (Wr. and St., 1889) | abyssal | + | – | – | – | – | – | + | + | – |
| *Keratoisis squarrosa* Kükenthal, 1919 | – | + | – | – | – | – | + | – | – | – |
| *Muricellisis echinata* Kükenthal, 1919 | deep bathyal | + | – | – | – | – | + | – | – | – |
| Pennatulacea | | | | | | | | | | |
| Suborder Sessiliflorae | | | | | | | | | | |
| Family Veretillidae | | | | | | | | | | |
| *Cavernularia elegans* Herklots, 1858 | – | + | – | – | – | – | + | – | – | |
| *Cavernularia habereri* Moroff, 1902 | littoral | + | – | – | – | – | – | + | – | – |
| *Cavernularia obesa* Milne-Edwards and Haime, 1850 | upper littoral | + | + | – | – | 1 | – | + | – | – |
| *Lituaria habereri* Balss, 1910 | deep littoral | + | – | – | – | – | + | – | – | – |
| Family Echinoptilidae | | | | | | | | | | |
| *Echinoptilum macintoshi* Hubrecht, 1885 | littoral | + | + | – | – | 1 | – | + | – | – |
| Family Kophobelemnidae | | | | | | | | | | |
| *Kophobelemnon stelliferum* (Müller, 1776) | u. lit.–abyss. | + | – | – | – | – | – | + | – | + |
| *Sclerobelemnon burgeri* (Herklots, 1858) | upper littoral | + | + | – | – | 1 | – | + | – | – |
| *Sclerobelemnon schmeltzii* Kölliker, 1872 | – | + | – | – | – | – | – | + | – | – |
| Family Funiculinidae | | | | | | | | | | |
| *Funiculina quadrangularis* (Pallas, 1766) | deep littoral | + | – | – | – | – | – | + | – | + |
| Family Protoptilidae | | | | | | | | | | |
| *Protoptilum orientale* Nutting, 1912 | upper bathyal | + | – | – | – | – | + | – | – | – |
| Family Stachyptilidae | | | | | | | | | | |
| *Stachyptilum macleari* Kölliker, 1880 | deep littoral | + | – | – | – | – | – | + | – | – |
| *Stachyptilum superbum* Studer, 1878 | d.lit.–d. bath. | + | – | – | – | – | – | + | – | – |
| *Stachyptilum dofleini* Balss, 1909 | deep littoral | – | + | + | – | 7 | + | – | – | – |
| *Stachyptilum* sp. | deep littoral | – | + | – | – | 3 | – | – | – | – |
| Family Scleroptilidae | | | | | | | | | | |
| *Scleroptilum grandiflorum* Kölliker, 1880 | deep littoral | – | + | + | – | 1 | – | – | – | + |
| Family Chunellidae | | | | | | | | | | |
| *Chunella hertwigi* (Balss, 1909) | deep littoral | + | – | – | – | – | + | – | – | – |
| *Chunella indica* (Th. and Hend., 1906) | d.lit.–u. bath. | + | + | – | – | 1 | – | + | – | – |
| Family Ombellulidae | | | | | | | | | | |
| *Ombellula huxleyi* Kölliker, 1880 | deep bathyal | + | – | – | – | – | + | – | – | – |
| *Ombellula lindahlii* Kölliker, 1875 | deep bathyal | + | – | – | – | – | + | – | – | – |
| Suborder Subselliflorae | | | | | | | | | | |
| Family Halipteridae | | | | | | | | | | |
| *Halipteris willemoesi* (Kölliker, 1870) | d.lit.–d.bath. | + | – | – | – | – | – | + | – | – |
| Family Virgulariidae | | | | | | | | | | |
| *Scytalium martensii* Kölliker, 1870 | upper bathyal | + | – | – | – | – | – | + | – | – |
| *Scytalium splendens* Thomson and Henderson, 1906 | – | + | – | – | – | – | – | – | – | – |
| *Virgularia abies* (Kölliker, 1870) | upper littoral | + | – | – | – | – | + | – | – | – |
| *Virgularia bromleyi* Kölliker, 1880 | bathyal | + | – | – | – | – | + | – | – | – |

Table 2. Continued.

| Taxon | depth range | R | TS | NRS | NRS | S | E | IP | NP | A |
|---|---|---|---|---|---|---|---|---|---|---|
| *Virgularia gustaviana* (Herklots, 1863) | littoral | + | + | – | – | 2 | – | + | – | – |
| *Virgularia rumphii* Kölliker, 1870 | upper littoral | – | + | + | – | 2 | – | + | – | – |
| *Virgularia* spp. | deep littoral | – | + | – | – | 3 | – | – | – | – |
| *Virgulariidae* sp. | deep littoral | – | + | – | – | 1 | – | – | – | – |
| Family Pennatulidae | | | | | | | | | | |
| *Pennatula fimbriatus* Herklots, 1858 | upper littoral | + | + | – | – | 1 | – | – | – | – |
| *Pennatula inermis* Nutting, 1912 | upper bathyal | + | – | – | – | – | + | – | – | – |
| *Pennatula murrayi* Kölliker, 1880 | deep littoral | + | – | – | – | – | – | + | – | + |
| *Pennatula naresi* Kölliker, 1880 | bathyal | + | – | – | – | – | + | – | – | – |
| *Pennatula phosphorea* Linnaeus, 1758 | deep littoral | + | – | – | – | – | – | + | – | + |
| *Pennatula rubescens* Nutting, 1912 | deep littoral | + | – | – | – | – | + | – | – | – |
| *Pennatula* sp. | upper bathyal | – | + | – | – | 2 | – | – | – | – |
| Family Pteroeididae | | | | | | | | | | |
| *Pteroeides bankanense* Bleeker, 1859 | – | + | – | – | – | – | – | + | – | – |
| *Pteroeides breviradiatum* Kölliker, 1869 | upper littoral | + | + | – | – | 1 | – | + | – | – |
| *Pteroeides dofleini* (Balss, 1910) | deep littoral | + | – | – | – | – | + | – | – | – |
| *Pteroeides sagamiense* Moroff, 1902 | littoral | + | + | – | – | 1 | + | – | – | – |
| *Pteroeides sparmannii* Kölliker, 1869 | upper littoral | + | + | – | – | 1 | – | + | – | – |
| *Pteroeides tenerum* Kölliker, 1869 | – | + | – | – | – | – | – | + | – | – |
| *Sarcophyllum bollonsi* Benham, 1906 | – | + | – | – | – | – | – | + | – | – |

*alcyonium* Milne-Edwards and Haime, 1850, and *Protodendron* Thomson and Dean, 1931. In total, at least 34 alcyonacean species were collected.

We determined from a search of the literature that 36 species and 13 families of pennatulaceans have been recorded in the Sagami Bay. We identified a total of 12 species in Sagami Bay and adjacent waters including 10 previously recorded species, a newly recorded species *Virgularia rumphii* and one newly recorded genus *Scleroptilum* Kölliker, 1874 (Table 2). In addition, we collected more than six unidentified species.

Analysis of the data compiled revealed that more than one third (260 of about 620 species in a previous study and 283 of about 640 in the present study) of the total number of Japanese octocoral species recorded have been collected from Sagami Bay and adjacent waters. It is worth mentioning that our study reported 15 first recorded species, two species that are most likely newly recorded, and six or more first recorded genera from Sagami Bay and adjacent waters (NRS in Table 2) as well as five first recorded species from Japanese waters (NRJ in Table 2). It should also be noted that sometimes when an octocoral has been reported as being recorded in Japan, it was actually recorded in Sagami Bay (Table 2, Fig. 2).

**Station distribution.**—The highest level of octocoral family biodiversity (11 families) was recorded at stations KS03-18 (157–172 m) and KS03-20 (200–211 m): eight families of gorgonian corals at station KS03-18 (157–172 m), four families of alcyonacean corals at KS02-35 (171–181 m), KS03-01 (78–81 m), KS03-20 (200–211 m), KS03-28 (166–168 m), and KS03-34 (179–182 m), and five families of pennatulaceans at CG04-01 (32–55 m). The stations with the highest octocoral species biodiversity were KS03-19 (170–176 m) for gorgonians (14 species), KS03-34 (179–182 m) for alcyonaceans (12 species), CG04-01 (32–55 m) for pennatulaceans (8 species), and KS03-19 (170–176 m) for all octocorals (25 species) (Table 1, Fig. 1).

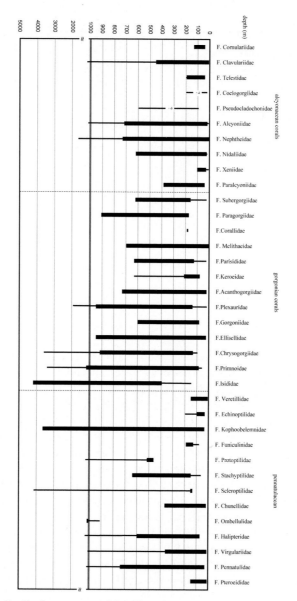

Figure 2. Depth distribution patterns of families of octocorals in Japan and adjacent waters and Sagami Bay modified from Matsumoto (2006). Source of data for Sagami Bay is in Table 2, and data for Japan and adjacent waters include Matsumoto (1997, 2004, 2005), and specimens from R/V TANSEI-MARU cruise between 1986–2005, KT98-14, KT02-03, KT04-06, KT05-30, R/V YAYOI of International Coastal Research Centre of University of Tokyo between 1998–2005, and the commercial fishing boats DAI-8-TACHI-MARU and KIRYO-MARU. The vertical line indicates the depth distribution record of Japanese and adjacent waters and bars indicate depth range at Sagami Bay.

**Depth distribution.**—From the literature, octocorals have been reported in Sagami Bay from upper littoral (0–40 m) to abyssal (> 3000 m) levels. Three abyssal octocoral species, *Keratoisis profunda*, *Kophobelemnon stelliferum*, and *Parastenella doederleini* have been observed previously (Table 2), and *P. doederleini* was also recorded at a depth

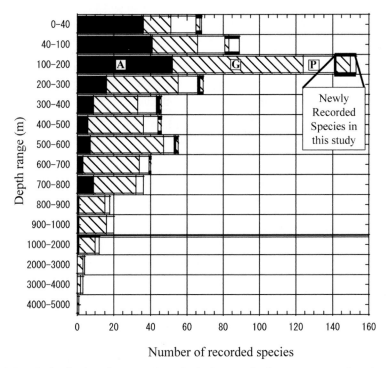

Figure 3. Depth distribution of octocoral species in Japan and adjacent waters and newly recorded species from the present study. Data are from Table 1 and Table 2. When the depth range in the literature was wide, the depth range was divided into 100 m sections and the occurrence at each depth range was computed. These wide distribution data usually included several specimen records from different depths. A: alcyonacean, G: gorgonian, P: pennatulacean. The bars on the right side of each depth range column indicate the newly recorded species.

of 400–550 m in Sagami Bay, which is shallower than abyssal. In the present study, we did not survey deeper than 2000 m (Table 1), and no abyssal species were collected, although old records of these abyssal species in this area suggest they had a shallow distribution.

Thirty-four families of octocorals have been observed between the depths of 100–200 m in Japan and adjacent waters, and 30 families of octocorals have been recorded between 100 m and 200 m from Sagami Bay and adjacent waters (Fig. 2). The highest degree of species diversity tended to occur between 100 m and 200 m (deep littoral) and 23% of all species and 37% of newly recorded species were observed within this depth range (Fig. 3). The second highest level of biodiversity was found between 40 m and 100 m (deep littoral).

## Discussion

**Horizontal distribution.**—The geographical distributions of the species of Sagami Bay and adjacent waters are presented in Table 2. Sagami Bay is also home to 146 coral species that are indigenous to the Indo-Pacific Oceans (subtropical-tropical water): 44 alcyonaceans, 79 gorgonians, and 23 pennatulaceans. Of the coral species recorded in Sagami Bay, 10 octocorals (one alcyonacean and nine gorgonians) also inhabit the northeast Pacific Ocean. The most common alcyonacean species is *Gersemia rubi-*

*formis*, while the common gorgonian species are *A. otsukai*, *Calyptrophora japonica*, *Fanellia tuberculata*, *Keratoisis paucispinosa*, *K. profunda*, *Paracalyptrophora kerberti*, *P. doederleini*, *Plumarella longispina*, and *Primnoa pacifica*. Furthermore, two alcyonacean coral genera and 18 gorgonian coral genera were recorded both in our study as well as in the northeast Pacific (Bayer, 1952a; Cimberg et al., 1981 cited in Krieger, 2001; McAllister and Alfonso, 2001; Cairns and Bayer, 2005; P. Etnoyer, Aquanautix Consulting, pers. comm.). Sagami Bay is also inhabited by 10 octocoral species from the Atlantic Ocean (three alcyonaceans, three gorgonians, and four pennatulaceans). Common species are the alcyonaceans *Alcyonium acaule*, *Daniela koreni*, and *G. rubiformis*, the gorgonians *Chrysogorgia agassizi*, *Chrysogorgia flexilis*, and *Thouarella hilgendorfi*, and the pennatulaceans *Funiculina quadrangularis*, *K. stelliferum*, *Pennatula murrayi*, and *Pennatula phosphorea*. Genera common to both Sagami Bay and the Atlantic Ocean include seven alcyonaceans, 23 gorgonians, and four pennatulaceans (Bayer, 1954; Breeze et al., 1997 cited in Gass and Willison, 2005; MacIsaac et al., 2001; Gass and Willison, 2005; Mortensen and Buhl-Mortensen, 2005; Watling and Auster, 2005). In contrast, 114 species of octocoral (32 alcyonaceans, 70 gorgonians, and 12 pennatulaceans) are endemic to Japan and adjacent waters, while 74 of these octocorals (17 alcyonaceans, 51 gorgonians, and six pennatulaceans) are only found in Sagami Bay and adjacent waters. Therefore, 56% of all octocoral species from Sagami Bay and adjacent waters are Indo-Pacific, 4% are northeast Pacific, 4% are Atlantic, and 44% are endemic to Japanese waters (Table 2). It is unclear whether the records in the literature reflect specimens that were correctly described or identified as not all materials have undergone the detailed revision. For example, *A. acaule* is described as a Mediterranean species (Ofwegen, L. P. van, Nationaal Natuurhistorisch Museum, Leiden, The Netherlands, pers. comm.). For this species and for several others in Table 2, we were unable to collect any new specimens or check the old material described in the literature. Therefore, these numbers of species may change slightly after the full revision of type and old specimens is completed.

With regard to the horizontal distributions, the primary question is why biodiversity and endemism of cold-water octocorals in Sagami Bay is so high. The second question is whether there is any relation or interchange between Sagami Bay and the surrounding areas, and how this interchange may have occurred. The high biodiversity and endemism around Sagami Bay is likely due to the co-occurrence of temperate, subtropical, and subarctic octocoral species. Other taxa inhabiting the bay, such as echinoderms, also have a high degree of biodiversity which has been explained in the same way (Takeda et al., 2006). The area is the biogeographic boundary where the warm Kuroshio current and the cold Oyashio current converge. The coexistence of warm- and cold-water currents near Sagami Bay may explain the mixture of taxa from warm and cold water. The distribution of the subtropical components (Indo-Pacific animals) is due to the major warm-water Kuroshio Current, and the distribution of subarctic components (cold-water corals) is due to the cold-water Oyashio Current. With regard to the second question, it has been suggested that the Aleutian Islands link shallow-water molluscan species between the northeast and northwest Pacific (Vermeij, 1990). However, whether or not the same is true for deep-water animals is not clear. The existence of related cold-water marine species between the North Pacific and Arctic-Atlantic may be explained by the trans-Arctic interchange after the submergence of the Bering Strait about 3.5 Ma during the early Pliocene (Vermeij, 1991). Although it is still not known whether the octocoral species of two oceans have been exchanged in the past, studies of other taxa have suggested the possibility of linkage between the North Pacific and Arctic-Atlantic basins (Vermeij, 1991; Collins et al., 1996). However, prior to further discussions on this topic, the related species need to be taxonomically revised.

**Vertical distribution.**—The deep littoral depth range (100–200 m) has the highest biodiversity both in Japanese waters and in Sagami Bay (Figs. 2, 3). With regard to vertical distributions, habitat complexity is the traditional explanation for the question "why a particular depth range has rich biomass and high species diversity?" however, this is not a sufficient explanation in this case. Instead, the high degree of octocoral biodiversity in the deep littoral zone may be due to the benthic-pelagic coupling, or the interaction between the benthic and pelagic ecosytems (Graf, 1992; Boero et al., 1996; Marcus and Boero, 1998; Raffaelli et al., 2003). In such ecotones, the transition zones between two adjacent ecosystems (Holland and Risser, 1991), the ecosystems interact with each other, resulting in a higher density of organisms and a greater number of species, due to the edge effect, than are found in either flanking community alone (Odum, 1971; Murcia, 1995). At 100–200 m in Sagami Bay, the pelagic and benthic ecosystems converge resulting in a high degree of biodiversity for both plankton and benthos in this ecotone zone.

## Conclusions

This study partially confirmed some of the octocoral fauna of Sagami Bay and adjacent waters. We succeeded in collecting 65 previously recorded species and probably 23 newly reported species. However, the fact that over 190 species recorded more than 100 yrs ago were not collected during our survey may be due to the fact that we did not sample some areas of Sagami Bay. One example of an area that was not included in our study is the Okinose Bank, which was well-known more than 100 yrs ago as an area with high biodiversity. It has now become difficult to survey these areas because of fishery rights and shipping traffic. In addition, particular sampling gears such as coral fishery nets and other devices that were used in the past, are now forbidden. Another possibility is that these species are now absent or more rare due to the environmental changes that have occurred over the past 100 yrs (Takeda et al., 2006). Sagami Bay is located very close to Tokyo, which has been extensively developed and is still growing, therefore, the ecosystem of Sagami Bay is threatened by human activity.

We are currently in the process of confirming specimens reported in the older literature (including type specimens) which are stored at various museums and universities. Once these old specimens are revised and recatalogued, we will have a better understanding of the current and past changes to the octocorals in Sagami Bay.

## Acknowledgments

We gratefully acknowledge the assistance of the crews of the R/V Shinyo-maru (KS03) of Tokyo University of Marine Science and Technology, R/V Tansei-maru (KT98-14, KT02-03, KT04-06, KT05-30), R/V Yayoi, and R/V Rinkai-Maru of University of Tokyo, R/V Tachibana-maru of Yokohama National University, as well as various commercial fishing boats (Chogoro-maru, Dai-3-Ido-Inkyo-maru, Hajime-maru, Daiyama-maru, Koshin-maru, Kiyomatsu-maru, Maruse-maru, Yohei-maru, Marutatsu-maru, Noboru-maru and Aoki-maru, Dai-8-Tachi-maru, Kiryo-maru). We also would like to thank K. Yanagi, H. Tachikawa, K. Tsuchiya, S. Ohta, T. Ishii, M. Nishida, the late E. Tsuchida, T. Oji, K. Kitazawa, I. Takeuchi, and N. Hirose for their help with collecting specimens. The following persons are thanked for comparison of material and fruitful discussion: B. Ruthensteiner, Zoologische Staatssammlung München, Germany; M. Grasshoff, Forschungsinstitut und Natur-museum Senckenberg, Frankfurt am Main, Germany; H. Ruhberg and P. Stiewe, Biozentrum Grindel und Zoologisches Museum, Hamburg, Germany; C. Lüter, Museum für naturkunde, Humboldt-Universität zu Berlin, Germany; L. P. van Ofwegen, Nationaal Natuurhistorisch Museum, Leiden, The Netherlands; R. van Soest, Institute for Biodiversity and Ecosystem Dynamics (Zoological Museum), University of Amsterdam, The Netherlands; the late F. M. Bayer and S. Cairns, National Museum of Natural History, Smithsonian Institution, Washington, D.C., USA; A. D. Johnston, Museum of Comparative Zoology, Harvard University, Cambridge, USA; S. Yamato, the Seto Marine Biological Laboratory, Kyoto University, Japan; R. Ueshima, The University Museum, University of Tokyo, Japan. The museum species list of the Northeast Pacific kindly provided by P. Etynor, Aquanautix, USA is also appreciated. Thanks also to Wakayama University, International Coastal Research Centre and the Department of Earth and Planetary Science of University of Tokyo, all of which provided facilities. We also thank S. McKay for revising the English of the manuscripts. The manuscript was greatly improved by the comments of L. P. van Ofwegen, S. D. Cairns, and an anonymous reviewer. A.K.M. was partly supported by the Sasakawa Scientific Research Grant from the Japan Society (No. 10-371) and by the Nippon Foundation Grant in the program HADEEP (Hadal Environmental Science/Education Program).

## Literature Cited

Balss, H. 1910. Japanische Pennatuliden. In: Doflein, F. (Ed.), Beitrage zur Naturgeschichte Ostasiens. Abh. Math.-phys. K. Bayer. Akad. Wiss., Suppl. 1: 1–106.

Bayer, F. M. 1951. Two new primnoid corals of the subfamily Calyptrophorinae (Coelenterata: Octocorallia). J. Wash. Acad. Sci. 41: 40–43.

_____. 1952. Two new species of *Arthrogorgia* (Gorgonacea: Primnoidae) from the Aleutian Islands region. Proc. Biol. Soc. Wash. 65: 63–70.

_____. 1954. Anthozoa: Alcyonaria. Fish. Gulf of Mexico; its origin, waters, and marine life. U.S. Department of the Interior, Fish and Wildlife Service. Fish. Bull. 89: 279–284

_____. 1956a. Octocorallia. Pages 167–231 in R. C. Moore, ed. Treatise on invertebrate palaeontology, part F, Coelenterata. University of Kansas Press, Lawrence.

_____. 1981. Key to the genera of Octocorallia exclusive of Pennatulacea (Coelenterata: Anthozoa), with diagnoses of new taxa. Proc. Biol. Soc. Wash. 94: 902–947.

_____. 1982. Some new and old species of the primnoid genus *Callogorgia* Gray, with a revalidation of the related genus *Fanellia* Gray (Coelenterata: Anthozoa). Proc. Biol. Soc. Wash. 95: 116–160.

_____ and M. Grasshoff. 1994. The genus group taxa f the family Ellisellidae, with clarification of the genera established by J. E. Gray (Cnidaria: Octocorallia). Senckenb. Biol. 74: 21–45.

Broch, H. 1935. Oktocorallen des nordlichsten Pazifischen Ozeans und ihre Beziehungen zur atlantischen Fauna. Avhandl. Norske Videnskaps Akad. Oslo (Matem.-Naturvid. Klasse) 1: 1–53.

Boero, F., G. Belmonte, G. Fanelli, S. Piraino, and F. Rubino. 1996. The continuity of living matter and the discontinuities of its constituents: do plankton and benthos really exist? Trends Ecol. Evol. 11: 177–180.

Cairns, S. D. and F. M. Bayer. 2005. A review of the genus *Primnoa* (Octocorallia: Gorgonacea: Primnoidae), with the description of two new species. Bull. Mar. Sci. 77: 225–256.

Collins, T. M., K. Frzer, A. R. Palmer, G. J. Vermeij, and W. M. Brown. 1996. Evolutionary history of northern hemisphere *Nucella* (Gastropoda, Muricidae): Molecular, morphological, ecological, and paleontological evidence. Evolution 50: 2287–2304.

Fabricius, K. and P. Alderslade. 2001. Soft corals and sea fans. A comprehensive guide to the tropical shallow water genera of the Central – West Pacific, the Indian Ocean and the Red Sea. (Australian Institute of Marine Sciences) Townsville, Australia, 215 p.

Fujita T. and H. Namikawa. 2003. Franz Doflein and Deep-water fauna of Sagami-Bay. Taxa 15: 1–12. (in Japanese)

Gass, S. E. and J. H. M. Willison. 2005. An assessment of the distribution of deep-sea corals in Atlantic Canada by using both scientific and local forms of knowledge. Pages 223–245 in A. Freiwald and J. M. Roberts, eds. Cold-water corals and ecosystems. Springer-Verlag, Berlin Heidelberg.

Graf, G. 1992. Benthic-pelagic coupling: a benthic view. Oceanogr. Mar. Biol. Annu. Rev. 30: 149–190.

Grasshoff, M. 1999. The shallow water gorgonians of New Caledonia and adjacent islands (Colenterata: Octocorallia). Senckenb. Biol.: 78: 1–121.

_____. 2000. The gorgonians of the Sinai coast and the Strait of Gubal, Red Sea. Cour. Forschungsinst. Senckenb. 224: 1–125.

_____ and G. Bargibant. 2001. Coral reef gorgonians of new caledonia/ Les Gorgones des recifs coralliens de nouvelle-caledonie. Paris. 336 p.

Holland, M. M. and P. G. Risser, 1991. The role of landscape boundaries in the management and restoration of changing environments – Introduction. Pages 1–7 in M. M. Holland, P. G. Risser, and R. J. Naiman, eds. Ecotones: The role of landscape boundaries in the management and restoration of changing environments. Chapman & Hall, New York.

Holm, O. 1894. Beitrage zur Kenntniss der Alcyonidengattung Spongodes Less. Zool. Jahrb., Abt. Syst. 8: 8–57.

Imahara, Y. 1996. Previously recorded octocorals from Japan and adjacent seas. Prec. Coral Octocoral Res. 4–5: 17–44.

_____. 2006. Preliminary report on the alcyonacean and pennatulacean octocorals collected by the Natural History Research of the Sagami Bay. Mem. Nat. Sci. Mus., Tokyo 40: 91–101

Isono, N. 1988. Misaki jikkenjo wo kyorai shita hito tachi - Nihon ni Okeru Doubutugaku no tanjou (The people related to Misaki Marine Biological Station – The birth of Zoology in Japan). Japan Scientific Societies Press, Tokyo, Japan. 230 p. (in Japanese).

Iwase, F. and A. K. Matsumoto. 2006. Preliminary list on gorgonian octocorals collected by the natural history research of the Sagami Bay. Mem. Nat. Sci. Mus., Tokyo 40: 79–89. (in Japanese with English abstract).

Japan Oceanographic Data Centre [Internet] Hydrographic and Oceanographic Department, Japan Coast Guard.4 October 2006. Available from: http://www.jodc.go.jp/

Kinoshita, K. 1908. Primnoidae von Japan. J. Coll. Sci. Imp. Univ. Tokyo 23: 1–74.

_____. 1909. Telestidae von Japan. Annot. Zool. Japon. 7: 113–123, pl. 3.

_____. 1913. Studien über einige Chrysogorgiiden Japans. J. Coll. Sci. Imp. Univ. Tokyo, 33: 1–47.

Kishinouye, K. 1903. Preliminary note on the Coralliidae of Japan. Zool. Anz. 26: 623–626.

Krieger, K. 2001. Coral (Primnoa) Impacted by Fishing Gear in the Gulf of Alaska. Pages 106–116 in J. H. M. Willison, J. Hall, S. E. Gass, E. L. R. Kenchington, M. Butler, and P. Doherty, eds. Proc. First Int. Symp. on Deep-sea Corals. Ecology Action Centre, Nova Scotia Museum, Halifax, Nova Scotia

Kükenthal, W. 1906. Japanische Alcyonaceen. Abh. K. Bayer. Akad. Wiss. 2, Kl. Suppl. 1(1): 1–86.

_____. 1909. Japanische Gorgoniden. II Teil. Die Familien der Plexauriden, Chrisogorgiiden und Melitodiden. Pages 1–78 in F. Doflein, ed. Beitrage zur Naturgeschichte Ostasiens. Abhandl. math.-phys. Klasse K. Bayer. Akad. Wissensch., Suppl.-Bd. 1(5).

_____ and H. Broch. 1991. Pennatulacea. Wissensch. Ergebn. deutschen Tiefsee-Expediton "Valdivia" 13(1) Lieferung 2: i-vi +113–576, pls. 13–29.

_____ and H. Gorzawsky, 1908b. Japanische Gorgoniden. I Teil. Die Familien der Primnoiden, Muricwiden und Acanthogorgiiden. In F. Doflein, ed. Beitrage zur Naturgeschichte Ostasiens. Abhandl. math.-phys. Klasse K. Bayer. Akad. Wissensch., Suppl.-Bd. 1: 1–71.

Kumano, M. 1937. Japanese Pennatulacea. Nat. Hist. Mag., 35: 246–256, pl. 11, (in Japanese).

MacIsaac, K., C. Bourbonnais, E. Kenchington, D. Gordon, Jr., and S. Gass. 2001. Observations on the occurrence and habitat preference of corals in Atlantic Canada. Pages 58–75 *in* J. H. M. Willson, J. Hall, S. E. Gass, E. L. R. Kenchington, M. Butler, and P. Doherty, eds. Proc. First Int. Symp. on Deep-sea Corals. Ecology Action Centre Nova Scotia Museum Halifax, Nova Scotia.

Marcus, N. H. and F. Boero. 1998. Minireview: the importance of benthic-pelagic coupling and the forgotten role of life cycles in coastal aquatic systems. Limnol. Oceanogr. 43: 763–768.

Matsumoto, A. K. 1997. Growth and Reproduction of *Melithaea flabellifera*. M.Sc. Thesis. University of Ochanomizu, Japan. (in Japanese)

——————. 2004. Heterogeneous and compensatory growth in *Melithaea flabellifera* (Octocorallia: Melithaeidae) in Japan. Hydrobiology 530/531: 389–397.

——————. 2005. Recent observations on the distribution of deep-sea coral communities on the Shiribeshi Seamount, Sea of Japan. Pages 345–356 *in* A. Freiwald and J. M. Roberts, eds. Cold-water corals and ecosystems. Springer-Verlag Berlin Heidelberg.

——————. 2006. Gorgonian corals as a calcium - carbonate producer in cold-waters. Ph.D. Thesis. University of Tokyo, Japan.

May, W. 1899. Beitrage zur Systematik und Chorologie der Alcyonaceen. Jena. Zeit. Naturw., 33 (N. F. 26): 1–180.

McAllister, D. E. and N. Alfonso. 2001. The distribution and conservation of deep-water corals on Canada's west coast. Pages 126–144 *in* J. H. M. Willson, J. Hall, S. E. Gass, E. L. R. Kenchington, M. Butler, and P. Doherty, eds. Proc. First Int. Symp. on Deep-sea Corals. Ecology Action Centre Nova Scotia Museum Halifax, Nova Scotia.

Moroff, T. 1902. Studien über Oktokorallien. I. Ueber die Pennatulaceen des Münchener Museums; II. Ueber einige neue Gorgonaceen aus Japan. Zool. Jahrb., Abt. Syst. 17: 363–409.

Mortensen, P. B. and L. Buhl-Mortensen. 2005. Deep-water corals and their habitats in the Gully, a submarine canyon off Atlantic Canada. Pages 247–277 *in* A. Freiwald and J. M. Roberts, eds. Cold-water corals and ecosystems. Springer-Verlag, Berlin Heidelberg.

Murcia, C. 1995. Edge effects in fragmented forests: Implications for conservation. Tree 10: 58–62.

Nutting, C. C. 1912. Description of the Alcyonaria collected by the U. S. Fish. Comm. Steamer "ALBATROSS", mainly in Japanese waters, during 1906. Proc. U. S. Nat. Mus. 43: 1–104.

Odum, E. P. 1971. Fundamentals in ecology (3th edition). Philadelphia, Saunders, 574 p.

Ofwegen, L. P. Van. 2005. A new genus of nephtheid soft corals (Octocorallia: alcyonacea: Nephtheidae) from the Indo-Pacific. Zool. Med. Leiden 79: 1–236.

Okubo, T. 1929. *Clavularia gotoi*, eine neue art von Alcyonaria aus Sagami Bai. Annot. Zool. Japon, 12: 47–58.

Raffaelli, D., E. Bell, G. Weithoff, A. Matsumoto, J. J. Cruz-Motta, P. Kershaw, R. Parker, D. Parry, and M. Jones. 2003. The ups and downs of Benthic ecology. Considerations of scale, heterogeneity and surveillance for Benthic-pelagic coupling. J. Exp. Mar. Biol. 285–286: 191–203.

Sanchez, J. A. 2005. Systematics of the bubblegum corals (Cnidaria: Octocorallia: Paragorgiidae) with description of new species from New Zealand and the Eastern Pacific. Zootaxa 1014: 1–72.

Suzuki, H. 1971. Notes on *Cornularia* (Stolonifera, Alcyonaria) found in the vicinity of the Manazuru Marine Biological Laboratory. Sci. Rept. Yokohama Nat. Univ. 2: 1–6.

Takeda, M., H. Namikawa, T. Kuramochi, H. Ono, M. Higuchi, and S. Matsumoto. 2006. Over view of the Survey Project "Study on Environmental Changes in the Sagami Sea and Adjacent Coastal Area with Time Serial Comparison of Fauna and Flora" I. Marine Organism (Brown Algae and Animals (Sponges - Annelids)). Mem. Nat. Sci. Mus., Tokyo, 40: 1–6. (in Japanese with English abstract).

Thomson, J. A. and N. I. Rennet. 1927. Report on Japanese pennatulids. J. Fac. Imp. Univ. Tokyo, Zool. 182: 115–143.

Utinomi, H. 1952. On a new deep-sea alcyonarian from Sagami Bay, *Carotalcyon sagamianum*, n. gen. et n. sp. Annot. Zool. Japon. 25: 441–446.

_____. 1955. On five stoloniferans from Sagami Bay, collected by His Majesty the Emperor of Japan. Jap. J. Zool. 11: 121–135.

_____. 1962. Preliminary list of octocorals of Sagmi Bay deposited in the Biological Laboratory of the Imperial Household. Publ. Seto Mar. Biol. Lab. 10: 105–108.

_____ and Y. Imahara. 1976. A new second species of dimorphic alcyonacean octocorals *Minabea* from the bays of Sagami and Suruga, with the emendation of generic diagnosis. Publ. Seto Mar. Biol. Lab. 23: 205–212.

Vermeij, G. J. 1990. Range limits and dispersal of molluscs in the Aleutian Islands, Alaska. The Veliger 33: 346–354.

_____. 1991. Anatomy of an invasion: the trans-Arctic interchange. Paleobiology 17: 281–307.

Watling, L. and P. J. Auster. 2005. Distribution of deep-water Alcyonacea off the Northeast Coast of the United States. Pages 279–296 *in* A. Freiwald and J. M. Roberts, eds. Cold-water corals and ecosystems. Springer-Verlag, Berlin Heidelberg.

Williams, G. 1995. Living genera of sea pens (Coelenterata: Pennatulacea): illustrated key and synopses. Zool. J. Linn. Soc. 113: 93–140.

_____. 1999. Index Pennatulacea. Annotated bibliography and indexes of the sea pens (Coelenterata: Octocorallia) of the world 1469–1999. Proc. Calif. Acad. Sci. 51: 19–103.

ADDRESSES: (A.K.M.) *Ocean Research Institute, The University of Tokyo, 1-15-1, Minamidai, Nakano-ku, Tokyo, 164-8639, Japan.* (F.I.) *Biological Institute of Kuroshio, 560-I, Nishidomari, Otsuki-cho, Kouchi, 788-0333, Japan.* (Y.I.) *Wakayama Prefectural Museum of Natural History, 370-1, Funo, Kainan-shi, Wakayama, 642-0001, Japan.* (H.N.) *Showa Memorial Institute, Tsukuba Research Center, National Science Museum of Tokyo, 4-1-1, Amakubo, Tsukuba-shi, Ibaraki, 305-0005, Japan.* CORRESPONDING AUTHOR: (A.K.M.) *E-mail: <amatsu@gorgonian.jp>.*

# Black corals (Cnidaria: Antipatharia) from Brazil: an overview

Livia de Laia Loiola

## Abstract

There are few records of black corals (Cnidaria: Antipatharia) from Brazil, where there are previous reports of only 18 species from the families Antipathidae, Myriopathidae, and Schizopathidae. Most of these records are from the deep-sea, especially from southwestern Atlantic seamounts and the Brazilian continental shelf margins. Most specimens were collected between 13° and 22°S, during a study to survey the living resources off Brazil (REVIZEE Program), carried out by the Brazilian government. This paper is an historical overview concerning the geographic and bathymetric distribution of black coral species reported off Brazil. The genus *Chrysopathes* is herein reported for the first time in the Atlantic Ocean and the family Aphanipathidae (Subfamilies Acanthopathinae and Aphanipathinae) is reported for the first time in the southwestern Atlantic.

Taxonomic studies of deep-sea black corals in Brazil were intensified during the 1990s, specifically due to material collected by the REVIZEE Program, an official project to survey the living resources of the Brazilian Economic Exclusive Zone. The Antipatharia specimens provided by this program, collected between 13° and 22°S, were deposited in the Cnidaria Collection of the Museu Nacional–Rio de Janeiro, and they have been examined during the last 6 yrs, especially those from the family Myriopathidae (see Loiola and Castro, 2005).

According to Castro et al. (2006), the greatest black coral richness off Brazil occurs in the area near the Cape of São Tomé (about 22°S), in depths between 100 and 500 m, with six species co-occurring there. This paper is an overview of the antipatharians reported off Brazil; the geographic and bathymetric distribution of the families are described and represented in tables and maps. Also, the first records of the genus *Chrysopathes* in the Atlantic Ocean, and of the family Aphanipathidae, both subfamilies Acanthopathinae and Aphanipathinae, in the southwestern Atlantic, are herein reported, and briefly described.

## Previous Records

Order Antipatharia

Family Antipathidae

The majority of black corals reported off Brazil are assigned to Antipathidae, represented by two genera: *Antipathes*, with three identified species and two different species yet to be identified at the specific level, and *Cirrhipathes*, with one identified species and

George, R. Y. and S. D. Cairns, eds. 2007. Conservation and adaptive management of seamount and deep-sea coral ecosystems. Rosenstiel School of Marine and Atmospheric Science, University of Miami.

Table 1. General information about Antipathidae species that occur off Brazil: species, localities, and references.

| Species | Localities | References |
|---|---|---|
| *Antipathes columnaris* (Duchassaing, 1870) | Northern Bahamas to Brazil (01°N); Caribbean | Opresko, 1974 |
| *Antipathes atlantica* Gray, 1857 | West Indies, Gulf of Mexico; Jamaica and Trinidad; Brazil (13°03′S–22°22′57″S) | Brook, 1889; Cairns et al., 1993; Warner, 1981; Loiola and Castro, 2001; Castro et al., 2006 |
| *Antipathes furcata* Gray, 1857 | Madeira; Caribbean; Barbados to Trinidad; Gulf of Mexico; Brazil (20°40′S–22°22′57″S) | Opresko, 1974; Cairns et al., 1993; Loiola and Castro, 2001; Castro et al., 2006 |
| *Antipathes* sp. 1 | Brazil (13°20.87′S–19°45′53″S) | Castro et al., 2006 |
| *Antipathes* sp. 2 | Brazil (18°S) | Castro, 1994 |
| *Cirrhipathes secchini* Echeverría, 2002 | Brazil (18°S) | Castro, 1994; Echeverría, 2002 |
| *Cirrhipathes* sp. 1 | Brazil (13°03′02″S) | Castro et al., 2006 |
| *Cirrhipathes* sp. 2 | Brazil (19°37.49′S–22°22′57″S) | Castro et al., 2006 |
| *Cirrhipathes* sp. 3 | Brazil (20°40′27″S) | Castro et al., 2006 |
| *Cirrhipathes* sp. 4 | Brazil (12°58.65′S–22°22′57″S) | Castro et al., 2006 |

four species unidentified. General information about records of Antipathidae species that occur in Brazil are shown in Table 1.

## Genus *Antipathes*

A single specimen of *Antipathes columnaris* was reported by Opresko (1974, see Echeverría and Castro, 1995), only from off the Amazon River mouth. *Antipathes atlantica* is the Antipathidae species reported with a wider latitudinal distribution off the eastern coast of Brazil, in depths between 50–300 m (Loiola and Castro, 2001; Castro et al., 2006). Also from this region, *Antipathes furcata* was reported at depths of 100–300 m (Loiola and Castro, 2001; Castro et al., 2006). Two unidentified *Antipathes* species were reported: one from deeper areas (66–390 m, see Castro et al., 2006), and another from shallow reef areas (up to 20 m), on the Abrolhos Bank (Castro, 1994). These two species have fan-shaped corallum, branched in a single plane. The geographic distribution of *Antipathes* species off Brazil, between 13° and 22°S, is shown in Figure 1.

## Genus *Cirrhipathes*

A species of *Cirrhipathes* was reported from shallow reef areas (10–25 m) on the Abrolhos Bank (Castro, 1994), and later described as a new species: *Cirrhipathes secchini*. Four unidentified species were reported by Castro et al. (2006) at depths of 50–706 m, off the Brazilian eastern coast. These four species differ with regard to the size and shape of their spines (C. B. Castro, Museu Nacional, Universidade Federal do Rio de Janeiro, pers. comm.). The geographic distribution of this genus off Brazil, between 13° and 22°S, is shown in Figure 2.

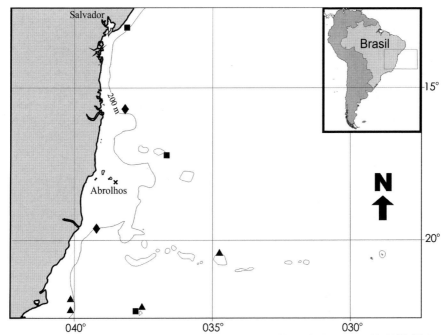

Figure 1. Distribution of Antipathidae—genus *Antipathes,* off Brazil, between 13°–22°S: ■ *Antipathes atlantica;* ▲ *Antipathes furcata;* ◆ *Antipathes* sp. 1; ✖ *Antipathes* sp. 2.

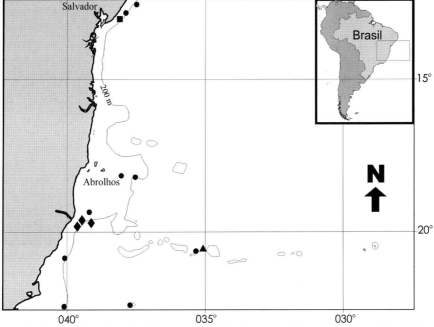

Figure 2. Distribution of Antipathidae—genus *Cirrhipathes,* off Brazil, between 13°–22° S: ■ *Cirrhipathes* sp. 1; ◆ *Cirrhipathes* sp. 2; *Cirrhipathes* sp. 3; ● *Cirrhipathes* sp. 4.

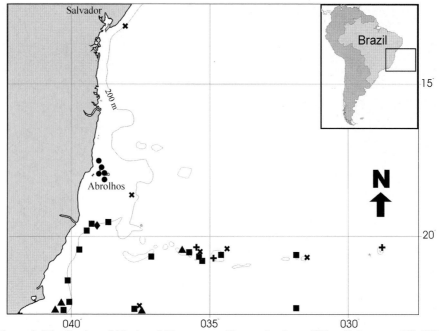

Figure 3. Distribution of Myriopathidae—genus *Tanacetipathes,* off Brazil, between 13°–22°S: ● *Tanacetipathes barbadensis* (Brook, 1889); ■ *Tanacetipathes tanacetum* (Pourtalès, 1880); ◆ *Tanacetipathes hirta* (Gray, 1857); ✖ *Tanacetipathes thamnea* (Warner, 1981); ▲ *Tanacetipathes longipinnula* n. sp.; ✚ *Tanacetipathes thalassoros* n. sp.

Family Myriopathidae

Records of Myriopathidae from the eastern coast of Brazil between 13° and 22°S (Fig. 3) include one *Plumapathes* species and six *Tanacetipathes* species (Loiola and Castro, 2005). General information about records of Myriopathidae species that occur in Brazil are shown in Table 2. *Plumapathes fernandezi* (Pourtalès, 1874) was reported by Echeverría (2002) as a unique specimen, collected from off Rio Grande do Sul State, depth unknown. *Tanacetipathes barbadensis* was reported by Loiola and Castro (2005) in shallow reefs (8–20 m) on the Abrolhos Bank. These specimens are mentioned as *Antipathes* sp. in Castro (1994), and the same specimens were erroneously identified as *Antipathes hirta* Gray, 1857 by Echeverría (2002). *Tanacetipathes hirta* (Gray, 1857) occurs at two points along the Brazilian eastern coast, between 100–417 m (Loiola and Castro, 2005; Castro et al., 2006). Pérez et al. (2005) described *Tanacetipathes paula* Pérez, Costa and Opresko, 2005 as a new species from the Archipelago of Saint Peter and Saint Paul, but this species was later considered by Loiola and Castro (2005) to be a synonym of *Tanacetipathes thamnea*. The Myriopathidae species with greater latitudinal distribution along the Brazilian eastern coast is *T. thamnea*, which occurs at depths of 53–558 m (Loiola and Castro, 2005; Castro et al., 2006).

*Tanacetipathes tanacetum* is the most abundant species along Brazilian seamounts and continental slope off the eastern coast. Opresko (1972) mentioned two records of this species (off the Parcel of Manoel Luís, and off the Atol das Rocas). Other records of *T. tanacetum* are from off southeastern Brazil, at depths ranging from 60 to 706 m (Echeverría and Castro, 1995; Loiola and Castro, 2005; Castro et al., 2006).

Table 2. General information about Myriopathidae species that occur off Brazil: species, localities, and references.

| Species | Localities | References |
|---|---|---|
| *Plumapathes fernandezi* (Pourtalès, 1874) | Eastern Pacific; Brazil (29°–34°S) | Opresko, 2001; Echeverría, 2002 |
| *Tanacetipathes barbadensis* (Brook, 1889) | Barbados; Trinidad; Brazil (18°S) | Brook, 1889; Warner, 1981; Loiola and Castro, 2005 |
| *Tanacetipathes hirta* (Gray, 1857) | Florida; Venezuela; Caribbean; Trinidad; Brazil (19°43.86′S–21°38′57″S) | Brook, 1889; Opresko, 1972 Warner, 1981; Loiola and Castro, 2005; Castro et al., 2006 |
| *Tanacetipathes tanacetum* (Pourtalès, 1880) | "Lesser Antilles"; Brazil (00°17′N; 03°50′S; 19°37.49′S–23°01′S) | Pourtalès, 1880; Brook, 1889; Opresko, 1972; Echeverría and Castro, 2005; Loiola and Castro, 2005; Castro et al., 2006 |
| *Tanacetipathes thamnea* (Warner, 1981) | Trinidad; Brazil (00°55′N; 13°06.79′S–20°40, 07′S) | Warner, 1981; Pérez el al., 2005; Loiola and Castro, 2005; Castro et al., 2006 |
| *Tanacetipathes longipinnula* Loiola and Castro, 2005 | Brazil (20°29.06′S–22°22′57″ S) | Loiola and Castro, 2005 Castro et al., 2006 |
| *Tanacetipathes thalassoros* Loiola and Castro, 2005 | Brazil (20°S–20°57′S) | Loiola and Castro, 2005 Castro et al., 2006 |

Two new species of *Tanacetipathes* were described by Loiola and Castro (2005), *Tanacetipathes longipinnula* and *Tanacetipathes thalassoros,* both from off the southeastern coast of Brazil; the first one at Almirante Saldanha Bank, from 50–300 m, and the second restricted to the Vitória Trindade seamounts chain, at depths of 50–100 m (Loiola and Castro, 2005).

Family Schizopathidae

The Schizopathidae from Brazil is represented by two specimens, collected off the Bahia State coast (Fig. 4), 2137 m, and identified as *Schizopathes affinis* (Loiola and Castro, 2001). This species represents the deepest record of a black coral off Brazil. General information about records of Schizopathidae species that occur in Brazil are shown in Table 3.

**New Records**

Family Aphanipathidae

Subfamily Acanthopathinae

*Acanthopathes humilis* (Pourtalès, 1867)

*Antipathes humilis* Pourtalès, 1867: 112; 1880: 118, pl. 3, figs. 18, 19, and 32.
*Aphanipathes humilis*: Brook, 1889: 131; Opresko, 1972: 994–999, fig. 10.
*Acanthopathes humilis*: Opresko, 2004: 230–232, fig. 12 a–b.

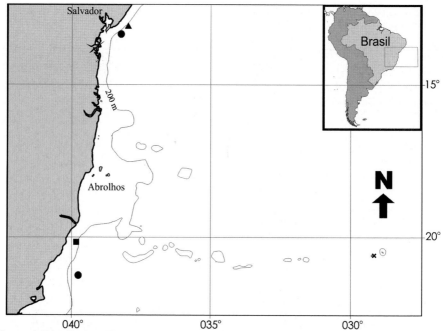

Figure 4. Distribution of Acanthopathidae, Cladopathidae, and Schizopathidae off Brazil, between 13°–22°S: ■ *Acanthopathes humilis*; ✖ *Aphanipathes* sp.; ● *Chrysopathes* sp.; ▲ *Schyzopathes affinis*.

*Material examined.*—Brazil, off Vitória, 20°10′S, 39°48′W, 315 m (MNRJ 5445: 1 colony).

*Brief description.*—Colony 8.0 cm tall, 14.0 cm wide. Stem small, 3.0 mm in diameter near the base, from where two lateral series of branches develop. Successive branches added laterally predominantly on the outer side of the lower branches, most of them being curved upwards near their point of insertion, and thus becoming parallel to the stem. Few smaller branchlets developed on the inner side of branches. Colony expands more laterally than vertically, resulting in a candelabrum-like shape. Branches up to the 14th order, lower order branches long (18–56 mm long, 0.7–1.5 mm in diameter), reaching the top of the colony; upper order branches short (4–23 mm, 0.5–0.7 mm in diameter); distance between points of insertion 2.5–20 mm (Fig. 5A). Spines anisomorphic, ornamented near the apex, cylindrical; hypostomal spines reduced in size (0.13–0.18 mm tall, 0.04–0.05 mm wide), circumpolypar spines larger (0.4–0.5 mm tall, 0.06–0.08 mm wide) than interpolypar (0.3–0.4 mm tall, 0.04–0.08 mm wide) and abpolypar spines (0.2–0.3 mm tall, 0.04–0.06 mm wide) (Fig. 5B–E). Spines on polypar side stand at 90°

Table 3. General information about Aphanipathidae, Cladopathidae, and Schizopathidae species that occur off Brazil: species, localities, and references.

| Species | Localities | References |
|---|---|---|
| *Acanthopathes humilis* (Pourtalès, 1867) | Western Atlantic; Brazil (20°10′S) | Opresko, 2004; present record |
| *Aphanipathes* sp. | Brazil (20°30′S) | Castro et al., 2006; present record |
| *Chrysopathes* sp. | Brazil (13°23′S; 22°24.655′S) | Present record |
| *Schizopathes affinis* Brook, 1889 | Cosmopolitan; Brazil (13°26.455′S) | Opresko, 1997 Loiola and Castro, 2001 Castro et al., 2006 |

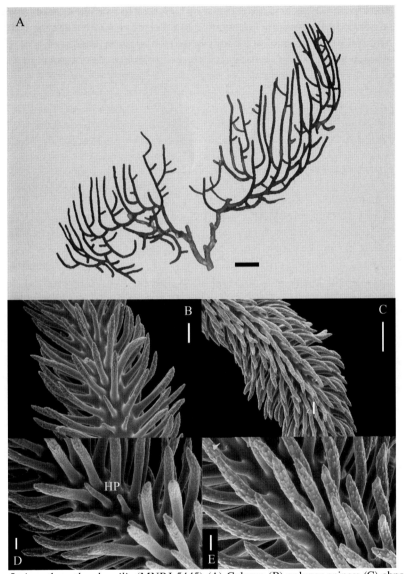

Figure 5. *Acanthopathes humilis* (MNRJ 5445) (A) Colony; (B) polypar spines; (C) abpolypar spines; (D) polypar spines of the middle region of a branch, HP–hipostomal spines; (E) abpolypar spines, closer view. Scale bars: A = 1.0 cm; B–C = 0.02 cm; D–E = 0.06 mm.

to branch or slightly curved upwards, some of them only with apexes curved, spines on abpolypar side extremely curved upwards. In thicker branches and in the stem, spines in 28–30 longitudinal rows, in narrower branches, 18–20 rows, 6–8 spines per mm in each row. Polyps badly preserved.

*Remarks*.—The Brazilian specimen greatly resembles the candelabrum variety of *A. humilis* described by Opresko (1972). There are small differences in spine size between the Brazilian and Opresko's (2004) specimens, but the planar branching pattern, the tendency of branchlets to develop on the convex side of the lower order branches, and the distinctly anisomorphic spines, characteristics of *A. humilis,* are found in the Brazilian specimen.

Aphanipathidae was recently established by Opresko (2004), and *A. humilis* was previously recorded in the western Atlantic from the Bahamas to Grenada (Opresko, 1972, 2004). This is the first record of this family in the southwestern Atlantic (Fig. 4). General information about records of Aphanipathidae species that occur in Brazil are shown in Table 3.

Subfamily Aphanipathinae

*Aphanipathes* sp.

*Aphanipathes* Brook, 1889: 121; Opresko, 2004: 212–214, fig. 1.

*Material examined.*—Brazil, off Trindade island, 20°30′S, 029°16′W, 360 m, RE-VIZEE Central V Sta. # 41 (MNRJ 4865: 1 colony).

*Brief description.*—Colony 8.0 cm tall, 11.0 cm wide, 6.0 cm thick. No distinct stem or basal plate, colony base fused to a scleractinian coral. Colony bushy, basal branches straight, nearly horizontal, distal branches 45–90° to basal ones. Branching up to the 5th order, lower order branches 30–41 mm long, 0.5–0.9 mm in diameter, upper order branches 8–25 mm long, 0.1–0.4 mm in diameter. Distance between points of insertion 2–10 mm (Fig. 6A). Spines triangular, conical, curved towards the branch apex, with acute apex; at midpoint of branches spines with big knob-like tubercles over apical half surface, at basal and distal ends of branches, spines less ornamented or smooth. No differences in size between polypar and abpolypar spines—0.10–0.14 mm tall, 0.05–0.09 mm wide at base. Spines arranged in 6–8 longitudinal rows on narrower branches, and in 8–10 rows on thicker branches, 4–5 spines per mm in each row (Fig. 6B–D). Polyps badly preserved.

*Remarks.*—This specimen was identified as *Aphanipathes* because of the bushy form of the corallum, with straight ascending branches, and the spines being conical with conical tubercles on the surface, like the four species assigned to this genus by Opresko (2004). *Aphanipathes sarothamnoides* Brook, 1889, the type species of this genus, has longer branchlets (5–10 cm) occurring farther apart from each other (1.5–2.5 cm), and taller spines (about 0.20 mm) (Brook, 1889; Opresko, 2004), than Brazilian specimen: branchlets 0.8–4.1 cm long, 0.2–1.0 cm apart, and spines 0.10–0.14 tall. *Aphanipathes salix* (Pourtalès, 1880) has branchlets 2–3 cm long, 0.3–2.0 cm apart, but the spines are taller than in Brazilian specimen—polypar 0.22 mm, and abpolypar 0.13 mm (Opresko, 1972). *Aphanipathes verticillata* Brook, 1889 differs from Brazilian specimens in that the spines are arranged in verticils (Brook, 1889; Opresko, 1972, 2004). *Aphanipathes pedata* (Gray, 1857) is branched strictly in one plane (Brook, 1889; Opresko, 1972), and according to the illustration given by Brook (1889: pl. 11, fig. 12), the spines are more extensively covered by tubercles than in the Brazilian specimen. Further studies of additional Brazilian specimens might confirm that this Brazilian specimen is a new species.

This subfamily is herein reported for the first time in the southwestern Atlantic, off Trindade Island, 360 m (Fig. 4). According to Brook (1889) and Opresko (2004), species of this genus have been reported from the West Indies (*A. pedata* and *A. salix*), south Pacific (*A. sarothamnoides*), and Indian Ocean (*A. verticillata*).

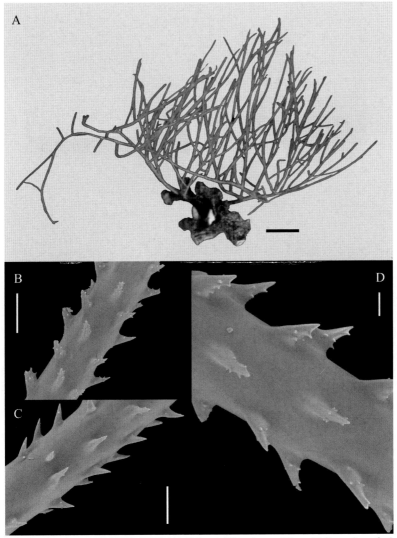

Figure 6. *Aphanipathes* sp. (MNRJ 4865) (A) Colony; (B) spines of the middle region of a branch; (C) spines of the basal region of a branch; and (D) spines, close view. Scale bars: A = 1.0 cm; B–C = 0.02 cm; D = 0.06 mm.

Family Cladopathidae

*Chrysopathes* sp.

*Chrysopathes* Opresko, 2003: 498; 515.

*Material examined.*—Brazil: off Salvador, 13°23'S, 038°37'W, 761 m, REVIZEE Bahia-2 Sta. # E0499 (MNRJ 4627: 1 colony); off Campos, 22°24.655'S, 039°55.413'W, 1130 m, BC-Sul-CENPES/UFRJ, (MNRJ 5150: 1 colony).

*Brief description.*—Colonies 10.0 (off Salvador) and 19.5 cm (off Campos) tall, 4.0 and 15.0 cm wide, respectively. Axis 0.5 and 1.0 mm in diameter near the base; basal plate, 10.5 and 4.0 mm in diameter, respectively. Branching lateral, up to the 2nd or-

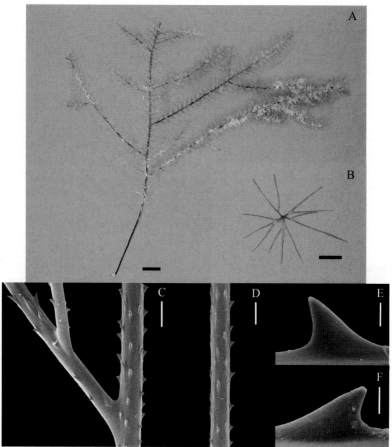

Figure 7. *Chrysopathes* sp. (MNRJ 5150) (A) Colony; (B) transversal view of a colony, cycle of pinnules; (C) spines of the basal region of a pinnule; (D) spines of the middle region of a pinnule; (E) polypar spines of the middle region of a pinnule, close view; and (F) abpolypar spines of the middle region of a pinnule, close view. Scale bars: A = 1.0 cm; B = 0.5 cm; C–D = 0.2 mm; E–F = 0.02 mm.

der, colonies fan shaped (Fig. 7A). Primary pinnules in six rows, in laterally alternating groups of three primaries each (Fig. 7B); polypar and abpolypar primaries 4.0–16.0 mm long; diameter of primaries 0.12–0.20 mm; distance between adjacent (in the same row) primaries 2.1–4.4 mm (Fig. 7A). Four to five primary pinnule cycles per cm of axis (Fig. 7B). Secondary pinnules only on the proximal half of primaries, on the abpolypar side or on a plane perpendicular to the plane defined by the primaries, either towards the distal or the proximal end of the colony, 0–2 per primary pinnules; 2–6 mm long, 0.08–0.16 mm in diameter; the secondary pinnules can occur on anterior, posterior and lateral primaries on all parts of the colony (Fig. 7B). Tertiary pinnules rarely present, only one per secondary, within a plane defined by the primaries (Fig. 7B). Spines conical, compressed, smooth; apices of polypar and abpolypar spines curved towards the distal end of the pinnules; furcated spines rarely present; six regular longitudinal rows (around the whole pinnule); polypar and abpolypar spines not differentiated in size, 0.02–0.09 mm tall, 0.03–0.07 mm wide at base; four to six spines per mm in a row; distance between adjacent spines in each row 0.14–0.34 mm (Fig. 7C–F). Polyps badly preserved.

*Remarks.*—Although the Brazilian specimens have a general corallum appearance similar to *Chrysopathes formosa* Opresko, 2003, there are some substantial differences in the subpinnulation pattern that indicate they are different species. Brazilian specimens can have single secondary pinnules, or subopposite pairs of secondaries per primary, and occasionally have tertiary pinnules; *C. formosa* has lateral primary pinnules with a single secondary, and tertiary pinnules are absent (Opresko, 2003). *Chrysopathes speciosa* Opresko, 2003 differs from Brazilian *Chrysopathes* also in the secondary pinnules pattern: up to four secondaries on some primary pinnules, with arrangements highly variable, either alternate, uniserial, subopposite, or irregular (Opresko, 2003). Further studies and comparisons with other Atlantic specimens from this family might indicate that these Brazilian specimens could be a new species, and that this genus could have a wider geographic distribution (D. M. Opresko, Oak Ridge National Laboratory, unpubl. data).

This is the first record of *Chrysopathes* in the Atlantic ocean, in two different Brazilian regions: off Salvador and off Campos (Fig. 4). Previously this genus was reported only in the eastern Pacific (Opresko, 2003). General information about records of Cladopathidae species that occur in Brazil are shown in Table 3.

## Acknowledgments

I thank the REVIZEE Program for providing the specimens, and Conselho Nacional de Desenvolvimento Científico e Tecnológico (CNPq) for grants and fellowships that aided the development of this study. Also, C. Castro (Museu Nacional, Universidade Federal do Rio de Janeiro) and D. Opresko (Oak Ridge National Laboratory) for helpful comments and suggestions throughout this study; M. Medeiros (Museu Nacional, Universidade Federal do Rio de Janeiro) for help in preparation of illustrations; and B. Vale (Departamento de Patologia, Fundação Oswaldo Cruz) for preparing the samples and images for SEM.

## Literature Cited

Brook, G. 1889. Report on the Antipatharia. Repts. Sci. Res. Voyage Challenger, Zool. 32: 1–222.

Cairns, S. D., D. M. Opresko, T. S. Hopkins, and W. W. Schroeder. 1993. New records of deepwater Cnidaria (Scleractinia and Antipatharia) from the Gulf of Mexico. Northeast Gulf Sci. 13: 1–11.

Castro, C. B. 1994. Corais do Sul da Bahia. Pages 161–176 *in* B. Hetzel and C. B. Castro, eds. Corais do Sul da Bahia. Nova Fronteira, Rio de Janeiro.

_____, D. O. Pires, M. S. Medeiros, L. L. Loiola, R. C. M. Arantes, C. M. Thiago, and E. Berman. 2006. Capítulo 4. Filo Cnidaria. Corais. Pages 147–192 *in* H. P. Lavrado and B. L. Ignácio, eds. Biodiversidade bentônica da região central da Zona Econômica Exclusiva Brasileira. Museu Nacional, Rio de Janeiro.

Duchassaing, P. 1870. Revue des Zoophytes et des Spongiaires des Antilles: 1–52, Paris.

Echeverría, C. A. 2002. Black corals (Cnidaria: Anthozoa: Antipatharia): first records and a new species from the Brazilian coast. Rev. Biol. Trop. 50: 1067–1077.

_____ and C. B. Castro. 1995. *Antipathes* (Cnidaria, Antipatharia) from southeastern Brazil. Bol. Mus. Nac., N. S., Zool. 364: 1–7.

Gray, J. E. 1857. Synopsis of the families and genera of axiferous zoophytes or barked corals. Proc. Zool. Soc. Lond. 25: 278–294.

Loiola, L. L. and C. B. Castro. 2001. Three new records of Antipatharia (Cnidaria) from Brazil, including the first record of a Schizopathidae. Bol. Mus. Nac., N. S., Zool. 455: 1–10.

_____ and _____. 2005. *Tanacetipathes* Opresko, 2001 (Cnidaria: Antipatharia: Myriopathidae) from Brazil, including two new species. Zootaxa 1081: 1–31.

Opresko, D. M. 1972. Redescriptions and reevaluation of the Antipatharians described by L. F. de Pourtalès. Bull. Mar. Sci. 22: 950–1017.

_____. 1974. A study of the classification of the Antipatharia (Coelenterata: Anthozoa) with redescriptions of eleven species. Ph.D. Diss, University of Miami, Coral Gables. 194 p.

_____. 1997. Review of the genus *Schizopathes* (Cnidaria: Antipatharia: Schizopathidae) with a description of a new species from the Indian Ocean. Proc. Biol. Soc. Wash. 110: 157–166.

_____. 2001. Revision of the Antipatharia (Cnidaria: Anthozoa). Part I. Establishment of a new family, Myriopathidae. Zool. Med. Leiden 75: 343–370.

_____. 2003. Revision of the Antipatharia (Cnidaria: Antipatharia). Part III. Cladopathidae. Zool. Med. Leiden 77: 495–536.

_____. 2004. Revision of the Antipatharia (Cnidaria: Antipatharia). Part IV. Aphanipathidae. Zool. Med. Leiden 78: 209–240.

Pérez, C. D., D. L. Costa, and D. M. Opresko. 2005. A new species of *Tanacetipathes* from Brazil, with a redescription of the type species T. tanacetum (Pourtalès) (Cnidaria, Anthozoa, Antipatharia). Zootaxa 890: 1–12.

Pourtalès, L. F. de. 1867. Contributions to the fauna of the Gulf Stream at great depths. Bull. Mus. Comp. Zool. 6: 103–120.

_____. 1874. Zoological results of the Hassler Expedition. Ill. Cat. Mus. Comp. Zool. 4: 29–54.

_____. 1880. Zoological results of the "Blake" expedition to the Caribbean Sea. Bull. Mus. Comp. Zool. 6: 113–118.

Warner, G. F. 1981. Species descriptions and ecological observations of black corals (Antipatharia) from Trinidad. Bull. Mar. Sci. 31: 147–163.

ADDRESS: *Projeto Coral Vivo, Arraial d'Ajuda Eco Parque, Estrada da Balsa, km 4,5, Arraial d'Ajuda, Porto Seguro, BA, Brazil 45816000. E-mail: <llloiola@yahoo.com.br>.*

# The azooxanthellate coral fauna of Brazil

DÉBORA DE OLIVEIRA PIRES

## Abstract

The azooxanthellate coral fauna diversity and distribution off Brazil are provided, with data both newly collected and compiled from the literature. The Brazilian coast is > 7000 km long and relatively little sampling has been conducted in the area, except during the last two decades. Knowledge of the azooxanthellate coral fauna of Brazil is fragmentory, despite recent contributions listing a total of 45 species. Here I report 56 species, including four new records, which is high compared to the 15 zooxanthellate coral species occurring in Brazil. At present, the ratio of azooxanthellate to zooxanthellate species is 4:1, contrasting with the ratio in the tropical-warm temperate western Atlantic (2:1) and the ratio worldwide (1:1). The species *Lophelia pertusa* (Linnaeus, 1758) (17–34°S) and *Solenosmilia variabilis* Duncan, 1873 (9–34°S) are the most dominant cold-water reef-building coral species in Brazil.

Pioneering studies on the azooxanthellate corals of the Brazilian coast have all been the result of foreign expeditions, such as the voyages of the HASSLER (1872), off Sergipe, Abrolhos, Bahia, and Cape Frio, Rio de Janeiro; H.M.S. CHALLENGER (1873) to St. Peter and St. Paul Rocks to Tristan da Cunha; the "U.S. Fish Commission Steamer ALBATROSS" (1887–1888) at 3–24°S off the coast; the Atlantis (1947–1948) to St. Peter and St. Paul Rocks, Rio de Janeiro; the CALYPSO (1961–1962), Ceará to Cape Frio; the OREGON (1963) off the Amazon; and the WALTHER HERWIG (1968), at 22–33°S off the coast. The material obtained by the Hassler was studied by Pourtalès (1874), who recorded the occurrence of seven species of azooxanthellate corals. Moseley, who participated in the circumnavigation expedition of the CHALLENGER (1872–1876), recorded five and seven species from the Brazilian waters in his studies of 1876 and 1881, respectively. Ridley (1881) described *Madracis brueggemanni* (Ridley, 1881) (= *Axohelia brueggemanni*) collected in Victoria Bank, off southeastern Brazil. Squires (1959) cited the occurrence of *Madrepora oculata* Linnaeus, 1758, in the north of Rio de Janeiro State, from collections of the ATLANTIS. However, Cairns (1979) did not recognize the identification of Squires as valid. Many specimens collected during the above expeditions were re-examined by Cairns (1979, 2000). Laborel, (1969, 1970) made a synthesis of the reef coral communities from the tropical coast of Brazil. He mentioned the existence of deep-sea coral reefs off São Paulo State and reported the occurrence of 14 azooxanthellate species, collected by the vessels CALYPSO, CANOPUS, and AKAROA.

By the end of the 1960s, five azooxanthellate coral species were sampled by the R/V PROF. W. BESNARD, from off Rio de Janeiro, São Paulo, and Rio Grande do Sul States (Tommasi, 1970). Leite and Tommasi (1976) described the distribution of *Cladocora debilis* Milne-Edwards and Haime, 1849 between Cape Frio (23°S) and Rio Grande do Sul State (34°35′S).

George, R. Y. and S. D. Cairns, eds. 2007. Conservation and adaptive management of seamount and deep-sea coral ecosystems. Rosenstiel School of Marine and Atmospheric Science, University of Miami.

The first extensive study on the azooxanthellate fauna from the Caribbean and adjacent waters was published by Cairns (1979), who recorded 25 species of azooxanthellate corals from Brazilian waters (> 200 m depth). This work included material collected during several expeditions, among them the American steamer ALBATROSS, the German ship WALTHER HERWIG, and the Brazilian R/V WLADIMIR BESNARD (1968–1969), and R/V ALMIRANTE SALDANHA (1972).

Fernandes and Young (1986) reported ten species of zooxanthellate and azooxanthellate corals occuring from Belmonte (Bahia) to the cape of São Tomé (Rio de Janeiro), collected by dredgings made by the Brazilian R/V ALMIRANTE CÂMARA. The French ship MARION DUFRESNE was used in 1987 to collect material from the Abrolhos Bank, in the Vitória-Trindade seamount chain, off Trindade island, and in Martim Vaz island. The data related to azooxanthellate corals (34 species/morpho-species) was provided by Zibrowius (1988).

In 1997 two studies were published that included azooxanthellate corals collected by R/V WLADIMIR BESNARD. Sumida and Pires-Vanin (1997) included seven species in their study on the composition and distribution of benthic communities off Ubatuba, São Paulo. and Pires (1997) included 15 species of azooxanthellate corals, also collected off São Paulo State, which extended the known geographical distribution of some species.

In his second monograph on azooxanthellate corals from the western Atlantic, Cairns (2000) included material collected shallower than 183 m and described a new species from off São Paulo, *Trochocyathus laboreli* Cairns, 2000. The introduction of non-indigenous corals of the genus *Tubastraea* in southeastern Brazil was reported by Castro and Pires (2001). Paula and Creed (2004) reviewed *Tubastraea coccinea* Lesson, 1829 and *Tubastraea tagusensis* Wells, 1982 from rocky shores of Ilha Grande Bay, southern Rio de Janeiro State.

Goff-Vitry et al. (2004) used Brazilian specimens of *Lophelia pertusa* (Linnaeus, 1758) and *Dendrophyllia alternata* Pourtalès, 1880, collected from the Campos basin, by Gardlines Surveys Ltd. and R/V W. BESNARD, respectively, in a molecular phylogenetic study. They found a high genetic differentiation between two samples of *L. pertusa*, one from Brazil and other from the northeast Atlantic, which indicates that these two populations have been isolated for a long time and could even represent two different species.

Three pockmarks located in Santos basin, at 700 m depth, were investigated by Sumida et al. (2004), who found four species of azooxanthellate corals—*Caryophyllia ambrosia caribbeana* Cairns, 1979, *Deltocyathus eccentricus* Cairns, 1979, *Deltocyathus italicus* (Michelotti, 1838), and *L. pertusa*. In 2005, a new genus and new species from southern Brazil was described: *Monohedotrochus capitolii* Kitahara and Cairns, 2005 collected by R/V ATLÂNTICO SUL ("Projeto Talude").

An increase in sampling occurred along the Brazilian coast during the last two decades, including governmental actions such as the project "Assessment of the Sustainable Yield of the Living Resources in the Exclusive Economic Zone–REVIZEE" (the most comprehensive biological survey ever conducted in the Brazilian continental platform and slope) and oil industry initiatives as the "Campos Basin Deep-sea Environmental Project/PETROBRAS". Among the material collected by the REVIZEE during benthos campaigns from the south, 22–34°S (Pires et al., 2004), 17 species of azooxanthellate corals were recorded. Twenty-five species were recorded in another study off eastern Brazil (13–23°S) (Castro et al., 2006b).

Here I summarize the present knowledge of the azooxanthellate coral fauna from Brazil, including some unpublished data, supply a list of taxa and indicate their latitudinal and bathymetric ranges. Results include four new records and 35 extensions of geographic and bathymetric distribution ranges.

## Methods

This synthesis was based on the review of the following literature: Laborel (1970); Cairns (1977a,b, 1979, 1982, 2000); Zibrowius (1980, 1988); Fernandes and Young (1986); Pires (1997); Sumida and Pires-Vanin (1997); Kitahara (2002); Goff-Vitry et al. (2004); Paula and Creed (2004); Sumida et al. (2004); Kitahara and Cairns (2005); and Castro et al. (2006a,b). It also includes an examination of material from the Cnidaria Collection of the Museu Nacional/Universidade Federal do Rio de Janeiro (MNRJ).

## Results and Discussion

Fifty-six species of azooxanthellate corals occur in Brazil (Table 1). Four of these species (marked with asterisks in Table 1) are new records and were deposited in MNRJ as follows:

*Caryophyllia paucipalata* Moseley, 1881. MNRJ 4451. 19°17′44″S, 037°57′13″W, 500 m, coll. R/V Astro Garoupa, Asessment of the Sustainable Yield of the Living Resources in the Exclusive Economic Zone Project–REVIZEE Central II, 22 Nov 1997.

*Flabellum* sp. MNRJ 5678. 22°16′S, 039°53′W (St. 16-1), 1059 m, coll. R/V Astro Garoupa, Campos Basin Deep-sea Environmental Project/CENPES, PETROBRAS, 22 Aug 2003.

*Flabellum* cf. *alabastrum* Moseley, 1876. MNRJ 6398. 19°50′–20°04′S, 038°21′–038°28′W, 669–686 m, coll. C. M. L. Silva, Ship Kayar I, "Programa Observadores de Bordo da Frota Arrendada", 14 Jun 2004.

*Schizocyathus fissilis* Pourtalès, 1874, MNRJ 2431. 24°42′5″S, 044°30′W, 320 m, 06 Dec 1988. MNRJ 2431. 25°16′6″S, 044°52′86″W, 258 m, 12 Dec 1988.

The coral fauna *sensu lato* from latitudes 13–23°S was evaluated by Castro et al. (2006b), who suggested that the present account is still incomplete. This conclusion is confirmed by the recent record of a new species by Kitahara and Cairns (2005), and by the new records presented herein.

The species richness presented here is high relative to the 15 species of reef-building corals occurring in Brazil. At present, the ratio of azooxanthellate to zooxanthellate species richness is approximately 4:1 (this study), contrasting with the ratio from the tropical-warm temperate western Atlantic (2:1) and the worldwide ratio (1:1) (Cairns, 2000).

Zibrowius (1988) provided a provisional list of azooxanthellate corals collected off the Brazilian coast by the cruise of the Marion Dufresne in 1987. The author provided a list of stations where corals occurred, to which additional locality data were added by Tavares (1999) for corals were collected from between 18°56′S, 037°49′W and 24°12′S, 042°16′W, and between 200–217 and 2370–2380 m depth. Zibrowius (1988) recorded four species, *Caryophyllia barbadensis* Cairns, 1979, *Concentrotheca laevigata* (Pourtalès, 1871), *Cyathoceras* sp., and *Deltocyathus agassizii* Pourtalès, 1867, which were not included among the 56 species in Table I because of the preliminary status of Zibrowius' identifications. The only data from Zibrowius (1988) used herein was the record of *Flabellum apertum* Moseley, 1876, for which the author provided a specific location.

Table 1. List of species of azooxanthellate corals from off Brazil, with their minimum and maximum latitude in Brazil and bathymetric ranges, and number of records (lots). Synthesis based on literature [Pourtalès (1874); Laborel (1970); Cairns (1977a,b, 1979, 2000); Fernandes and Young (1986); Zibrowius (1980, 1988); Pires (1997); Sumida and Pires-Vanin (1997); Kitahara (2002); Goff-Vitry et al. (2004); Paula and Creed (2004); Sumida et al. (2004); Kitahara and Cairns (2005); Castro et al. (2006a,b)] and on material from the Cnidaria Collection of the Museu Nacional/UFRJ. * marked in species, latitudes, or depths denotes new records and geographic or bathymetric extensions, respectively.

| Species | Min. Latitude | Max. Latitude | Depth Min. | Depth Max. | N. records |
|---|---|---|---|---|---|
| *Madracis asperula* Milne Edwards and Haime, 1849 | 03°20'S | 22°40'S | 24 | 110 | 11 |
| *Madracis brueggemanni* (Ridley, 1881) | 01°24'S | 22°00'S | 45 | 110 | 16 |
| *Madracis pharensis* forma *pharensis* (Heller, 1868) | 03°20'S | 21°48'S | – | – | 2 |
| *Fungiacyathus symmetricus* (Pourtalès, 1871) | 09°01'S | 27°38'S | 46 | 250 | 4 |
| *Fungiacyathus crispus* (Pourtalès, 1871) | off the Amazon | off the Amazon | – | – | 1 |
| *Bathelia candida* Moseley, 1881 | off Rio Grande do Sul | off Rio Grande do Sul | – | – | 1 |
| *Madrepora oculata* Linnaeus, 1758 | 09°01'S | 30°03'S | 370 | 759* | 10 |
| *Astrangia solitaria* (Lesueur, 1817) | Fernando de Noronha | southern Bahia | 0 | – | many |
| *Astrangia rathbuni* Vaughan, 1906 | southern Bahia | southern Brazil | 0 | 90 | many |
| *Caryophyllia ambrosia caribbeana* Cairns, 1979 | 13°22'S | 32°50'S | 274* | 1,326* | 23 |
| *Caryophyllia berteriana* Duchassaing, 1850 | 20°40'S | 31°20'S | 250 | 800* | 9 |
| *Caryophyllia paucipalata** Moseley, 1881 | 19°17'S* | 19°17'S* | 500* | 500* | 1* |
| *Caryophyllia scobinosa* Alcock, 1902 | off Rio Grande do Sul | off Rio Grande do Sul | – | – | 1 |
| *Cladocora debilis* Milne-Edwards and Haime, 1849 | 19°43'S* | 34°25'S | 46 | 438 | 73 |
| *Premocyathus cornuformis* (Pourtalès, 1868) | 09°01'S | 24°35'S | 46 | 600 | 2 |
| *Coenocyathus parvulus* (Cairns, 1979) | Cumuruxatiba, Bahia | 24°20'S | 50.6 | 130 | 3 |
| *Monohedotrochus capitolii* Kitahara and Cairns, 2005 | 28°43'S | 33°02'S | 150 | 460 | 8 |
| *Trochocyathus rawsonii* Pourtalès, 1874 | 00°18'S | 0°18'S | – | – | 1 |
| *Trochocyathus laboreli* Cairns, 2000 | 15°54'S | 33°45'S | 125 | 390 | 16 |
| *Paracyathus pulchellus* (Philippi, 1842) | 00°18'N | 33°37'S | 223 | 310 | 3 |
| *Deltocyathus calcar* Pourtalès, 1874 | 10°44'S | 25°53'S | 91 | 540 | 26 |
| *Deltocyathus italicus* (Michelotti, 1838) | 03°20'S | 29°29'S* | 500 | 2,050 | 25 |
| *Deltocyathus eccentricus* Cairns, 1979 | 15°35'S | 29°29'S* | 247 | 700 | 11 |
| *Deltocyathus halianthus* (Lindström, 1877) | 23°22'S | 24°14'S | 46 | 55 | 3 |
| *Deltocyathus moseleyi* Cairns, 1979 | 20°40'S | 20°40'S | 500 | 500 | 1 |

Table 1. Continued.

| Species | Min. Latitude | Max. Latitude | Depth Min. | Depth Max. | N. records |
|---|---|---|---|---|---|
| Deltocyathus pourtalesi Cairns, 1979 | 24°09′S | 24°09′ | 600 | 600 | 1 |
| Stephanocyathus diadema (Moseley, 1876) | 08°37′S | 25°53′S* | 1,234 | 2,212* | 6 |
| Stephanocyathus paliferus Cairns, 1977 | 0°18′N | 22°24′S* | 274 | 1,649* | 2 |
| Desmophyllum dianthus (Esper, 1794) | 22°13′* | 28°35′S* | 768* | 1,127* | 5 |
| Lophelia pertusa (Linnaeus, 1758) | 17°25′S | 34°50′S* | 272* | 1,152* | 33 |
| Dasmosmilia lymani (Pourtalès, 1871) | 03°20′S | 27°58′S | 86* | 320 | 5 |
| Dasmosmilia variegata (Pourtalès, 1871) | off Maranhão | 24°25′S | 180* | 320 | 5 |
| Solenosmilia variabilis Duncan, 1873 | 03°20′S | 34°33′S* | 46 | 1,157* | 30 |
| Phyllangia americana americana Milne Edwards and Haime, 1849 | northern Brazil | São Paulo | 0 | 180* | many |
| Rhizosmilia maculata (Pourtalès, 1874) | 03°20′S | 20°51′S | 8 | 65.2 | 12 |
| Phacelocyathus flos (Pourtalès, 1878) | 09°01′S | 09°01′S | 560 | 560 | 1 |
| Anomocora fecunda (Pourtalès, 1871) | 0°18′S | St. Peter and St. Paul rocks | 182 | 182 | 2 |
| Pourtalosmilia conferta Cairns, 1978 | St. Peter and St. Paul rocks | 25°05′S | – | – | 2 |
| Deltocyathoides stimpsonii (Pourtalès, 1871) | off Maranhão | off Maranhão | – | – | 1 |
| Sphenotrochus auritus Pourtalès, 1874 | 01°12′S | 34°35′S | 15 | 82* | 23 |
| Flabellum floridanum Cairns, 1991 | off São Paulo | off São Paulo | – | – | 1 |
| Flabellum apertum Moseley, 1876 | 18°58′S | 34°23′S | 400 | 900 | 5 |
| Flabellum cf. alabastrum* Moseley, 1876 | 19°50′S* | 20°04′S* | 666* | 686* | 2* |
| Flabellum sp. | 22°16′S* | 22°16′S* | 1,059* | 1,059* | 1* |
| Placotrochides frustum Cairns, 1979 | 03°22′S | 03°22′S | 763 | 763 | 1 |
| Javania cailleti (Duchassaing and Michelotti, 1864) | 17°04′S* | 33°42′S | 107* | 250* | 8 |
| Polymyces fragilis (Pourtalès, 1868) | 22°39′S* | 32°24′S | 180 | 650* | 7 |
| Schizocyathus fissilis* Pourtalès, 1874 | 24°42′S* | 25°16′S* | 258* | 320* | 2* |
| Stenocyathus vermiformis (Pourtalès, 1868) | St. Peter and St. Paul rocks | 24°36′S* | 180* | 650* | 9 |
| Balanophyllia dineta Cairns, 1977 | 04°27′N | 04°27′N | 116 | 116 | 1 |
| Dendrophyllia alternata Pourtalès, 1880 | 23°49′S* | 30°02′S | 277 | 530 | 9 |
| Rhizopsammia goesi (Lindström, 1877) | 17°00′S | 17°00′S | 18 | 18 | 1 |
| Cladopsammia manuelensis (Chevalier, 1966) | 27°51′S | 33°42′S | 78 | 320 | 16 |
| Enallopsammia rostrata (Pourtalès, 1878) | 20°28′S* | 30°03′S | 270* | 1,332* | 14 |
| Tubastraea coccinea Lesson, 1829 | 22°58′S | 23°20′S | 0 | 5 | many |
| Tubastraea tagusensis Wells, 1982 | 22°58′S | 23°20′S | 0 | 5 | many |
| Total number of species: 56 | | | | | |

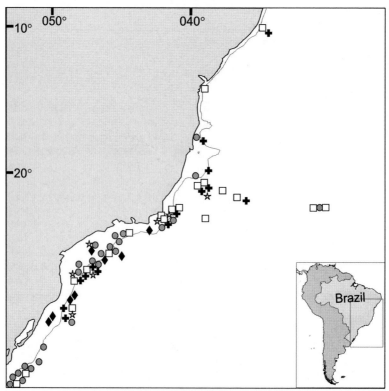

Figure 1. Records of the primary deep-sea scleractinian reef-building species along the eastern Brazilian coast. Circles = *Lophelia pertusa*; squares = *Solenosmilia variabilis*; stars = *Enallopsammia rostrata*; + marks = *Madrepora oculata*; lozenges = *Dendrophyllia alternata*.

Examination of the records of deep-sea reef-building coral species along the eastern Brazilian coast (Fig. 1) revealed extensive and almost continuous distribution ranges. Results indicate that *L. pertusa* and *Solenosmilia variabilis* Duncan, 1873 are the two major cold-water reef builders in Brazil.

## Acknowledgments

I thank C. Castro and M. Medeiros ("Museu Nacional/Universidade Federal do Rio de Janeiro") for suggestions that improved the manuscript and for helping with the figure, respectively. Thanks to S. Cairns (Smithsonian Institution) and B. Hoeksema (Nationaal Natuurhistorisch Museum Naturalis) who made valuable comments on the manuscript. Thanks to the Asessment of the Sustainable Yield of the Living Resources in the Exclusive Economic Zone Project–REVIZEE, Campos Basin Deep-sea Environmental Project/CENPES, PETROBRAS and P. Pezzuto ("Universidade do Vale do Itajaí", "Programa Observadores de Bordo da Frota Arrendada") for collecting and donating material to the Museu Nacional. I thank "Conselho Nacional de Desenvolvimento Científico e Tecnológico (CNPq)" for funding. Thanks to George Institution for Biodiversity and Sustainability (GIBS) and "Coordenação de Aperfeiçoamento de Pessoal de Nível Superior (CAPES)" for travel grants to attend the 3ISDSC.

# Literature Cited

Cairns, S. D. 1977a. A revision of the Recent species of *Stephanocyathus* (Anthozoa: Scleractinia) in the western Atlantic, with descriptions of two new species. Bull. Mar. Sci. 27: 729–739.

_____. 1977b. Stony corals. I. Caryophylliina and Dendrophylliina (Anthozoa: Scleractinia). Mem. Hourglass Cruises 3: 27 p.

_____. 1979. The deep-water Scleractinia of the Caribbean Sea and adjacent waters. Stud. Fauna Curaçao 57: 1–341.

_____. 1982. Antarctic and subantarctic scleractinia. Antarctic Res. Ser. 34(1): 74 p.

_____. 2000. A revision of the shallow-water azooxanthellate Scleractinia of the Western Atlantic. Stud. Fauna Curaçao 75: 1–240.

Castro, C. B. and D. O. Pires. 2001. Brazilian coral reefs: what we already know and what is still missing. Bull. Mar. Sci. 69: 357–371.

_____, B. Segal, D. O. Pires, and M. S. Medeiros 2006a. Distribution and diversity of coral communities in the Abrolhos Reef Complex, Brazil. Pages 19–39 *in* G. F. Dutra, G. Allen, T. Werner, and S. A. McKenna, eds. A rapid marine biodiversity assessment of the Abrolhos Bank, Bahia, Brazil. RAP Bulletin of Biological Assessment 38.

_____, D. O. Pires, M. S. Medeiros, L. L. Loiola, R. C. M. Arantes, C. M. Thiago, and E. Berman. 2006b. Cnidaria: Corais. Pages 147–192 *in* H. P. Lavrado and B. L. Ignácio, eds. Biodiversidade bêntica da costa central brasileira. (Série Livros n. 18). Museu Nacional, Rio de Janeiro.

Fernandes, A. C. S. and P. S. Young. 1986. Corais coletados durante a "Operação Geomar X" em junho de 1978 (Coelenterata, Anthozoa, Scleractinia). Publções avulsas Mus. nac. 66: 23–31.

Goff-Vitry, M. C., A. D. Rogers, and D. Baglow. 2004. A deep-sea slant on the molecular phylogeny of the Scleractinia. Mol. Phylogen. Evol. 30: 167–177.

Kitahara, M. V. unpubl. data. Sistemática dos corais de profundidade do sul do Brasil. Universidade do Vale do Itajaí, 134 p.

_____ and S. D. Cairns. 2005. *Monohedotrochus capitolii*, a new genus and species of solitary azooxanthellate coral (Scleractinia, Caryophylliidae) from southern Brazil. Zool. Med. Leiden 79-2: 117–123.

Laborel, J. 1969. Madreporaires et hydrocorallaiaires récifaux des côtes Brasiliennes. Systématic, écologie, répartition verticale et geographique. Ann. Inst. Océanogr., Paris 47: 171–229.

_____. 1970. Les peuplements de madréporaires des cotes tropicales du Brésil. Ann. Univ. Abidjan (E) 2(3): 1–260.

Leite, C. F. and L. R. Tommasi. 1976. Distribuição de *Cladocora debilis* Meth, 1849 (Faviidae, Anthozoa, Cnidaria), ao sul de Cabo Frio (23°S). Bolm Inst. Oceanogr., S. Paulo 25: 101–112.

Moseley, H. N. 1876. Preliminary report to Professor Wyville Thomson, F. R. S., Director of the Civilian Scientific Staff, on the true corals dredged by H. M. S. "CHALLENGER" in deep water between the dates Dec. 30th; 1870, and August 31st, 1875. Proc. R. Soc. Lond. 24: 544–569.

_____. 1881. Report on certain Hydroid, Alcyonarian, and Madreporarian corals procured during the voyage of the H. M. S. "CHALLENGER", in the years 1873–1876. Part 3. On the deep-sea Madreporaria. Rep. Sci. Res. Challenger (Zool.) 2: 127–208.

Paula, A. F. P. and J. C. Creed. 2004. Two species of the coral *Tubastraea* (Cnidaria, Scleractinia) in Brazil: a case of accidental introduction. Bull. Mar. Sci. 74: 175–183.

Pires, D. O. 1997. Cnidae of Scleractinia. Proc. Biol. Soc. Wash. 110: 167–185.

_____, C. B. Castro, M. S. Medeiros, and C. M. Thiago. 2004. Anthozoa. Pages 71–73 *in* A. C. Z. Amaral and C. L. B Rossi-Wongtschowski, eds. Biodiversidade Bentônica da Região Sudeste-Sul do Brasil–Plataforma Externa e Talude Superior. Série Documentos REVIZEE: Score Sul. 216 pp, Instituto Oceanográfico, USP, São Paulo.

Pourtalès, L. F. 1874. Zoological results of the "Hassler" expedition. Deep-sea corals. Mem. Mus. Comp. Zool. Harv. 4: 33–50.

Ridley, S. O. 1881. Account of the zoological collections made during the survey of H. M. S. "ALERT" in the Straits of Magellan and on the coast of Patagonia. X. Coelenterata. Proc. Zool. Soc. Lond. 1881: 101–107.

Squires, D. F. 1959. Deep sea corals collected by the Lamont Geological Observatory. 1 Atlantic corals. Am. Mus. Novit. 1965: 1–42.

Sumida, P. Y. G. and A. M. S. Pires-Vanin. 1997. Benthic associations of the shelfbreak and upper slope off Ubatuba-SP, Southeastern Brazil. Estuar. Coast. Shelf Sci. 44: 779–784.

_____, M. Y. Yoshinaga, L. A. S. Madureira, and M. Hovland. 2004. Seabed pockmarks associated with deepwater corals off SE Brazilian continental slope, Santos Basin. Mar. Geol. 207: 159–167.

Tavares, M. 1999. The cruise of the Marion Dufresne off the Brazilian coast: account of the scientific results and list of stations. Zoosystema 21: 597–605.

Tommasi, L. R. 1970. Nota sobre os fundos detríticos do circalitoral inferior da plataforma continental brasileira ao sul do Cabo Frio (RJ). Bolm Inst. Oceanogr., S. Paulo 18: 55–62.

Zibrowius, H. 1980. Les scléractiniaires de la Mediterranée et de l'Atlantique nord-oriental. Mém. Inst. Océanogr. 11: 1–284.

_____. 1988. Les coraux Stylasteridae et Scleractinia. Pages 132–136 in A. Guille and J. M. Ramos, eds. Les rapports des campagnes à la mer MD55/Brésil à bord du "MARION DUFRESNE" 6 may–2 june, 1987. Terres Australes et Antarctiques Françaises.

ADDRESS: *Museu Nacional, Universidade Federal do Rio de Janeiro, Departamento de Invertebrados, Quinta da Boa Vista, s/n°, Rio de Janeiro, RJ, 20940-040, Brazil. E-mail: <debora. pires@coralvivo.org.br>.*

# Azooxanthellate *Madracis* coral communities off San Bernardo and Rosario Islands (Colombian Caribbean)

Nadiezhda Santodomingo, Javier Reyes, Adriana Gracia, Andrés Martínez, Germán Ojeda, and Carolina García

## Abstract

Azooxanthellate habitat-forming corals develop in deep waters adjacent to shallow fringing coral reefs off San Bernardo and Rosario Islands (Colombian Caribbean). This study was carried out to characterize biological and geological features of the continental margin where these azooxanthellate coral communities flourish. The principal habitat-forming corals species found were *Madracis myriaster* (Milne-Edwards and Haime, 1849) and other branching *Madracis* species. These communities rest on sandy mud bottoms over the shelf break, in depths ranging from 120–180 m. *Madrepora* sp., antipatharians, and gorgonians were collected directly attached to adjacent limestone hardgrounds. The azooxanthellate coral habitats were found on areas of irregular topography (channels, small mounds) and nearby sites with evidence of benthic mud-gas seepage from beneath the seafloor. Irregular topography and gas seeps might be important factors contributing to the settlement and accumulation of coral communities, but the mechanisms involved are not fully understood. Questions remain pertaining to the possible linkage between shallow- and deep-water corals in the Caribbean region.

Azooxanthellate coral reef ecosystems are receiving increasing attention worldwide because they are known to be important biodiversity hotspots of high biological and socioeconomic value (Wilson, 2001; Freiwald et al., 2004). Studies on the taxonomy and distribution of azooxanthellate corals in the Tropical Western Atlantic (TWA) have been carried out by Cairns (1979, 2000) and the inventories for this region are well established. According to Cairns (1979, 2000), the azooxanthellate corals *Lophelia pertusa* (Linnaeus, 1758) and *Madrepora oculata*, identified as the main deep-water reef-building species (Freiwald et al., 2002; Hovland et al., 2002), are widespread throughout TWA. The majority of those records are from Florida, the Bahamas, and to a lesser extent, from locations off the Caribbean coast of Central and South America (Cairns, 1979, 2000; Freiwald et al., 2004). *Lophelia pertusa*, which has been known to exist in the Western Atlantic from Nova Scotia to Brazil at depths of 146–1200 m, has been recently found in the Gulf of Mexico associated with hydrocarbon and other fluid seepages (Schroeder, 2002; Schroeder et al., 2005). *Madrepora oculata* is known to be present from Georgia (USA) to Brazil in depths from 144 to 391 m (Cairns, 1979). *Enallopsammia profunda* (Pourtalès, 1867), a Western Atlantic endemic species is known as a common reef-building coral from South Carolina to southwestern Florida slope, ranging from 490 to 900 m depth (Reed et al., 2006). Also, the apozooxanthelate coral *Oculina varicosa* Lesueur, 1820 build coral reefs in depths from 70 to 100 m off Florida coast (Reed, 2002a,b).

George, R. Y. and S. D. Cairns, eds. 2007. Conservation and adaptive management of seamount and deep-sea coral ecosystems. Rosenstiel School of Marine and Atmospheric Science, University of Miami.

Azooxanthellate corals in the Colombian Caribbean were found in 1998 during three exploratory INVEMAR-Macrofauna expedition surveys from 20 to 500 m depth. Inventories have since increased from 40 species (Reyes, 2000) to 65 species. Twenty-two of these are first records for Colombia (Lattig and Reyes, 2001; Reyes and Santodomingo, 2005), one of them is a new species (Lattig and Cairns, 2000), and two are possibly new species. After the INVEMAR-Macrofauna surveys, only *L. pertusa* fossil fragments have been collected off northern Colombia's Guajira Peninsula, whereas live *M. oculata* has been recorded off the central western Colombian coast (Cairns, 1979; Reyes et al., 2005); however, to date, neither *Madrepora* nor *Lophelia* coral mounds have been recorded in the Colombian Caribbean region. Instead, highly diverse azooxanthellate coral communities formed by *Madracis myriaster*, *Eguchipsammia* spp., and *Anomocora fecunda* were discovered on the shelf break (Reyes et al., 2005). Such findings suggested the possibility that these species were involved in a coral thicket development similar to those described for *Madracis* and *Dendrophyllia* fossil communities in the Niger Delta in the eastern Atlantic (Allen and Wells, 1962); or even a coral mound development similar to those described for *Lophelia* and *Madrepora* from high latitudes (Freiwald et al., 2002).

During 2005, a new expedition to the area of San Bernardo–Rosario Islands (MARCORAL) was undertaken with the purpose of (1) confirming the presence of azooxanthellate coral communities; (2) characterizing the bathymetry and seafloor relief on which these communities develop; (3) sampling the substrate around known habitats; and (4) imaging the shallow sub-bottom to understand possible controls of the underlying geology. Here we report preliminary results of this expedition.

## Methods

The "MARCORAL" expedition was carried out on board R/V ANCON (April 22–May 2, 2005). The sampling area was designed to extend beyond the sites of previous *Madracis* records off San Bernardo–Rosario Islands during the "MACROFAUNA" projects (Fig. 1); therefore, the area was divided in two sectors: sector A, 14 × 4 km (9°51.089′N, 76°11.137′W; 9°44.218′N, 76°15.052′W) and sector B, 4 × 4 km (9°54.013′N, 76°11.332′W; 9°51.706′N, 76°9.085′W) (Fig. 1).

Bathymetry, bottom and sub-bottom reflectivity data were acquired by using a single-beam echosounder (Knudsen 320 BR; 12-200 kHz frequencies) at a maximum surveying speed of 5 kt. In the Sector A, the 17 echosounder lines were acquired along the NE-SW direction, each of which was 14 km long and spaced at 250 m (Fig. 2). In addition, eight lines perpendicular to the principal ones (4 km long and spaced 2 km) were acquired. In Sector B, 17 N-S tracklines (4 km length) and five perpendicular tracklines were collected. In total, 406 km of tracklines were surveyed covering an area of approximately 72 km². Depth soundings were processed and gridded using commercial software to obtain bathymetric maps and Digital Elevation Models (DEM).

Seafloor sampling was undertaken using a Van Veen grab (60 l, 0.03 m²), a heavy-chained rocky dredge (1 × 0.4 m opening; 1.5 kt for 5 min), and an epibenthic trawl net (9 × 1 m opening; 3 kt for 10 min). Sampling stations are shown in Figure 2. Sediment facies were characterized through standard grain size analysis and determination of calcium carbonate content (%CaCO₃). Our information was further integrated with ancillary oceanographic and geologic data (CIOH, 1985; Ingeominas, 1997) using commercial Geographic Information System software. Rock samples were rinsed, labeled, and dry-stored. Biological samples were sorted by major taxa and preserved in 70%–95% ethanol. Taxonomic identifications were obtained at the lowest level possible. Voucher specimens were archived at the Museo de Historia Natural Marina de Colombia (MHNMC) in Santa Marta, Colombia.

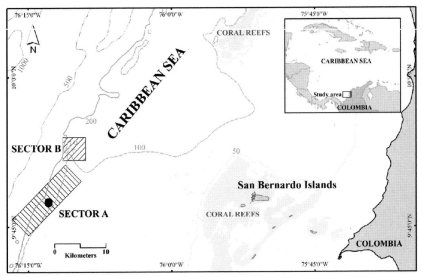

Figure 1. Study area off of the coral reef complex at San Bernardo Islands. Areas inside rectangles were characterized during MARCORAL expedition (Sectors A and B). Black dot indicates the previous *Madracis* spp. record of MACROFAUNA expeditions.

## Results

Coral specimens were collected in all of the 40 effective grabs, one epibenthic trawl, and four rocky dredge samples. In total, 40 azooxanthellate corals were identified to species level and another three to genus level (Appendix 1). Other major macrobenthic invertebrates were collected including molluscs, bryozoans, echinoderms, and crustaceans, but their taxonomic identification are still in progress and will be treated separately.

*Madracis myriaster* was the most abundant coral species (Fig. 3A–B); *Madracis asperula* and *Madracis brueggemanni* were commonly found among *M. myriaster* fragments. Due to difficulties in sorting, all specimens belonging to these three species were identified as azooxanthellate (non-symbiotic) branching *Madracis* spp.; the absence of symbiotic algae was established by the inspection of some pieces of tissue under light microscopy, verifying the classification of Cairns et al. (1999) and Best (2001).

Dead *Helioseris cucullata* (Ellis and Solander, 1786) coral fragments were usually collected among these samples (Fig. 3C). Five other colonial branching Scleractinia were abundant: *Eguchipsammia cornucopia*, *Anomocora fecunda*, *Coenosmillia arbuscula*, *Thalamophyllia riisei*, and *M. oculata*. In addition, the solitary corals *Javania cailleti*, *Caryophyllia berteriana*, and *Oxysmilia rotundifolia* were collected (Fig. 3D). Their abundance and corallum growth pattern suggest that the latter eight azooxanthellate coral species and *Madracis* spp. could also be considered habitat-forming species.

Two different seabed types were associated with habitat-forming species, sandy mud with branching *Madracis* communities; and hardgrounds (mainly limestone outcrops) on which octocorals, antipatharians, and *Madrepora* were found to settle. Additionally, sponge communities were found growing associated with rich *Halimeda* debris sediments on the continental shelf.

**Branching *Madracis* communities.**—Sampling grabs 3, 12, and 28 in sector A contained a significant abundance of azooxanthellate coral habitat-forming species, repre-

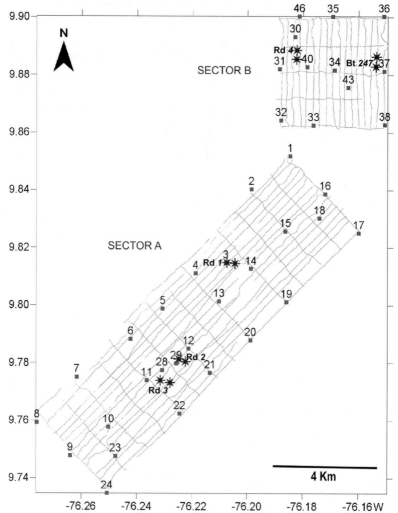

Figure 2. Map of the position of profiling grids and seabed sampling places carried out in MAR-CORAL expedition. Dashed lines = sectors A and B echosounder tracklines; black squares with station number (1–46) = van Veen grab stations; black stars (Bt 247) = epibenthic trawling station; black stars (Rd1–4) = heavy rock dredge.

senting about 65%–75% of the dry weight from the 2 mm sieved fraction. Corals were also abundant in heavy chain rock dredge (Rd) samples Rd2 and Rd3 (Fig. 2), which displayed a mixture of clays and some hardgrounds (limestone) fragments. Branching *Madracis* spp. were the dominant species at these sites; other scleractinian corals such as *A. fecunda*, *C. arbuscula*, *T. riisei*, *E. cornucopia*, and *J. cailleti* were also common, but altogether were less frequent than *Madracis*. Solitary polyps of *J. cailleti* were often found to settle on skeletons of the colonial species mentioned above. Samples collected at these stations were an assemblage of living corals (with polyps), recently dead corals (white skeletons), bioeroded or coral skeletons exposed to seabed erosion (Fe-Mn mineralization), and fossil coral debris.

*Madracis* coral communities occurred on the shelf-break, ranging in depth of 120–180 m depth (Fig. 4), where the seafloor was identified as a lithobioclastic sandy mud facies

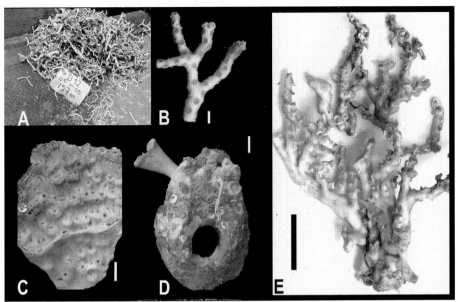

Figure 3. Coral species collected off San Bernardo Islands; (A) sample of *Madracis* spp. community, Sector A, grab 12 (157 m depth); (B) *Madracis myriaster* fragment (grab 12), scale bar 5 mm; (C) recently dead *Helioseris cucullata* skeleton, Sector A, rocky dredge 2 (151–123 m depth), scale bar 1 cm; (D) *Oxysmilia rotundifolia* attached to chimney-shaped mudstone, Sector A, rocky dredge 3 (154–117 m depth), scale bar 1 cm; (E) *Madrepora oculata*, Sector A, rocky dredge 1 (147–108 m depth), scale bar 10 cm.

(25%–45% sand; 26%–36% carbonates); water temperature in this range of depth is estimated from 18 to 21 °C (*sensu* World Ocean Database 2001 from NODC, 2001). Echosounder profiles showed that seabed morphology of the shelf break where these coral samples were taken is irregular and characterized by low mounds (10–20 m high above background depth) with variable dimensions (50–100 m wide). Channels were imaged across the continental shelf break (110 m depth) and into the continental slope beyond 300 m depth (Fig. 4). Coral thickets may correspond to a diffuse reflector signal on echosounder data as showed in the profile (grab D12, Fig. 5), and were often located on channel flanks. Such seabed morphology could be derived from sediment gravity flows, which are common turbidity deposits at those beds, or might be an artifact of the bathymetry data interpolation due to sparse bathymetric sampling (250 m spaced tracklines).

Dead *Agaricia* and *Helioseris* (Fig. 3) coral skeletons were commonly found among the samples. These genera are zooxanthellate coral dwellers of fore-back reef and deep-reef slope. Their well preserved skeletons suggest either in situ deposition or a short transport distance; some specimens displayed still white skeletons, suggesting that corals had died recently.

**Hard-bottom communities.**—Hard-ground outcrops provided settling substrate to several anthozoans, e.g., antipatharians of the genus *Stichopathes*, *Antipathes*, and *Tanacetipathes*, and some octocorals such as *Swiftia* sp., *Scleracis* sp., and *Verrucella* sp., among others. Moreover, scleractinians such as *Polycyathus mayae* and *Madracis* recruits were found inside rock crevices, while *O. rotundifolia*, *J. cailleti*, and *C. berteriana* settled on rock surfaces. A colony of living *M. oculata* (Fig. 3E) was gathered

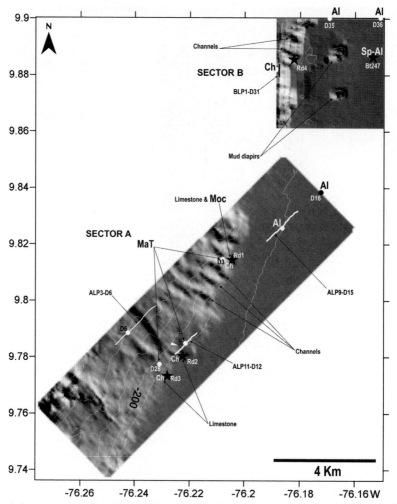

Figure 4. Sectors A and B shaded relief model based on bathymetric grid (Gaussian model, 130 m variogram lag distance) indicating relevant seabed features. Continental shelf is relatively flat, except at central Sector B where two mud diapirs arose from 110 to 80 m depth, inside a bowl-shape depression. Continental break (approximately at 110 m depth) following N-S direction in Sector B and NE-SW direction in Sector A. Continental slope inclination decreased from North to South, and the number of small gravity flow sediments channels-ridges (aligned mounds?) also decreased. Dots indicate grab stations (e.g., D16: grab sample number); light and dark lines indicate echosounder tracklines (e.g., ALP9) and the nearest grab station (e.g., D15). Stars indicate benthic trawls (Bt247) or heavy rock dredge stations (Rd1–4). Light arrow indicates previous *Madracis* spp. records. *Madracis* thicket (MaT); calcareous algae records (Al); sponge-algae community (Sp–Al); *Madrepora oculata* (Moc); columnar limestones pipe-shaped records (Ch).

in dredge 1. These hard substrates were also colonized by other invertebrates such as serpulids, vermetids, brachiopods, and some solitary caryophyllid corals. The occurrence of invertebrates and living calcareous red algae on rock surfaces suggested that hardgrounds remain as exposed croppings on the seabed for long periods of time. All rocky dredge samples were collected near sites where branching *Madracis* corals were obtained (Fig. 2).

Hardgrounds were classified as four types: (1) rocks composed of coarse sand grains in a carbonate matrix (Fig. 6A); (2) reef coral limestones with evidence of karstic dissolution

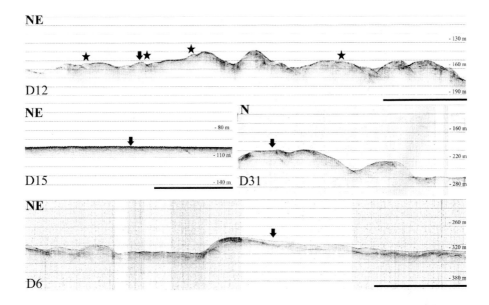

Figure 5. Knudsen echosounder profiles (12 kHz) at selected points off the San Bernardo Islands showing different seabed configurations. D12 tracklines (ALP11–D12) segment, diffuse reflectors indicate *Madracis* thickets (MaT) occurrences, black stars; coralline "mounds" are low relief up 10 m high, instead of non coralline "mounds" that reach 30 m high. D15 trackline (ALP9-D15) aspect of the continental shelf trackline records where calcareous algae were found. D31 trackline (BLP1-D31) black arrow indicates the grab D31, where chimney shaped mudstones were found, along with strong reflector; to the South, mud mounds show diffuse reflectors. D6 trackline (ALP3-D6) indicates muddy depositional laminate facies of the deep slope. Dashed lines are each 20 m depth. Scale bar 300 m, D15 length is equal to 300 m. Arrows indicate approximately grab position. Stars indicate *Madracis* thickets.

(Fig. 6B); (3) reef coral and red calcareous algae conglomerates; and (4) columnar limestone composed of carbonate mud of columnar shape (resembling mud pipes) (Fig. 6C).

Rock cuts made through columnar limestone showed concentric rings, perhaps related to different accretion episodes. These rings may be indicative of sequential gas/mud discharge events (grab 31, Figs. 4, 6C) as has been described for samples derived from mud volcanoes in the Gulf of Cadiz (Magalhães et al., 2005), as a result of bacterial carbonate precipitation (Jørgensen, 1992; Rejas et al., 2005). As shown by Vernette et al. (1992), the continental shelf of the south Caribbean is pierced by mud diapirs and mud volcanoes. Two conspicuous domes (30 m high) imaged in depths of 110–80 m on sector B, are not far (nearest sample Rd4 was undertaken at a distance of 1.5 km, Fig. 4) from the site where the columnar limestone were collected, supporting their mud diapiric origin; the faults system along this accretionary wedge is described by Flinch (2003). An alternative hypothesis for the origin of these pipe-shaped rocks could be diagenetic, related to lithification of endobenthic animal burrows (Macintyre and Milliman, 1970; Curran and Martin, 2003). The external surfaces of the columnar limestone were covered by serpulid worms and other scleractinians, e.g., *O. rotundifolia* (Fig. 3D).

**Continental shelf communities.**—Unlike the muddy bottoms below 280 m depth (grab 6, Fig. 5), sediments from the continental shelf above 110 m depth were classified as biolithoclastic muddy sand facies (grab 15, Fig. 5). The seafloor of the continental shelf is gently sloping (0–2°; Fig. 4). Above this platform, on the northeastern part of sector B, a diverse sponge community composed by species from the genus *Xetospongia, Ircinia, Halicometes,* and *Topsentia,* among others, thrives on sediments with abundant *Halimeda* debris (Figs. 4, 5). Crabs were plentiful in the sponges, e.g., *Portunus spinicarpus* (Stimpson, 1871) and *Iliacantha subglobosa* Stimpson, 1871, as well as ophiurans (e.g., *Ophiomusium testudo* Lyman, 1875) and crinoids *Comactinia meridionalis* L. Agassiz, 1865.

Green algae specimens of the genus *Rhipiliopsis,* which are the deepest living green algae in the world (Littler and Littler, 2000), were collected on sandy bottoms of the continental shelf (101 m deep; grab 1, Fig. 2). To our knowledge, these are the first records of these green fleshy algae in depths over 100 m in Colombian Caribbean waters.

## Discussion

*Madracis* species are common hermatypic zooxanthellate corals in tropical shallow reefs in the Caribbean sea (Zlatarski and Martínez, 1982; Prahl and Erhardt, 1985; Guzmán and Guevara, 1998; Veron, 2000); however, some species of this cosmopolitan coral genus can adapt to deep waters because they are non-symbiotic (azooxanthellate) or exhibit a facultative symbiotic life strategy (Best, 2001). During three INVEMAR-Macrofauna expeditions in the Colombian Caribbean, the remarkable abundance of some branching *Madracis* species (e.g., *M. myriaster* and *M. asperula*) was documented, implying that they are capable of building coral assemblages below 70 m depth (Reyes et al., 2005).

Fossil *Madracis-Dendrophyllia* assemblages found in tropical waters off the Niger Delta were dated from late Pleistocene and earlier Holocene age (Allen and Wells, 1962), and they have a similar species composition to our living Colombian communities, i.e., *Madracis* spp., *Paracyathus pulchellus, Cladocora debilis* and *Dendrophyllia* (= *Eguchipsammia*). According to Allen and Wells (1962), living *Madracis* communities from the Colombian Caribbean could share some other characteristics with fossil assemblages: they are located in a constant range depth, on similar substrate (soft bottoms), and they probably accumulated at temperatures < 20 °C. Thus, the environmental conditions where Colombian *Madracis* communities develop today may resemble the hypothetical environment where African *Madracis-Dendrophyllia* thrived during Holocene. Our finding of living *Madracis* communities in Colombian Caribbean provides new information about the history and evolution of coral ecosystems in tropical areas, and raises new questions about a possible wider zonation of corals in tropical waters, comprising these *Madracis* assemblages as an "intermediate" community between shallow coral reefs and deep-water coral reefs.

Reef-building and habitat-forming corals (*L. pertusa* and *M. oculata*) have been recorded mainly in hard bottoms of steep slopes (Mortensen et al., 1995, 2001; Freiwald et al., 1999, 2002; Masson et al., 2003; Wheeler et al., 2005), and rarely in non-rocky bottoms with relatively smooth slopes (Remia and Taviani, 2005; Taviani et al., 2005). The branching *Madracis* spp. communities in the Colombian Caribbean were collected on sandy mud bottoms, but around sites of irregular seabed topography such as channels and small mounds. This rough seascape may provide a favorable hydrodynamic regime where corals can settle and thrive, as proposed earlier by Mortensen et al. (2001). The oceanographic conditions of this area are largely influenced by the Panama-Colombia Countercurrent, which begins as a superficial current off the Panama coast and strengthens eastward at around 100–200 m depth in the Colombian off-shore. This flow, which

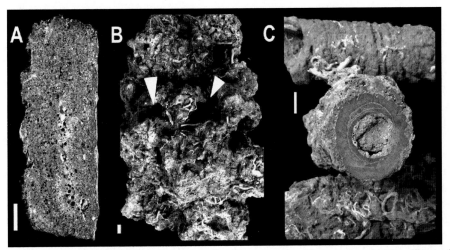

Figure 6. Limestones collected off San Bernardo; (A) packstone with coarse sand grains; (B) coral reef limestone, white arrows show evidence of karstic dissolution crevices; and (C) pipe-shaped columnar limestone. Scale bar 1 cm.

is opposite to the Caribbean Current, is a semi-continuous flow along the entire southern boundary of the Caribbean, and is associated with offshore cyclonic eddies (Andrade and Barton, 2000; Andrade et al., 2003). As a result, the interactions between strong currents and seafloor topography on the Colombian shelf-break may control the distribution of *Madracis* spp., under oceanographic conditions comparable to those observed in areas where *L. pertusa* thrives (Freiwald et al., 1997, 1999; Hovland et al., 2002; Roberts et al., 2003). No oceanographic data are available to verify if a local input of nutrients through the seabed may exist at or near the sites where *Madracis* communities are growing.

Fluid seepage from beneath the seafloor is another factor related to the occurrences of *L. pertusa* at high latitudes (Henriet et al., 1998; Hovland and Risk, 2003; Gomes et al., 2004; Schroeder et al., 2005), and may have an impact on the existence of azooxanthellate coral communities in the study area. Offshore of the Colombian continental margin includes an "accretionary wedge" where gas actively escapes from submarine mud-volcanoes of large dimensions (Vernette et al., 1992; Flinch, 2003). Two large domes, suggestive of mud diapirs shown in sector B, in addition to the columnar limestone samples collected, are evidence of possible mud volcanism and active gas seepage. *Oxysmilia rotundifolia* was the only coral found growing directly attached to the columnar limestone, suggesting that for this species at least, there might be an ecological linkage with mud diapirism and gas discharge. It is noteworthy that some of the principal habitat-forming species *Madracis* spp. were common in (> 500 fragments), but not found directly settled on, sediment samples from the hard substrates that contained the columnar limestones.

The linkage between coral communities, mud domes, and gas seepage has been recognized previously on the extensive shallow coral reefs off Rosario and San Bernardo Islands, and surrounding areas, where gas seepage is active today, such as in Salmedina and Burbujas "bubbles" shoals (Díaz et al., 1996, 2000).

According to Vernette et al. (1983), during the last Holocene regression approximately 18000 yrs ago, the Colombian Caribbean coastline was some 120 m below the current sea level, and therefore shallow coral reefs developed at that time were drowned once sea level rose back to its present level; a similar phenomenon has been observed offshore of Barbados in the West Indies, where there is evidence that the sea level was 121 ± 5 m below present during the last glacial maximum (Fairbanks, 1989). The carbonate rocks collected off Columbia between 120 and 180 m depth provide direct evidence of the existence of a calcareous platform (Vernette et al., 1983), and exhibit signals of karstic dissolution in shallow environments or sub-aereally exposed to rainfall. Moreover, the dead fragments of *Agaricia* sp. and *H. cucullata*, support the idea that these drowned shallow coral reefs may have provided the substrate that *Madrepora* spp., antipatharians, and gorgonians are colonizing today, while the adjacent soft muddy bottoms are the substrate for the development of *Madracis* spp. communities.

To summarize, favorable conditions for the establishment of remarkable coral communities exist in this sector of the Colombian continental margin. Corals are distributed from shallow to deeper waters along extensive fringing coral reefs surrounding Rosario and San Bernardo–Rosario Islands (Díaz et al., 2000). *Madracis* spp. and calcareous red algae are developing in "intermediate-depth" waters, and even the growth of *Madrepora* spp. could be successful at deeper waters such as is indicated by the specimens recorded down to 1050 m (Cairns, 1979, 2000). Finally, it is important to emphasize that *Madrepora* and *Lophelia*, as well as *Madracis* spp. have a wide Caribbean distribution (Cairns, 1979, 2000; Reyes et al., 2005) and although no visible threats appear to affect these ecosystems, over the last few years shrimp fisheries and the petroleum industry have started expansion plans that may eventually endanger deep-sea coral communities.

## Acknowledgments

We would like to thank G. Bedoya, J. R. Peláez, and E. Londoño (EAFIT University) for their handling of the technical aspects of the geophysical survey, as well as the crew of the B/I AN-CÓN for their cooperation aboard the ship. Thanks to J. Quintero, I. Daniel, N. Cruz, P. Flórez, A. Polanco, M. Ruíz, G. Navas; N. Rangel, G. Borrero, and the Taxonomy Research Group of INVEMAR for their collaboration and good fellowship, and to S. Cairns, J. Reed, and an anonymous reviewer for their valuable comments and advice on improving the content of this paper. This project was funded by Colciencias (Project Code 2115-09-16649). Contribution No. 951 of INVEMAR.

## Literature Cited

Allen, J. R. L. and J. W. Wells. 1962. Holocene coral banks and subsidence in the Niger Delta. J. Geol. 70: 381–397.

Andrade, C. A. and E. D. Barton. 2000. Eddy development and motion in the Caribbean Sea. J. Geophys. Res. 105: 26191–26201.

_____, _____, and C. N. K. Mooers. 2003. Evidence for an eastward flow along the Central and South American Caribbean Coast. J. Geophys. Res. 108: 1–11.

Best, M. B. 2001. Some notes on the terms "deep-sea ahermatypic" and "azooxanthellate" illustrated by the coral genus *Madracis*. Pages 19–29 *in* J. H. Martin Willison, J. Hall, S. E. Gass, E. L. R. Kenchington, M. Butler and P. Doherty, eds. Proc. First Int. Symp. on Deep-sea Corals, Ecology Action Centre Nova Scotia Museum, Halifax, Nova Scotia.

Cairns, S. D. 1979. The Deep-water Scleractinia of the Caribbean Sea and adjacent waters. Stud. Fauna Curaçao 57: 1–341.

_____. 2000. A revision of the shallow-water azooxanthellate Scleractinia of the western Atlantic. Studies Nat. Hist. Caribbean Region 75: 1–231.

_____, B. W. Hoeksema, and J. Van der Land. 1999. List of extant stony corals. Atoll Res. Bull. 459: 13–46

CIOH. 1985. Carta Sedimentológica Golfo del Darién a Punta Canoas. Centro de Investigaciones Oceanográficas e Hidrográficas. Servicio Hidrográfico de Colombia. Sección de Cartografía. 1:300.000. Catálogo de Cartografía Náutica. Dirección General y Marítima.

Curran, H. A. and A. J. Martin. 2003. Complex decapod burrows and ecological relationships in modern and Pleistocene intertidal carbonate environments, San Salvador Island, Bahamas. Palaeo 192: 229–245.

Díaz, J. M., J. A. Sánchez, and G. Díaz-Pulido. 1996. Geomorfología y formaciones arrecifales recientes de Isla Fuerte y Bajo Bushnell, Plataforma continental del Caribe Colombiano. Bol. Invest. Mar. Cost. 25: 87–105.

_____, L. M. Barrios, M. H. Cendales, J. Garzón-Ferreira, J. Geister, M. López-Victoria, G. H. Ospina, F. Parra-Velandia, J. Pinzón, B. Vargas-Rangel, F. A. Zapata, and S. Zea. 2000. Áreas coralinas de Colombia. Instituto de Investigaciones Marinas y Costeras "José Benito Vives de Andreis" INVEMAR. Serie de Publicaciones Especiales, No. 5. Santa Marta. 175 p.

Fairbanks, R. G. 1989. A 17,000-year glacio-eustatic sea level record: influence of glacial melting rates on the Younger Dryas event and deep-ocean circulation. Nature 342: 637–642.

Flinch, J. F. 2003. Structural evolution of the Sinu-Lower Magdalena area (Northern Colombia). Pages 776–796 *in* C. Bartolini, R. T. Buffler, and J. Blickwede, eds. The Circum-Gulf of Mexico and the Caribbean: Hydrocarbon habitats, basin formation, and plate tectonics. AAPG Memoir 79.

Freiwald, A., R. Henrich, and J. Pätzold. 1997. Anatomy of a deep-water coral reef mound from Stjernsund. West Finnmark, northern Norway. SEPM Spec. Publ. 56: 141–161.

_____, J. Wilson, and R. Henrich. 1999. Grounding icebergs shape deep water coral reefs. Sediment. Geol. 125: 1–8.

_____, J. H. Fosså, A. Grehan, T. Koslow, and J. M. Roberts. 2004. Cold-water Coral Reefs. UNEP-WCMC, Cambridge, UK. 84 p.

_____, V. Hühnerbach, B. Lindberg, J. B. Wilson, and J. Campbell. 2002. The Sula reef complex, Norwegian shelf. Facies 47: 179–200.

Gomes, P. Y., M. Y. Yoshinaga, L. A. Saint-Pastous Madureira, and M. Hovland. 2004. Seabed pockmarks associated with deepwater corals off SE Brazilian continental slope, Santos Basin. Mar. Geol. 207: 159–167.

Guzmán, H. M. and C. A. Guevara. 1998. Arrecifes coralinos de Bocas del Toro, Panamá: 1. Distribución, estructura y estado de conservación de los arrecifes continentales de la Laguna de Chiriquí y la Bahía Almirante. Rev. Biol. Trop. 46: 601–623.

Henriet, J. P., B. De Mol, S. Pillen, M. Vanneste, D. Van Rooij, W. Versteeg, P. F. Croker, P. M. Shannon, V. Unnithan, S. Bouriak, and P. Chachkine. ( Porcupine-Belgica 97 Shipboard Party). 1998. Gas hydrate crystals may help build reefs. Nature 391: 648–649.

Hovland, M. and M. Risk. 2003. Do Norwegian deep-water coral reefs rely on seeping fluids? Mar. Geol. 198: 83–96.

_____, S. Vasshus, A. Indreeide, L. Austdal, and Ø. Nilsen. 2002. Mapping and imaging deep-sea coral reefs off Norway, 1982–2000. Hydrobiologia 471: 13–17.

Ingeominas. 1997. Atlas geológico digital de Colombia, Escala 1/500.000: Ministerio de Minas y Energía, República de Colombia, Sheets 1, 3, 4, and 6.

Jørgensen, N. O. 1992. Methane-derived carbonate cementation of marine sediments from the Kattegat, Denmark: Geochemical and geological evidence. Mar. Geol. 103: 1–13

Lattig, P. and S. D. Cairns. 2000. A New Species of *Tethocyathus* (Scleractinia: Caryophylliidae), a Trans-Isthmian Azooxanthellate species. Proc. Biol. Soc. Wash. 113: 590–595.

_____ and J. Reyes. 2001. Nueve primeros registros de corales azooxanthelados (Anthozoa: Scleractinia) del Caribe colombiano (200–500 m). Bol. Invest. Mar. Cost. 30: 19–38.

Littler, D. S. and M. M. Littler. 2000. Caribbean reef plants. An identification guide to the reef plants of the Caribbean, Bahamas, Florida and Gulf of Mexico. Offshore Graphics, Washington. 542 p.

Macintyre, I. G. and J. D. Milliman. 1970. Physiographic features on the outer shelf and upper slope, Atlantic Continental Margin. Southeastern United States. Geol. Soc. Am. Bull. 81: 2577–2597.

Magalhães, V. H., L. M. Pinheiro, M. K. Ivanov, V. Díaz-del-Río, L. Somoza, and J. Gardner. 2005. Distribution of Mud volcanism and control of seafloor seepage related authigenic carbonates in the Gulf of Cadiz. Geophys. Res. (Abstract) 7: 5589.

Masson, D. G., B. J. Bett, D. S. M. Billett, C. L. Jacobs, A. J. Wheeler, and R. B. Wynn. 2003. The origin of deep-water, coral-topped mounds in the northern Rockall Trough, Northeast Atl. Mar. Geol. 194: 159–180.

Mortensen, P. B., M. Hovland, T. Brattegard, and R. Farestveit. 1995. Deep water bioherms of the scleractinian coral *Lophelia pertusa* at 64°N on the Norwegian shelf: Structure and associated megafauna. Sarsia 80: 145–158.

_____, _____, J. H. Fossa, and D. M. Furevik. 2001. Distribution, abundance and size of *Lophelia pertusa* coral reefs in mid-Norway in relation to seabed characteristics. J. Mar. Biol. Assoc. UK 81: 581–597.

NODC National Oceanographic Data Center. 2001. World Ocean Database 2001. Silver Spring, MD: National Oceanographic and Atmospheric Administration. 29 September 2003; July 2005. Available from: http://www.nodc.noaa.gov/OC5/WOD01/data2001.html.

Prahl, H. and H. Erhardt. 1985. Colombia, Corales y Arrecifes Coralinos. FEN, Bogotá. 295 p.

Reed, J. K. 2002a. Deep-water *Oculina* coral reefs of Florida: biology, impacts, and management. Hydrobiologia 471: 43–55.

_____. 2002b. Comparison of deep-water coral reefs and lithoherms off southeastern USA. Hydrobiologia 471: 57–69.

_____, D. C. Weaver, and S. A. Pomponi. 2006. Habitat and Fauna of Deep-Water *Lophelia pertusa* Coral Reefs of the Southeastern U.S.: Blake Plateau, Straits of Florida, and Gulf of Mexico. Bull. Mar. Sci. 78: 343–375.

Rejas, M., C. Taberner, S. Schouten, M. de Baas, P. Mata, J. M. de Gilbert, V. Díaz del Río, and L. Somoza. 2005. Anaerobic methane oxidation and carbonate precipitation in the carbonate crusts from The Gulf of Cádiz. Geophys, Res. (Abstract) 7: 3144.

Remia, A. and M. Taviani. 2005. Shallow-buried Pleistocene *Madrepora*-dominated coral mounds on a muddy continental slope, Tuscan Archipelago, NE Tyrrhenian Sea. Facies 50: 419–425.

Reyes, J. 2000. Lista de los Corales (Cnidaria: Anthozoa: Scleractinia) de Colombia. Biota Colombiana 1: 164–176.

_____ and N. Santodomingo. 2005. Azooxanthellate coral biodiversity in the Southern Caribbean. 3rd Int. Symp. on Deep-Sea Corals. Science and Management. University of Miami. RSMAS. University of Florida. Abstract #0507, page 150.

_____, _____, A. Gracia, G. Borrero-Pérez, G. Navas, L. M. Mejía-Ladino, A. Bermúdez, and M. Benavides. 2005. Southern Caribbean azooxanthellate coral communities off Colombia. Pages 309–330 *in* A. Freiwald and J. M. Roberts eds. Cold-water corals and ecosystems. Springer-Verlag Berlin Heidelberg.

Roberts, J. M., D. Long, J. B. Wilson, P. B. Mortensen, and J. D. Gage. 2003. The cold-water coral *Lophelia pertusa* (Scleractinia) and enigmatic seabed mounds along the north-east Atlantic margin: are they related? Mar. Pollut. Bull. 46: 7–20.

Schroeder, W. W. 2002. Observations of *Lophelia pertusa* and the surficial geology at a deep-water site in the northeastern Gulf of Mexico. Hydrobiologia 471: 29–33.

_____, S. D. Brooke, J. B. Olson, B. Phaneuf, J. J. McDonough III, and P. Etnoyer. 2005. Ocurrence of deep-water *Lophelia pertusa* and *Madrepora oculata* in the Gulf of Mexico. 297–307 *in* A. Freiwald and J. M. Roberts, eds. Cold-water corals and ecosystems. Springer-Verlag. Berlin, Heidelberg.

Taviani, M., A. Remia, C. Corselli, A. Freiwald, E. Malinverno, F. Mastrototaro, A. Savini, and A. Tursi. 2005. First geo-marine survey of living cold-water *Lophelia* reefs in the Ionian Sea (Mediterranean basin). Facies 50: 409–417.

Vernette, G., S. H. de Martínez, J. O. Martínez, and C. Parada. 1983. Sediment characteristic and depositional processes on the Colombian continental shelf within the Caribbean sea (from the Magdalena River to the Morrosquillo Gulf). Pages 303–317 *in* 10ª Conferencia Geológica del Caribe, Cartagena-Colombia.

_____, A. Mauffret, C. Bobier, L. Briceño, and J. Gayet. 1992. Mud diapirism, fan sedimentation and strike-slip faulting, Caribbean Colombian margin. Tectonophysics 202: 335–349.

Veron, J. E. N. 2000. Corals of the world. Vols. 1-3. Australian Institute of Marine Science, Townsville, Australia. 463 p.

Wheeler, A. J., M. Kozachenko, A. Beyer, A. Foubert, V. Huvenne, M. Klages, D. G. Masson, K. O. Le-Roy, and J. Thiede. 2005. Sedimentary processes and carbonate mounds in the Belgica Mound province, Porcupine Seabight, NE Atlantic. Pages 571–603 *in* A. Freiwald and J. M. Roberts, eds. Cold-water corals and ecosystems. Springer-Verlag. Berlin.

Wilson, J. B. 2001. *Lophelia* 1700 to 2000 and beyond. Pages 1–5 *in* J. H. M. Willison, J. Hall, S.E. Gass, E. L. R. Kenchington, M. Butler, and P. Doherty, eds. Proc. First Int. Symp. on Deep-sea Corals, Ecology Action Centre Nova Scotia Museum, Halifax, Nova Scotia.

Zlatarski, V. N. and N. Martínez. 1982. Les Scléractiniaires de Cuba. Editions de l'Academie Bulgare des Sciences, Sofia. 471 p.

ADDRESSES: (N.S., J.R., A.G., A.M., G.O., C.G.) *Instituto de Investigaciones Marinas y Costeras, Cerro de Punta Betín, Sociedad Portuaria, Santa Marta, Colombia.* CORRESPONDING AUTHOR: (N.S.) *Telephone: +57-54-211380, Fax: +57-54-312986, E-mail: <nadiaks@invemar.org.co>.*

Appendix 1. List of azooxanthellate corals living in *Madracis* spp. communities and hardgrounds (Colombian Caribbean).

---

Order Scleractinia Bourne, 1900
 Family Pocilloporidae Gray, 1840
  Genus *Madracis* Milne-Edwards and Haime, 1849
   *Madracis asperula* Milne-Edwards and Haime, 1849
   *Madracis brueggemanni* (Ridley, 1881)
   *Madracis myriaster* (Milne-Edwards and Haime, 1849)
   *Madracis pharensis* (Heller, 1868)
 Family Oculinidae Gray, 1847
  Genus *Madrepora* Linnaeus, 1758
   *Madrepora carolina* (Pourtalès, 1871)
   *Madrepora oculata* (Linnaeus, 1758)
 Family Caryophylliidae Gray, 1847
  Genus *Anomocora* Studer, 1878
   *Anomocora fecunda* (Pourtalès, 1871)
   *Anomocora marchadi* (Chevalier, 1966)
   *Anomocora prolifera* (Pourtalès, 1871)
  Genus *Caryophyllia* Lamarck, 1801
   *Caryophyllia ambrosia caribbeana* Cairns, 1979
   *Caryophyllia barbadensis* Cairns, 1979
   *Caryophyllia berteriana* Duchassaing, 1850
   *Caryophyllia* sp.
  Genus *Cladocora* Ehrenberg C. G., 1834
   *Cladocora debilis* Milne-Edwards and Haime, 1849
  Genus *Coenocyathus* Milne-Edwards and Haime, 1848
   *Coenocyathus parvulus* (Cairns, 1979)
  Genus *Coenosmilia* Pourtalès, 1874
   *Coenosmilia arbuscula* Pourtalès, 1874
  Genus *Deltocyathus* Milne-Edwards and Haime, 1848
   *Deltocyathus calcar* Pourtalès, 1874
   *Deltocyathus eccentricus* Cairns, 1979
   *Deltocyathus* sp. cf. *italicus* (Michelotti, 1838)
  Genus *Oxysmilia* Duchassaing, 1870
   *Oxysmilia rotundifolia* (Milne-Edwards and Haime, 1848)
  Genus *Paracyathus* Milne-Edwards and Haime, 1848
   *Paracyathus pulchellus* (Philippi, 1842)
  Genus *Phacelocyathus* Cairns, 1979
   *Phacelocyathus flos* (Pourtalès, 1878)
  Genus *Polycyathus* Duncan, 1876
   *Phacelocyathus mayae* Cairns, 2000
  Genus *Stephanocyathus* Seguenza, 1864
   *Stephanocyathus paliferus* Cairns, 1977
  Genus *Tethocyathus* Kühn, 1933
   *Tethocyathus variabilis* Cairns, 1979
  Genus *Thalamophyllia* Duchassaing, 1870
   *Thalamophyllia riisei* (Duchassaing and Michelotti, 1864)

---

Appendix 1. Continued.

---

Genus *Trochocyathus* Milne-Edwards and Haime, 1848
   *Trochocyathus* sp. cf. *Trochocyathus faciatus* Cairns, 1979
   *Trochocyathus rawsonii* Pourtalès, 1874
Family Flabellidae Bourne, 1905
  Genus *Flabellum* Lesson, 1831
   *Flabellum moseleyi* Pourtalès, 1880
  Genus *Javania* Duncan, 1876
   *Javania cailleti* (Duchassaing and Michelotti, 1864)
  Genus *Polymyces* (Pourtalès, 1868)
   *Polymyces fragilis* Pourtalès, 1868
Family Guyniidae Hickson, 1910
  Genus *Guynia* Duncan, 1873
   *Guynia annulata* Duncan, 1872
  Genus *Schizocyathus* Pourtalès, 1874
   *Schizocyathus fissilis* Pourtalès, 1874
Family Turbinoliidae Milne-Edwards and Haime, 1848
  Genus *Sphenotrochus* Milne-Edwards and Haime, 1848
   *Sphenotrochus auritus* Pourtalès, 1874
   *Sphenotrochus lindstroemi* Cairns, 2000
Family Gardineriidae Stolarski, 1996
  Genus *Gardineria* Vaughan, 1907
   *Gardineria minor* Wells, 1973
Family Dendrophylliidae Gray, 1847
  Genus *Balanophyllia* Searles Wood, 1844
   *Balanophyllia caribbeana* Cairns, 1977
   *Balanophyllia cyathoides* (Pourtalès, 1871)
   *Balanophyllia palifera* Pourtalès, 1878
   *Balanophyllia wellsi* Cairns, 1977
   *Balanophyllia* sp.
  Genus *Eguchipsammia* Cairns, 1994
   *Eguchipsammia cornucopia* (Pourtalès, 1871)
   *Eguchipsammia* sp.

# Distribution of deep-sea corals in the Newfoundland and Labrador region, Northwest Atlantic Ocean

Vonda E. Wareham and Evan N. Edinger

## Abstract

Deep-sea corals were mapped using incidental by-catch samples from stock assessment surveys and fisheries observations. Thirteen alcyonaceans, two antipatharians, four solitary scleractinians, and 11 pennatulaceans were recorded. Corals were broadly distributed along the continental shelf edge and slope, with most species found deeper than 200 m; only nephtheid soft corals were found on the shelf. Large branching corals with robust skeletons included *Paragorgia arborea* (Linnaeus, 1758), *Primnoa resedaeformis* (Gunnerus, 1763), *Keratoisis ornata* (Verrill, 1878), *Acanthogorgia armata* (Verrill, 1878), *Paramuricea* spp., and two antipatharians. Coral distributions were highly clustered, with most co-occurring with other species. Scientific survey data delineated two broad coral species richness hotspots: southwest Grand Bank (16 spp.) and an area of the Labrador slope between Makkovik Bank and Belle Isle Bank (14 spp.). Fisheries observations indicated abundant or diverse corals off southeast Baffin Island, Cape Chidley, Labrador, Tobin's Point, and the Flemish Cap. Corals on the Flemish Cap comprised exclusively soft coral, sea pens, and solitary scleractinians. Most coral-rich areas were suggested in earlier research based on stock assessment surveys or Local Environmental Knowledge (LEK). Currently there are no conservation measures in place to protect deep-sea coral in this region.

Deep-sea corals can be found worldwide (Freiwald and Roberts, 2005), but until recently data on their distributions in the Northwest Atlantic Ocean were limited (Collins, 1884; Deichmann, 1936; Miner, 1950; Nesis, 1963; Tendal, 1992, 2004; Breeze et al., 1997; MacIsaac et al., 2001; Gass and Willison, 2005; Watling and Auster, 2005; Mortensen et al., 2006). Within the last decade, knowledge of corals in eastern Canada has increased dramatically, with attention being primarily focused on the Scotian Shelf (Breeze et al., 1997; MacIsaac et al., 2001). The continental shelf and slope of Newfoundland and Labrador have received very little attention (Nesis, 1963; Tendal, 1992; Gass and Willison, 2005). This study presents deep-sea coral distribution patterns and diversity data in an area that has, up to now, received only exploratory treatment.

In Atlantic Canada, deep-sea corals have been found at depths > 200 m primarily along the continental shelf edge and continental slope, particularly near submarine canyons or saddles where the shelf has been incised (Breeze et al., 1997; MacIsaac et al., 2001; Gass and Willison, 2005; Mortensen and Buhl-Mortensen, 2005b). Such bathymetric features are considered good habitat for corals because they are associated with strong currents that winnow away fine sediment, exposing harder substrates, and provide a reliable source of fine particulate organic matter for suspension feeding corals (Hecker et al., 1980; Harding, 1998). Conversely, increased sedimentation can be hazardous to

George, R. Y. and S. D. Cairns, eds. 2007. Conservation and adaptive management of seamount and deep-sea coral ecosystems. Rosenstiel School of Marine and Atmospheric Science, University of Miami.

corals, congesting polyps and inhibiting feeding processes (Hecker et al., 1980). Hard substrates are believed to be important, especially to larger gorgonian corals such as *Primnoa resedaeformis* and *Paragorgia arborea*; two species that are typically found attached to cobbles, boulders, or bedrock (Mortensen and Buhl-Mortensen, 2004, 2005a; Mortensen et al., 2006). Other corals have a calcareous holdfast for anchoring in soft sediment (e.g., *Acanella arbuscula*). Many solitary Scleractinians such as *Flabellum alabastrum* simply lie on the ocean floor, while others, e.g., *Desmophyllum dianthus* usually retain a holdfast.

Deep-sea corals are slow-growing and long-lived (Lazier et al., 1999; Andrews et al., 2002; Risk et al., 2002; Roark et al., 2005; Sherwood et al., 2006), and can reach heights up to 2.5 m (Miner, 1950; Tendal, 1992; Breeze et al., 1997; Mortensen and Buhl-Mortensen, 2005). Large corals increase complexity of benthic environments through their arboreal growth and robust skeletons (Krieger and Wing, 2002); in turn, these structures can create habitat for other benthic organisms during some stages of their life history (Auster, 2005). Large sessile organisms, however, are more susceptible to anthropogenic disturbances, especially fishing gear in contact with the ocean floor (Watling and Norse, 1998; Krieger, 2001; Fosså et al., 2002; Hall-Spencer et al., 2002; Thrush and Dayton, 2002; Anderson and Clark, 2003). The Northeast Newfoundland Shelf, southern Labrador Shelf, and the Grand Banks of Newfoundland have been subject to intense bottom trawling, (Kulka and Pitcher, 2002). As shelf stocks were depleted, fishing effort shifted to deeper waters on the slope, with potentially severe consequences for deep-water ecosystems (Koslow et al., 2000).

The goal of this study is to map the distribution and diversity of deep-sea corals off the coasts of Newfoundland, Labrador, and southeast Baffin Island using incidental by-catch from scientific surveys and fisheries observations aboard commercial vessels. General information on the distribution patterns of deep-sea corals and their diversity in the region is limited, and the extent to which they have been impacted by fishing activities in the Newfoundland and Labrador region is unknown. This study presents distributions and depth ranges of individual coral species and highlights areas with high diversity and abundance of deep-sea corals in waters off Newfoundland and Labrador, Canada. The study area in question encompasses the continental shelf and slope of the Grand Banks of Newfoundland, Northeast Newfoundland Shelf, the Flemish Cap, Labrador Shelf, Davis Strait, and Baffin Basin. It is our hope that this work will add to the growing information on deep-sea corals in Atlantic Canada (Cairns and Chapman, 2001; Gass and Willison, 2005; Mortensen et al., 2006) and provide necessary data to help conserve coral habitat in the Northwest Atlantic.

## Materials and Methods

Coral data was gathered opportunistically from three sources, commencing in 2002 up to and including May 2006. The Canadian Department of Fisheries and Oceans (DFO) Multispecies Stock Assessment Surveys covered central and southern Labrador as well as northeast to southwest Newfoundland, but excluded the Gulf of St. Lawrence. The Northern Shrimp Stock Assessment Survey, co-sponsored by the Northern Shrimp Research Foundation and DFO, covered the southeast Baffin Island and northern Labrador. Observations from the Fisheries Observer Program (FOP) were the third source of data, and covered a broader area, extending from Baffin Basin to the Grand Banks and the Flemish Cap. Each data source encompassed different management zones and incorporated slightly different sampling techniques. Therefore, each source is discussed separately.

**Multispecies stock assessment surveys and the northern shrimp survey.**—DFO multispecies stock assessment surveys (2002–2006) consisted of an annual spring and fall survey aboard the Canadian Coast Guard Ship (CCGS) WILFRED TEMPLEMAN and the CCGS TELEOST. Survey tows followed a stratified random survey design and covered NAFO (Northwest Atlantic Fisheries Organization) divisions 2HJ (southern Labrador), and 3KLMNOP (northeast-southern Newfoundland, and The Grand Banks). The CCGS WILFRED TEMPLEMAN conducted shallow water tows < 700 m, while the CCGS TE-LEOST conducted both shallow and deep water tows < 1500 m. Research Vessels (RV) used in the study were equipped with a Campelen 1800 shrimp trawl with rockhopper footgear: tight rubber disks (102 × 35 cm diameter) with spacers along the footrope. The 16.9 m wide net had four panels constructed of polyethylene twine: wing panel 80 mm mesh size, the square and first belly 60 mm mesh size, the second belly and cod end 40 mm mesh size with a 12.7 mm liner in the cod end (cf. McCallum and Walsh, 1996). Tow duration was 15 min at 3 kt (± 1 kt); average tow length was 1.4 km (0.79 nmi), and tows were conducted along a consistent depth contour. The area swept was calculated by multiplying net wing span by tow distance for an average swept area per tow of 0.025 km$^2$ (0.073 nmi$^2$). The total area that can be surveyed by DFO per year for all NAFO divisions in the region was 690,676 km$^2$. However, not all NAFO divisions were surveyed each year: divisions 2J and 3KLNOP were surveyed in each year from 2002–2005, but not consistently; divisions 0B and 2G were surveyed in 2005; division 2H was surveyed in 2004. Each catch was sampled for fish species, and sub-sampled for invertebrate by-catch. Individual coral species were assigned a numerical species code with the exception of pennatulaceans, which were grouped. All suspected corals were assigned a species code, bagged, labelled with locator number, and frozen. DFO technicians and scientific crew were provided with coral identification guides produced by the Bedford Institute of Oceanography and by the authors. Training workshops for DFO crew and fisheries observers were organized in 2004 and 2005 by the authors. All specimens were forwarded to Memorial University (MUN) and identified by the authors using gross morphology (shape, size, hardness, color, and presence/absence of horny axis), polyp morphology, and, in some cases, sclerite description using Scanning Electron Microscopy. Photographs of coral species are presented in Figure 1.

The Northern Shrimp Survey was conducted from late July to early September 2005 on the Fisheries Products International vessel MV CAPE BALLARD. This was the first of five annual research deployments scheduled for the Artic Region. Survey tow methods were standardized with DFO surveys and followed a random stratified sampling program at depths of 100–750 m in NAFO divisions 0B (southeast Baffin Island) and 2G (northern Labrador). All research surveys were standardized with the assumption that each catch was searched meticulously; therefore any coral caught within a given set was assumed to be almost always recovered by the RV technicians, and lack of corals in a given set was interpreted as absence of coral in that sample.

**Fisheries observer program.**—Fisheries observers are deployed aboard Canadian and Foreign fishing vessels and are responsible for monitoring compliance to fisheries regulations and for the collection of scientific and technical data related to fishing operations (Kulka and Firth, 1987). Fisheries managers and scientists use these data to manage and study fisheries. Observers are deployed in most fisheries in the region covering a broad array of depths extending from Baffin Basin to southern Newfoundland and the Flemish Cap. Prior to collecting corals, all observers were equipped with the same identification guides as the RV technicians, and participated in coral identification workshops in 2004 and 2005 organized by the authors. Coral data were collected between April 2004 and May 2006. Observer coverage at sea varied between 0–100% depending on the fishery, quota allocation, gear type and NAFO division. Sampling protocol required each observer to submit at least one sample of each coral species encountered on each trip, and to record all other occurrences of each coral species on set/catch data sheets. Samples and records were tracked to assess accuracy of data from each observer. Coral distribution data from fisheries observers presented here include identified samples and records that could be compared with an identified sample previously submitted by the individual observer reporting the record. Data from fisheries observers had several limitations. First, distribution data from observers were biased by fishing effort. Second, unlike the RV surveys, observer coral data were not standardized for variations in tow length, gear type, and search time. Finally, observers may not have had sufficient search time to locate all corals within a catch, especially in high volume fisheries such as shrimp. Given these limitations, observer data were treated as presence, but not absence, of coral species.

**Definition of coral species richness hotspots and abundance peaks.**—Coral species richness hotspots were identified qualitatively in the scientific survey data as areas with higher species richness per tow than surrounding sets. Observer data were not used to identify coral species richness hotspots, but were used to describe the range of individual species, and to characterize coral distributions in areas not covered by the scientific surveys.

*Mapping of deep-sea coral.*—Data from research surveys (Multispecies and Northern Shrimp Surveys), and samples and records from observers were combined into a master database and mapped in MapInfo Professional 8.0 Software. Bathymetry data were obtained from the General Bathymetric Chart of the Oceans (GEBCO, 2003). Distribution maps were verified visually for accuracy and cross checked with data points plotted on Canadian Hydrographic Service bathymetry charts. Any discrepancies were investigated and adjusted appropriately.

**Results**

**Multispecies stock assessment surveys and northern shrimp survey.**—Thirty-five research surveys carried out between December 2002 and January 2006 were explored: 2002 (1), 2003 (2), 2004 (10), 2005 (19), and 2006 (3). The 34 multispecies surveys yielded 1968 tows from NAFO divisions 2HJ and 3KLMNOP. The one Northern Shrimp Survey yielded 227 tows from NAFO divisions 0B and 2G. The total area swept was 52.75 $km^2$, or approximately 0.00685% of the survey area. NAFO divisions 3Pn (2002 only) and 3Ps (2005 only) were the only divisions within the region in which surveys covered > 0.01% of area within the time frame of this study. One area was not adequately surveyed, a small area adjacent to Hudson Strait in the Labrador Sea (~61°20′N, 62°30′W). It was excluded because of the high probability of gear damage due to rough substrates and the reported concentrations of large gorgonians (e.g., *P. arborea*, *P. resedaeformis*) found in this particular area (D. Orr, DFO, pers. comm.). In total, 976 coral specimens were collected from 622 scientific survey sets that captured at least one coral specimen per set. See Table 1 for a summary of each species frequency, mean depth, and range.

**Fisheries observer program.**—From January 2004 to May 2006 fisheries observers documented 1304 coral occurrences from nine of the 25 directed fisheries operating in the region: 397 occurrences were from submitted samples and 907 were from records only (Table 1). The Greenland halibut (*Reinhardtius hippoglossoides*) fishery had the highest frequency of coral by-catch (n = 677) and fished the deepest depths (average depths 889–1070 m depending on gear type) on the continental slope (Table 2). Fishing effort was concentrated in deep waters off the southeastern slope of the Baffin Island Shelf, the southeast Labrador slope, the Northeast Newfoundland Shelf, and in a deep water trough of the Northeast Newfoundland Shelf called Funk Island Deep (Fig. 2A). Mobile gear used in this fishery included Otter trawls (1 net) and twin trawls (2 nets), with the main difference between these being the number of nets hauled per vessel. Fixed gear was also used, benthic longline and gillnet, but mainly off the southwest Grand Bank, in areas unsuitable for trawling.

The northern shrimp (*Pandalus borealis* and *Pandalus montagui*) fishery had the second highest frequency of coral by-catch (n = 226) with effort concentrated on the Labrador Shelf edge (average depths 349–415 m depending on gear type; Table 2). Three trawl types were utilized: shrimp, twin, and triple trawls. Mandatory Nordmore Grate bycatch reduction devices (22–28 mm bar spacing) were used in conjunction with each gear type to help reduce by-catch of mobile species. The grate allows shrimp to pass through and into the net, while oversized by-catch are redirected out through an exit door in the top panel of the net.

Both the shrimp and Greenland halibut fisheries deployed multiple gear types but only the latter used both mobile and fixed gear classes. Overall, mobile gear captured 943 coral occurrences (samples and records), compared to 363 occurrences by fixed gear (Table 3). The Otter Trawl had the highest frequency of coral by-catch (n = 636) of all gear types. Other fisheries in the region captured corals as well and are summarized in Table 2. Further analysis of coral by-catch patterns among directed fisheries and gear types will be published separately.

**Deep-sea coral distribution and diversity patterns.**—Twenty-eight deep-sea coral species were identified in the region, with two additional forms represented that have not been identified to species level. In total, there were 13 alcyonaceans, two antipatharians, four solitary scleractinians, and 11 pennatulaceans (Table 1; Figs. 1, 2).

The order Alcyonacea was subdivided into three informal groups: soft corals with polyps contained in massive bodies, gorgonians with a consolidated axis, and gorgonians without a consolidated axis (Bayer, 1981). See Figure 2A for alcyonacean distributions.

Soft corals consisted of one alcyoniid (*Anthomastus grandiflorus*) and at least two nephtheids (*Gersemia rubiformis*; *Capnella florida*), however, a third nephtheid species was suspected. Because of uncertainty in identifying nephtheid corals to species, all nephtheids were mapped as a single group.

Nephtheids (n = 898) had the highest frequency of all species documented. *Gersemia rubiformis* (Fig. 1X), was the only species in the study that was consistently distributed on the continental shelf (n = 308). Depth for this species ranged between 47–1249 m with average depths < 174 m. Individual colonies were < 5 cm high (when frozen and con-

Table 1. Summary of coral occurrences by species from DFO Multispecies Stock Assessment Surveys (RV) 2003–2006, the Northern Shrimp Stock Assessment Survey (RV) 2005, records (fisheries observations with no samples; FOP$_r$), and samples (fisheries observations with samples; FOP$_s$) from fisheries observations April 2004–January 2006. (?) = species suspected but not confirmed: Nephtheid soft corals are represented by *Capnella florida*, *Gersemia rubiformis*, and at least one other form of nephtheid, which has yet to be identified. *Paramuricea placomus* and *Paramuricea grandis* may both be represented, spicule mounts suggest presence of both species, but some specimens are not yet determined. *Stauropathes arctica* is confirmed, but the identity of a second growth form of antipatharian assigned *Bathypathes* sp., has not yet been determined, partly due to uncertainty in the taxonomy of this group. Sea pens 4 and 12 are two distinct forms yet to be identified; sea pen spp. refers to pennatulaceans too poorly preserved to be identifiable beyond order.

| Order/Family/Species | Coral frequencies (#) | | | | Mean depth and range (m) | | |
| --- | --- | --- | --- | --- | --- | --- | --- |
| | RV | FOP$_s$ | FOP$_r$ | Total | RV | FOP$_s$ | FOP$_r$ |
| Alcyonacea | | | | | | | |
| Nephtheidae | | | | | | | |
| *Capnella florida* (Verrill, 1869) | 284 | 118 | 97 | 499 | 444 (47–1,404) | 615 (230–1,287) | 552 (269–1,087) |
| *Gersemia rubiformis* (Ehrenberg, 1834) | 166 | 31 | 111 | 308 | 174 (41–722) | 286 (51–1,249) | 547 (56–1,249) |
| Nephtheids (?) | 88 | 3 | – | 91 | 253 (67–1,398) | 717 (399–1,135) | – |
| Alcyoniidae | | | | | | | |
| *Anthomastus grandiflorus* (Verrill, 1878) | 49 | 15 | 1 | 65 | 913 (171–1,404) | 834 (302–1,277) | 821 |
| Primnoidae | | | | | | | |
| *Primnoa resedaeformis* (Gunnerus, 1763) | 11 | 6 | 11 | 28 | 402 (162–676) | 559 (357–1,157) | 432 (380–592) |
| Paragorgiidae | | | | | | | |
| *Paragorgia arborea* (L.) | 8 | 10 | 9 | 27 | 573 (370–846) | 719 (448–1,277) | 481 (402–576) |
| Anthothelidae | | | | | | | |
| *Anthothela grandiflora* (Sars, 1856) | 2 | – | – | 2 | 723 (528–918) | – | – |
| Isididae | | | | | | | |
| *Keratoisis ornata* (Verrill, 1878) | 6 | 8 | 16 | 30 | 491 (195–664) | 786 (302–1,100) | 879 (403–1,262) |
| *Acanella arbuscula* (Johnson, 1862) | 81 | 50 | 187 | 318 | 822 (154–1,433) | 827 (344–1,277) | 893 (302–1,244) |
| Acanthogorgiidae | | | | | | | |
| *Acanthogorgia armata* (Verrill, 1878) | 30 | 19 | 25 | 74 | 873 (171–1,415) | 819 (278–1,260) | 513 (302–1,207) |

Table 1. Continued.

| Order/Family/Species | Coral frequencies (#) | | | | Mean depth and range (m) | | |
|---|---|---|---|---|---|---|---|
| | RV | FOP$_s$ | FOP$_r$ | Total | RV | FOP$_s$ | FOP$_r$ |
| Chrysogorgiidae | | | | | | | |
| *Radicipes gracilis* (Verrill, 1884) | 11 | 4 | 13 | 28 | 1052 (785–1,337) | 997 (384–1,491) | 981 (419–1,207) |
| Plexauridae | | | | | | | |
| *Paramuricea grandis* (Verrill, 1883) | 22 | 16 | 11 | 49 | 810 (152–1,415) | 800 (457–1,193) | 594 (402–773) |
| *Paramuricea placomus* (L.) | 22 | 16 | 11 | 49 | 810 (152–1,415) | 800 (457–1,193) | 594 (402–773) |
| Antipatharia | | | | | | | |
| Schizopathidae | | | | | | | |
| *Stauropathes arctica* (Lütken, 1871) | 2 | 14 | 21 | 37 | 1,075 (1,013–1,136) | 1,027 (745–1,287) | 1,069 (872–1,228) |
| *Bathypathes* sp. (Brooke, 1889) (?) | 2 | 14 | 21 | 37 | 1,075 (1,013–1,136) | 1,027 (745–1,287) | 1,069 (872–1,228) |
| Scleractinia | | | | | | | |
| Flabellidae | | | | | | | |
| *Flabellum alabastrum* (Moseley, 1873) | 53 | 19 | 69 | 141 | 819 (218–1,433) | 629 (353–1,135) | 833 (339–1,200) |
| *Dasmosmilia lymani* (Pourtalès, 1871) | 1 | – | – | 1 | 457 | – | – |
| Caryophlliidae | | | | | | | |
| *Desmophyllum dianthus* (Esper, 1794) | – | 2 | – | 2 | – | 883 (713–1,052) | – |
| *Vaughanella margaritata* (Jourdan, 1895) | 1 | 3 | – | 4 | 1,320 | 1,199 (1,163–1,252) | – |
| Pennatulacea | | | | | | | |
| Virgulariidae | | | | | | | |
| *Halipteris finmarchia* (Sars, 1851) | 29 | 11 | 336 | 577 | 749 (113–1,433) | 745 (344–1,028) | 699 (182–1,244) |
| Pennatulidae | | | | | | | |
| *Pennatula grandis* (Ehrenberg, 1834) | 16 | 16 | 336 | 577 | 876 (488–1,404) | 633 (320–1,018) | 699 (182–1,244) |
| *Pennatula phosphorea* (L.) | 36 | 9 | 336 | 577 | 823 (96–1,345) | 713 (146–1,223) | 699 (182–1,244) |

Table 1. Continued.

| Order/Family/Species | Coral frequencies (#) | | | | Mean depth and range (m) | | |
|---|---|---|---|---|---|---|---|
| | RV | $FOP_s$ | $FOP_r$ | Total | RV | $FOP_s$ | $FOP_r$ |
| *Pennatula aculeata* (Koren and Danielsen, 1858) | 1 | – | 336 | 577 | 229 | – | 699 (182–1,244) |
| Funiculinidae | | | | | | | |
| *Funiculina quadrangularis* (Pallas, 1766) | 18 | – | 336 | 577 | 1,018 (346–1,433) | – | 699 (182–1,244) |
| Umbellulidae | | | | | | | |
| *Umbellula lindahli* (Kölliker, 1875) | 4 | 3 | 336 | 577 | 463 (145–620) | 870 (723–984) | 699 (182–1,244) |
| Protoptilidae | | | | | | | |
| *Distichoptilum gracile* (Verrill, 1882) | 2 | 2 | 336 | 577 | 921 (794–1,048) | 750 (346–1,154) | 699 (182–1,244) |
| Anthoptilidae | | | | | | | |
| *Anthoptilum grandiflorum* (Verrill, 1879) | 47 | 19 | 336 | 577 | 796 (171–1,433) | 833 (320–1,072) | 699 (182–1,244) |
| Kophobelemnidae | | | | | | | |
| *Kophobelemnon stelliferum* (Müller, 1776) | 1 | 1 | 336 | 577 | 1,235 | 1,154 | 699 (182–1,244) |
| sea pen sp. 4 | 2 | – | 336 | 577 | 647 (493–801) | – | 699 (182–1,244) |
| sea pen sp. 12 | 2 | – | 336 | 577 | 1,224 (1,135–1,313 ) | – | 699 (182–1,244) |
| sea pen spp. | 3 | 18 | 336 | 577 | 775 (212–1,200) | 697 (277–1,063) | 699 (182–1,244) |
| Total coral specimens | 976 | 397 | 907 | 2,281 | | | |
| Total sets sampled | 2,192 | 45,566 | | 47,758 | | | |
| Total sets with at least one coral specimen | 622 | 864 | | 1,477 | | | |
| Total sets without coral | 1,581 | 44,702 | | 46,281 | | | |

Table 2. Summary of coral by-catch frequencies by target fishery, gear type, and average depths fished; data from fisheries observations documented between April 2004 and January 2006.

| Target species | Gear types (fixed) | (mobile) | Average depth fished (m) | Corals captured (frequency) |
|---|---|---|---|---|
| Skate (*Raja* spp.) | Gillnet | – | 439 | 2 |
| White hake (*Urophycis tenuis* Mitchill, 1814) | Gillnet | – | 218 | 6 |
| Redfish (*Sebastes* sp.) | – | Otter trawl | 447 | 189 |
| Yellowtail flounder (*Limanda ferruginea* Storer, 1839) | – | Otter trawl | 165 | 18 |
| Atlantic halibut (*Hippoglossus hippoglossus* Linnaeus) | Longline | – | 867 | 56 |
| Snow or queen crab (*Chionoecetes opilio* Fabricius, 1788) | Crab pot | – | 399 | 128 |
| Angler, common monkfish (*Lophius americanus* Valenciennes, 1837) | Gillnet | – | 331 | 2 |
| Greenland halibut (*Reinhardtius hippoglossoides* Walbaum, 1792) | – | Otter trawl | 891 | 429 |
|  | Gillnet | – | 995 | 140 |
|  | Longline | – | 1,070 | 27 |
|  | – | Twin trawl | 889 | 81 |
| Shrimp (*Pandalus borealis* Kroyer, 1838; *Pandalus montagui* Leach, 1814) | – | Shrimp trawl | 415 | 102 |
|  | – | Twin trawl | 349 | 111 |
|  | – | Triple trawl | 365 | 13 |
| Total |  |  |  | 1,304 |

Table 3. Summary of coral by-catch frequencies by species, target fishery, gear class, and gear type; data from fisheries observations documented between April 2004 and January 2006.

| | Gear types per coral frequency | | | | | | | |
| | Fixed gear | | | Mobile gear | | | | |
| Coral | Crab Pot | Gillnet | Longline | Otter Trawl | Shrimp Trawl | Twin Trawl | Triple Trawl | Total |
|---|---|---|---|---|---|---|---|---|
| *Capnella florida* | 32 | 4 | 2 | 74 | 63 | 40 | – | 215 |
| *Gersemia rubiformis* | 72 | 21 | 1 | 26 | 16 | 6 | – | 142 |
| Nephtheid | – | 1 | – | 1 | 1 | – | – | 3 |
| *Anthomastus grandiflorus* | – | 2 | 8 | 4 | 1 | 1 | – | 16 |
| *Primnoa resedaeformis* | – | 1 | – | 2 | 14 | – | – | 17 |
| *Paragorgia arborea* | – | 2 | 1 | 10 | 6 | – | – | 19 |
| *Keratoisis ornata* | – | – | 21 | 3 | – | – | – | 24 |
| *Acanella arbuscula* | 14 | 44 | 7 | 100 | – | 72 | – | 237 |
| *Acanthogorgia armata* | 2 | 3 | 12 | 27 | – | – | – | 44 |
| *Radicipes gracilis* | – | 6 | 9 | 2 | – | – | – | 17 |
| *Paramuricea* spp. | 7 | 1 | 1 | 18 | – | – | – | 27 |
| Antipatharia | – | 14 | 1 | 17 | – | 3 | – | 35 |
| *Flabellum alabastrum* | – | 6 | 3 | 56 | – | 22 | 1 | 88 |
| *Desmophyllum dianthus* | – | 1 | 1 | – | – | – | – | 2 |
| *Vaughanella margaritata* | – | 3 | – | – | – | – | – | 3 |
| Pennatulacea | 1 | 41 | 16 | 296 | 1 | 48 | 12 | 415 |
| Total | 128 | 150 | 85 | 636 | 102 | 192 | 13 | 1,304 |
| Gear total | | 363 | | | | 943 | | |

tracted), had a wide range of colour variations, and were frequently observed attached to pebbles, broken shells, and live gastropods.

*Capnella florida* (Fig. 1Y) was mostly found in deeper waters on the continental shelf edge and slope (n = 499). However some samples were captured in shallower waters on the shelf. Depth for this species ranged between 47–1404 m with average depths > 444 m. Individual samples were massive bodied colonies < 15 cm high, with multiple branches that terminated in clusters of non-retractable polyps. Colonies were mostly black with some variations of brown and beige. Attached substrates consisted of mostly rock and gravel, but some colonies were observed attached to live gastropods and sponges.

*Anthomastus grandiflorus* (Fig. 1W) was found only on the shelf edge and slope of the southern Labrador Shelf, Northeast Newfoundland Shelf, and the Grand Banks of Newfoundland (n = 65). The depth of this species ranged between 171–1404 m with average depths > 821 m. Most individual colonies, when contracted, were < 5 cm high characterized by a capitulum that had long polyp tubes and large polyps extended from the cap. Individual colonies were supported by a sterile basal stalk that was observed attached to pebble and cobble substrates. One juvenile specimen was observed attached to a *Keratoisis ornata* stem.

Seven species of gorgonian with a consolidated axis were recorded: *Acanella arbuscula*, *Acanthogorgia armata*, *Paramuricea* spp. (*Paramuricea placomus* and *Paramuricea grandis*), *P. resedaeformis*, *K. ornata*, and *Radicipes gracilis*. Most of these have large (> 30 cm) fan-like skeletons, with the exception of *A. arbuscula* and *R. gracilis*, which are usually smaller.

*Acanella arbuscula* (Fig. 1M) had the highest frequency of all gorgonians (n = 318) with concentrations on the shelf edge and slope of southeast Baffin Island, Hawke Chan-

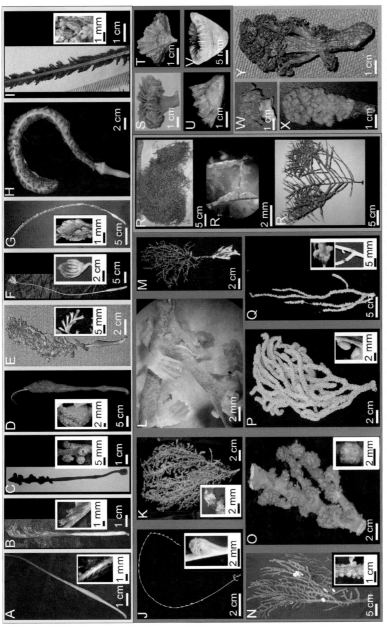

Figure. 1. Deep-sea coral specimens collected off Newfoundland, Labrador, and Baffin Island. Order Pennatulacea: (A) *Distichoptilum gracile*; (B) *Pennatula phosphora*; (C) *Kophobelemnon stelliferum*; (D) *Pennatula aculeata*; (E) *Pennatula grandis*; (F) *Umbellula lindahli*; (G) *Halipteris finmarchia*; (H) *Anthoptilum grandiflorum*; (I) *Funiculina quadrangularis*. Order Alcyonacea: (J) *Radicipes gracilis*; (K) *Acanthogorgia armata*; (M) *Acanella arbuscula*; (N) *Paramuricea* spp.(?): (P) *Primnoa resedaeformis*; (Q) *Keratoisis ornata*; (L) *Anthothela grandiflora*; (O) *Paragorgia arborea*; (W) *Anthomastus grandiflorus*; (X) *Gersemia rubiformis*; (Y) *Capnella florida*. (R) Order Antipatharia (?). Order Scleractinia: (S) *Desmophyllum dianthus*; (T) *Flabellum alabastrum*; (U) *Dasmosmilia lymani*; (V) *Vaughanella margaritata.*\* (?) = species not confirmed.

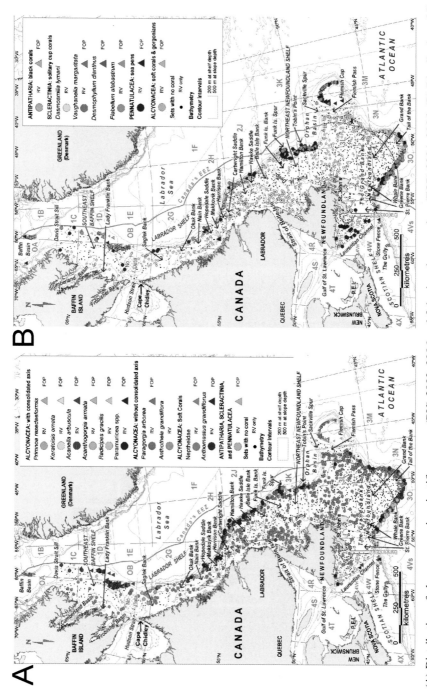

Figure 2. (A) Distribution of alcyonaceans in scientific surveys (RV) 2003–2006 and fisheries observer data (FOP) 2004–2006. (B) Distribution of Pennatulaceans, solitary scleractinians, and antipatharians in scientific surveys (RV) 2003–2006 and fisheries observer data (FOP) 2004–2006.

nel, Funk Island Spur, and southwest Grand Bank (Fig. 2A). Depth of this species ranged between 154–1433 m with average depths > 822 m. Individual colonies were usually < 15 cm high, red, bush-like, and supported by a distinctly banded stem and calcareous root-like base. Polyps were located at opposite angles on brittle segmented branches. Samples were usually damaged and captured in multiples with several tows acquiring 50–100 individual colonies per tow.

*Acanthogorgia armata* (Fig. 1K) had the second highest frequency of this group (n = 74) with concentrations off: southeast Baffin Island, Hawke Channel, Tobin's Point, and the southwest Grand Bank. Depth for this species ranged between 171–1415 m with average depths > 513 m. Individual colonies were < 50 cm in height, characterized by dense yellow-beige branches with long narrow polyps that have crown like tips. Many samples were covered with juvenile gooseneck barnacles (*Lepas* sp. Linnaeus) and Icelandic scallops (*Chlamys islandica* Müller, 1776). As well, two samples were attached to two separate *K. ornata* colonies.

*Paramuricea* samples are believed to be *P. grandis* based on spicule analysis, but *P. placomus* may also be present. When grouped, *Paraumuricea* spp. had the third highest occurrence of this group (n = 49), with most of these concentrated on the continental slope off the Labrador Shelf—Northeast Newfoundland Shelf, and the continental slope off the southwest Grand Bank. One sample was captured as far north as the Hudson Strait and a second sample was captured as far south as the Tail of the Bank. Depth for this genus ranged between 152–1415 m, with most samples recovered at average depths > 594 m. Individual colonies were flexible and fan shaped in one plane, and ranged between 20–85 cm in height. Polyps were short, round, and compressed to the branch. All specimens were black with the exception of one vivid orange sample photographed ~1 hr after capture (Fig. 1N). Samples were seldom observed with a substrate attached even though holdfasts were present. Several samples were attached to cobbles and one sample (50 cm tall fan) was firmly attached to a large subfossil *K. ornata* base. The lifespan of this *K. ornata* colony was likely to be several centuries, based on known growth rates of *Keratoisis* sp. from the family Isididae (Roark et al., 2005).

*Primnoa resedaeformis* (Fig. 1P), for the most part, were not widely distributed (n = 28). Most were found off Saglek Bank, with five other samples documented on the north end of Hamilton Bank on the Labrador Shelf. Depth for this species ranged between 162–1157 m with average depths > 402 m. Two samples on Saglek Bank were captured at depths of 162 m and 165 m; the remaining colonies occurred at slope depths down to 1157 m. Individual colonies were < 35 cm in height and were characterized by dense downward-directed yellow or pink polyps covering a rigid dark brown skeleton. Cross-sections of the stem revealed concentric growth rings alternating between calcite and gorgonin (Andrews et al., 2002; Sherwood et al., 2005). Four subfossil *P. resedaeformis* skeletons were submitted, but not mapped, the largest being ~ 35 cm from the base to the truncated tips of the branches.

*Keratoisis ornata* (Fig. 1Q) was only found on the southwest Grand Bank (n = 30) between 43°52′N–45°00′N and 52°10′W–55°35′W. Depth of this species ranged between 195–1262 m with average depths > 491 m. Individual colonies were < 50 cm with the majority of the samples submitted being either fragments of the original colony or large masses that were severely damaged in trawls. The skeletons are rigid and characterized by thick white calcified branches with proteinaceous internodes. Polyps ranged in density from thick mats to sparse polyps on a predominantly bare skeleton. One rare intact sample, captured by longline gear, had many associated species encrusting on the axis or attached among the branches. The notable species were the solitary scleractinian *D. dianthus*, gorgonian *A. armata*, and numerous juvenile Icelandic scallops and gooseneck barnacles.

*Radicipes gracilis* (Fig. 1J) were distributed on the southwest Grand Bank only (n = 28). Depth of this species ranged between 384–1419 m with average depths > 981 m. Individual colonies were < 80 cm in height and consisted of a single coiled or twisted iridescent axis. Polyps were sparse but evenly spaced on the stem and the entire colony was supported by a calcareous root-like holdfast.

The final group belonging to the alcyonaceans are the gorgonians that lack a consolidated axis. Two species were documented: *P. arborea* and *Anthothela grandiflora*. *Paragorgia arborea* (Fig. 1O) (n = 27) were clustered in an area adjacent to the Hudson Strait and off the northeast shelf of Hamilton Bank. The remainder were sporadically distributed off Cape Chidley, Funk Island Spur, and the Grand Bank slope. Depth of this species ranged between 370–1277 m with average depths > 481 m. Most samples were small, fragmented pieces (< 25 cm); no whole intact colonies were collected. Polyps were usually retracted inside bulbous branch tips. Samples were either red, yellow-beige or salmon colour. One particular set from the Northern Shrimp Survey captured 50 kg of *P. arborea* from an area adjacent to the Hudson Strait (61°22′N, 61°10′W). The subsample was forwarded to DFO and consisted of a large laterally compressed midsection of the main stem (25 cm in length and 11 cm in diameter) and several branch tips. *Anthothela grandiflora* (Fig. 1L) had only two small samples submitted from RV surveys; one sample from off Cape Chidley, Labrador at 528 m and a second sample from the Funk Island Spur at 918 m. Both samples were identified using polyp morphology and sclerite descriptions (cf. Verseveldt, 1940; Miner, 1950).

The order Antipatharia (Fig. $1R_{1,2}$) was represented by two species, *Stauropathes arctica*, and probably *Bathypathes* sp. (S. France, Univ. of Louisiana, pers. comm.). See Figure 2B for antipatharian distributions. The majority of the samples were from observers with only two samples from scientific surveys. Antipatharians were widely distributed on the continental slope in deep waters from Baffin Basin to southwest Grand Bank (n = 37). Three clusters emerged: southeast Baffin Basin (North of Davis Strait sill), southeast Baffin shelf (south of Davis Strait sill), and on the southwest slope of the Orphan Basin (Tobin's Point area). Depth of this species ranged between 745–1287 m with average depths > 1027 m. Two growth forms were observed among samples: *S. arctica* (Pax, 1932; Opresko, 2002) with an open-branched skeleton (Fig. $1R_2$), and a compressed "tumbleweed-like" form referred to as *Bathypathes* sp. (Fig. 1R). A 45 cm long × 30 cm wide × 15 cm high specimen of the tumbleweed form, captured at 1013 m off the northwest Flemish Cap, was the largest antipatharian recorded.

The order Scleractinia was represented by four solitary cup corals: *F. alabastrum*, *D. dianthus*, *Vaughanella margaritata*, and *Dasmosmilia lymani*. Cup corals were distributed along the continental shelf edge and slope with concentrations off the southwest Grand Bank and southeast Baffin Island. See Figure 2B for scleractinian distributions.

*Flabellum alabastrum* (Fig. 1T) distributions were clustered off the Southeast Baffin Shelf, the Flemish Cap, and southwest Grand Bank (n = 141). To a lesser extent, other occurrences were noted off the slope of the Northeast Newfoundland Shelf and Labrador Shelf. Depth ranged between 218–1433 m with average depths > 629 m. *Flabellum alabastrum* samples were identified by the corallum and compressed calice (cf. Cairns, 1981). *Vaughanella margaritata* samples (n = 4) were documented off the Southeast Baffin Shelf and east of the Hopedale Saddle. Depth of this species ranged between

1163–1320 m with average depths > 1199 m. Three samples with multiple specimens of living and subfossil *V. margaritata* per sample, were captured by benthic gillnets targeting Greenland halibut off the Southeast Baffin Shelf. One specimen, missing a holdfast, was captured in a RV survey east of the Hopedale Saddle (Fig. 1V).

*Desmophyllum dianthus* (Fig. 1S) samples were submitted only by observers (n = 2). Both samples were from the southeast Grand Bank; one sample was captured at 1052 m by benthic gillnet gear and the second sample, which was attached to a *K. ornata* colony, was captured at 713 m by longline gear. In addition one subfossil specimen, not included in the dataset, was captured by a gillnet at 1125 m off the Southeast Baffin Shelf. *Dasmosmilia lymani* (Fig. 1U) was only documented once off the southwest slope of Grand Bank (44°52′N, 54°23′W) at 457 m. All scleractinians were identified by gross morphology of the corallum, and confirmed by S. Cairns, Smithsonian Institution.

The order Pennatulacea (Fig. 1A–I) was represented by 11 types of sea pen. Nine sea pens were identified to species level, and two have yet to be determined. See Figure 2B for pennatulacean distributions. Pennatulaceans were distributed along the edge of the continental shelf east of Baffin Basin, off southeast Baffin Island, Tobin's Point, the Flemish Cap, and the southwest Grand Bank (n = 577). The greatest diversity of sea pens was found near the southwest Grand Bank. Species were found at depths between 96–1433 m. Individual colonies varied in size between 10–80 cm. Specimens were identified by peduncle, rachis, and polyp morphology (cf. Williams, 1995, 1999). *Anthoptilum grandiflorum* (Fig. 1H) and *Pennatula phosphora* (Fig. 1B) were the most abundant followed by *Halipteris finmarchia*, *Funiculina quadrangularis*, and *Pennatula grandis* (Table 1). Numerous samples of *H. finmarchia* were observed with commensal sea anemones *Stephanauge nexilis* (Verrill, 1883) firmly attached to the rachis (cf. Miner, 1950). Samples that were damaged beyond identification to genus, were grouped as "sea pen spp.", as were observer records of sea pens.

**Coral species richness hotspots.**—Based on maps of scientific survey data, two coral species richness hotspots were identified (Fig. 3). The first hotspot (Fig. 4A) was situated on the Labrador continental slope between Makkovik Bank (55°30′N, 57°05′W) and Belle Isle Bank (52°00′N, 51°00′W). The second hotspot (Fig. 4B) was located off southwest Grand Bank and Tail of the Bank (~ 42°50′–45°10′N, 49°00′–55°00′W). Both areas had higher species richness per tow than sets surrounding them and higher frequencies of scientific survey stations containing corals than surrounding areas. Southwest Grand Bank and Tail of the Bank had the greatest species richness (16 spp.), with nine alcyonaceans, five pennatulaceans, and two scleractinians recorded. Makkovik Bank-Belle Isle Bank hotspot (14 spp.) had seven alcyonaceans, four pennatulaceans, two scleractinians, and one antipatharian recorded. Species richness per tow based on scientific survey data ranged between 0 and 11 species per set with only two sets capturing nine or more species.

**Discussion**

Distribution maps presented contribute to the growing knowledge of deep-sea coral distribution and diversity off Newfoundland, Labrador, and southeast Baffin Island. Thirteen alcyonaceans, two antipatharians, four solitary scleractinians, and 11 pennatulaceans were documented. Corals were more widely distributed on the continental edge and slope than previously thought, but only nephtheids were found on top of the shelf.

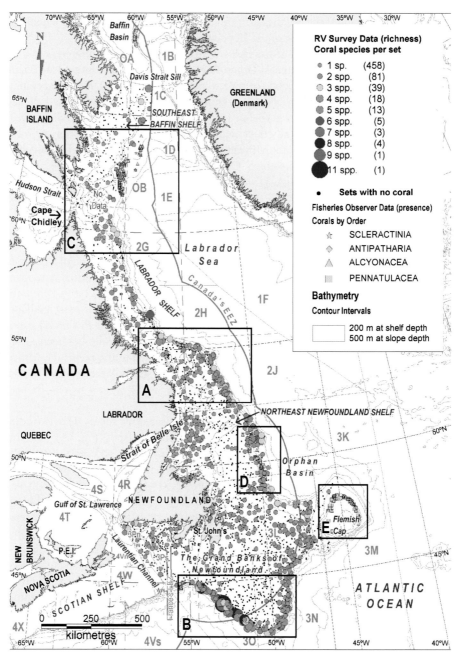

Figure 3. Coral species richness per set in scientific survey data. Most speciose areas were (A) Makkovik Bank to Belle Isle Bank, and (B) southwest Grand Bank. Distribution of coral rich areas from DFO survey data and fisheries observer data: (C) Southeast Baffin Shelf to Cape Chidley, (D) Funk Island Spur to Tobin's Point, and (E) Flemish Cap.

Only ahermatic corals were identified with no occurrence of the reef building *Lophelia pertusa* (Linnaeus, 1758), as reported at the Stone Fence, Scotian margin (Gass and Willison, 2005; Mortensen et al., 2006). The present study documented hundreds more unique records than previously known, and with much more complete and systematic data coverage than was previously available. Many of the coral-rich areas in this study were identified previously by Gass and Willison (2005). Their results partially overlapped with the findings presented in this study off Cape Chidley, Northeast Newfoundland Shelf, and the Grand Banks of Newfoundland.

**Distribution of hotspots.**—Coral species richness hotspots were identified in two distinct locations. The hotspot on the southwest Grand Bank and Tail of the Bank had the greatest species richness in the study with 16 coral species documented. The topography of this area is complex with steep slopes, and numerous canyons. The area is most likely influenced by warm Labrador slope water (cf. Haedrich and Gagnon, 1991). Previous reports from fisheries observers and local fishermen indicated the presence of corals in this region, but very little data from scientific surveys had been documented (Gass and Willison, 2005; Mortensen et al., 2006). The second hotspot extended along the continental shelf edge and slope of the Labrador Shelf from Makkovik Bank to Belle Isle Bank. It spanned the greatest area and included 14 coral species. Most corals were concentrated on the shelf edge and slope with some neptheid soft corals on the bank tops. *Acanella arbuscula* and soft corals were the most abundant species, as both dominated the Funk Island Spur along with several species of sea pens (i.e., *A. grandiflorum*, *P. phosphorea*, and *P. grandis*). Rare species were documented in this area, with one occurrence of *A. grandiflora* off the Funk Island Spur at 918 m and one occurrence of *V. margaritata* at 1320 m off Hopedale Saddle. Jourdan (1895) reported *V. margaritata* (= Caryophyllia) was common off the south side of the Flemish Cap at 1267 m, but only one sample was collected from the Labrador slope and several samples east of the Hudson Strait.

Topography of the southern Labrador Shelf encompasses five banks (> 200 m), three saddles (> 500 m), and two shelves. The Labrador Current flows along the edge of the Labrador shelf and branches at Hamilton Bank (Lazier and Wright, 1993). The main stream continues along the edge and slope of the Newfoundland Shelf, while the secondary branch of the Labrador Current continues inshore along the coasts of Labrador and Newfoundland (Lazier and Wright, 1993). Gass and Willison (2005) also made reference to corals on the southern Labrador slope but did not identify it as a biodiversity hotspot. They documented sporadic coral occurrences off Harrison Bank using fisheries observer reports, off Hamilton Bank using fishers' LEK, off the slope Northeast Newfoundland Shelf using DFO surveys and fishers' LEK, and off the south slope of Orphan Basin using fishers' LEK. Gass and Willison (2005) documented *P. arborea* near Tobin's Point, southwest Grand Bank, "trees" on the southern margin of the Orphan Basin, and *A. arbuscula* and "trees" on the southern Labrador slope. By contrast, *P. arborea* was relatively rare in the current study. Scarcity of *P. arborea* in the current study may reflect depletion of this large species by fisheries damage.

**Other areas of interest.**—When observer data were mapped in conjunction with survey data three additional areas of interest were identified: the area from southeast Baffin Shelf to Cape Chidley (Fig. 4C), Funk Island Spur to Tobin's Point (Fig. 4D), and the north side of the Flemish Cap (Fig. 4E). The coral clusters identified at Cape Chidley

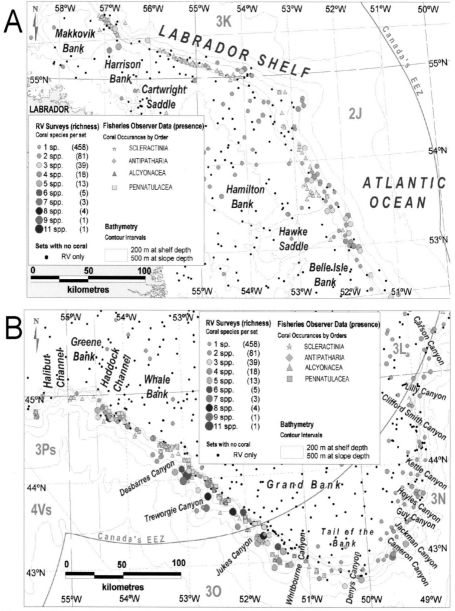

Figure 4. (A) Coral species richness per set in scientific survey data and coral occurrences by Order from fisheries observer data for Makkovik Bank-Belle Isle Bank Hotspot. (B) Coral species richness per set in scientific survey data and coral occurrences by Order from fisheries observer data for southwest Grand Bank and Tail of the Bank.

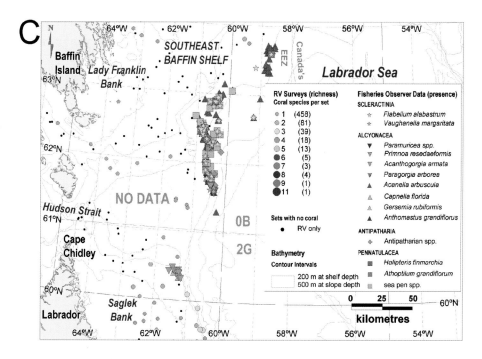

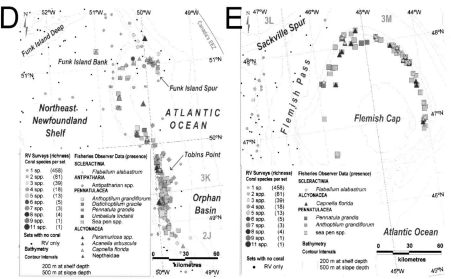

Figure 4. Continued. (C) Coral species richness per set in scientific survey data and coral occurrences by species from fisheries observer data for Southeast Baffin Shelf-Cape Chidley area. (D) Coral species richness per set in scientific survey data and coral occurrences by species from fisheries observer data for Funk Island Spur-Tobin's Point. (E) Coral species richness per set in scientific survey data and coral occurrences by species from fisheries observer data for the Flemish Cap area.

and Southeast Baffin Shelf were dominated by gorgonians, especially *Primnoa resedae-formis*, which was most abundant off Cape Chidley, based on only one year of survey data. This area was not identified as a biodiversity hotspot in the current analysis, but recent unpublished data and past reports suggest high coral abundance and intermediate coral biodiversity. High densities of *P. resedaeformis* were documented off Cape Chidley by MacIsaac et al. (2001) and Gass and Willison (2005). The area immediately east of Hudson Strait (~61°20′N, 62°30′W) was not surveyed, nor were observer data available. Nonetheless, the largest samples submitted for this study were collected from the outer edges of this area with many large gorgonians documented, primarily *P. resedaeformis* and *P. arborea* (Gass and Willison, 2005). 2006 Northern Shrimp Survey sets within this area recovered up to 500 kg of coral, mostly *P. resedaeformis* and *P. arborea* in a single tow (Wareham and Edinger, unpubl. data). The Hudson Strait region is influenced by strong currents and high nutrient flows from both the Labrador Current and Arctic waters from the Hudson Strait (Drinkwater and Harding, 2001). Observer samples and records of *A. arbuscula* were most numerous off Southeast Baffin Shelf. This area is intensively fished for Greenland halibut and northern shrimp. Coral records from surveys, observer reports, and fishers' LEK were previously documented off southeast Baffin Island (Davis Strait) and Cape Chidley (MacIsaac et al., 2001; Gass and Willison, 2005).

Tobin's Point, off the northeast Newfoundland Shelf is intensively fished for Greenland halibut, shrimp, and snow crab (*Chionecetes opilio*). The coral species reported there were dominated by *C. florida*, sea pens, and antipatharians. The Flemish Cap had mostly the neptheid *C. florida*, sea pens, the scleractinian *F. alabastrum*, and one antipatharian. Numerous juvenile *A. grandiflorus* (n = 541) were documented on the northeast side of the Flemish Cap but were not documented within the time frame of this study (Wareham and Edinger, unpubl. data). The Flemish Cap was not covered by Canadian scientific surveys, so coral distributions were derived only from observer data. The clustering of corals on the north side of the Flemish Cap may be an artifact of fishing effort, which was concentrated on the smooth north side of the cap. The south side is deeply incised by canyons, making it difficult terrain for trawling; the south side of the cap may contain suitable habitat for a variety of corals, but has had relatively little sampling effort to date. In general, the top of banks had the lowest coral diversity with only nephtheids present. Most corals are probably incapable of colonizing the top of banks due to limited hard substrates and cold temperatures (cf. Mortensen et al., 2006).

**Substrates of coral biodiversity hotspots and other areas of interest.**—Information on surficial geology has been sparse and limited to Soviet Fishing Investigations by Litvin and Rvachev (1963), and highly generalized maps of surficial geology of the continental margin of Eastern Canada focusing on the bank tops (Fader and Miller, 1986; Piper et al., 1988). Substrates for each hotspot and other areas of interest are discussed separately.

The Grand Banks of Newfoundland (Grand, Whale, Green, and St. Pierre Banks) are relatively shallow and are heavily influenced by wave action. Sand dominates the bank top, with gravel, shell beds, and muddy-sand patches throughout (Fader and Miller, 1986). The edge and slope are a veneer of adlophous sands and gravels. Substrates on the slope progressively change with depth from sand-mud to mud (Litvin and Rvachev, 1963; Piper et al., 1988).

The Flemish Cap and Flemish Pass are located in international waters just east of Grand Bank. Flemish Cap is a dome-shaped plateau ranging from 150 to 350 m deep.

Cap substrates are dominated by sand and shell beds, which change to muddy sand-sandy mud and boulders on the slope; mud is predominant at slope depths > 1000 m (Litvin and Rvachev, 1963). The Flemish Pass is a 1200 m deep trough that separates the Flemish Cap from the Grand Bank. Flemish Pass is strongly influenced by the Labrador Current. Substrates in the Flemish Pass consist mostly of sandy mud with some pebbles and stones (Litvin and Rvachev, 1963). The Northeast Newfoundland Shelf includes Funk Island Bank, Funk Island Spur, and Tobin's Point. Substrates abruptly change from sand on top of the Funk Island Bank at ~ 300 m, to sandy-mud on the slope at ~ 500 m, to mud off Funk Island Spur at > 1000 m (Litvin and Rvachev, 1963). The Labrador Shelf extends along the entire coastline of Labrador, with the widest section off Hamilton Bank. Transverse troughs up to 600 m deep divide the shelf into banks (Piper et al., 1988). Sand substrates dominate southern bank tops with scattered pebbles and gravel; slope composition changes rapidly with depth from muddy-sand to sandy-mud to mud. Hawke Saddle, located south of Hamilton Bank, has a mud substrate at 500 m but changes to sandy-mud on the saddle slope towards the shelf edge (Litvin and Rvachev, 1963). There is little information available on slope substrates north of Harrison Bank on the Labrador Shelf and Baffin Island Shelf.

**Comparison with local ecological knowledge (LEK).**—Many of the coral areas identified by fishers' LEK (Gass and Willison, 2005) were confirmed with scientific survey and observer data in the current study. Nonetheless, several important differences emerged. First, the current data suggest that there is much more continuous coral habitat along the southern Labrador slope, with a wider variety of corals than indicated from LEK. Second, the southwest Grand Bank and the Tail of the Bank hotspot, identified in the current study as an area of high species richness and coral record density, was much less prominent in LEK data or in previously available scientific survey data. These discrepancies between studies may largely be a result of more complete scientific survey sampling efforts throughout the Newfoundland and Labrador region. In the current study, the lack of large gorgonian records on the southeast Grand Bank, and the relative scarcity of *P. arborea* samples may reflect loss of corals due to trawling impacts. Evidence of deleterious effects on deep-sea corals by mobile fishing gear (e.g., trawls) has been published in detail (Watling and Norse, 1998; Auster and Langton, 1999; Fosså et al., 2002; Hall-Spencer et al., 2002; Anderson and Clark, 2003), mostly focusing on the effects of trawling on deep-sea scleractinians, with limited attention to impacts on deep-sea gorgonians (Krieger, 2001; Mortensen et al., 2005, Stone, 2006). In the current study, mobile gears captured more corals than fixed gears, and in general, covered larger areas. The duration of trawl tows ranged between 1 and 10 hrs per tow, making the precision of coral localities from observer data highly variable. Although the deep-sea coral clusters recognized by fishers have persisted despite a long history of intensive deep-water trawl fishing in the region (cf. Kulka and Pitcher, 2002), there is little information on changes in abundances of deep-sea corals through time (Gass and Willison, 2005).

**Associated invertebrate diversity.**—A variety of sessile invertebrates were observed growing commensally on deep-sea corals in this study. Although trawl samples tend to underestimate associated invertebrate diversity of corals (Buhl-Mortensen and Mortensen, 2005), samples from fixed gear (e.g., longline and gillnet) have contributed many intact coral assemblages, or groups of coral living together in what appears to be commensal relationships. For example, two colonies of *A. armata* were found to be attached to two separate *K. ornata* colonies, one of which also included the scleractinian *D. dianthus*. Another *K. ornata* sample had juvenile colonies of *A. grandiflorus* and *A. armata* attached. Many observations of gooseneck barnacles, Icelandic scallops, sea anemones, and echinoderms, all juveniles, were attached to *K. ornata*. Many nephtheid soft corals were observed attached to living gastropods. These observations suggest that hard substrates may be limited in some areas, and emphasize the important contribution that large corals can make towards creating and structuring deep-sea habitat, including habitat for other deep-sea corals. The nature of associations between corals and fish are difficult to determine in trawl survey data because fish and corals may have co-occurred in the same habitat without any direct biological interaction (Edinger et al., 2007).

**Limitations and conservation implications.**—The findings reported here complement earlier work (Gass and Willison, 2005), and provide specific information on deep-sea coral distribution and diversity in the region. However, caution must be exercised when interpreting these results for three reasons. First, the data resulted from only 3 yrs of sampling, with only 1 yr of scientific sampling in northern Labrador and the Davis Strait, and with sampling gaps in scientific data. Second, the distribution data from fisheries observers were biased by fishing effort. Finally, point maps imply a greater degree of sampling area coverage than the area actually surveyed in the present work. Coral conservation is a fairly new concept in Eastern Canada. Three areas with unique features including very high densities of corals or unique species occurrences were established on the Scotian margin to help protect corals: the Northeast Channel Coral Conservation Area (2002), the Stone Fence *Lophelia* reef fisheries closure (2004), and The Gully Marine Protected Area (2004; Breeze and Fenton, 2007). Corals in Newfoundland and Labrador waters are generally widespread along the continental edge and slope. Hence a network of representative areas would be the most appropriate conservation approach (cf. Fernandes et al., 2005).

## Acknowledgments

We thank all the fisheries observers, DFO technicians, and Canadian Coast Guard personnel that helped contribute to the study. J. Firth, P. Veitch, D. Orr, and K. Gilkinson (DFO) made invaluable contributions and provided moral support. We thank the Northern Shrimp Research Foundation for coral data from the Northern Labrador and Baffin Island regions. R. Devillers (MUN) and T. Bowdring (DFO) assisted in generating maps. S. Gass provided guidance and raw data on prior records including LEK records. DFO Science division provided logistical support. P. Etnoyer, S. France, K. Gilkinson, R. Haedrich, O. Sherwood provided useful critiques of the manuscript. Study supported by the Department of Fisheries and Oceans, Oceans and Habitat Division, NSERC Discovery grant to E. E., and by Memorial University. Special thanks to our families and friends for their ongoing support, and never-ending patience.

## Literature Cited

Anderson, O. F. and M. R. Clark. 2003. Analysis of by-catch in the fishery for orange roughy *Hoplostenthus atlanticus*, on the south Tasman Rise. Mar. Freshwat. Res. 54: 643–652.

Andrews, A. H., E. E. Cordes, M. M. Mahoney, K. Munk, K. H. Coale, G. M. Cailliet, and J. Heifetz. 2002. Age, growth, and radiometric age validation of a deep-sea, habitat forming gorgonian (*Primnoa resedaeformis*) from the Gulf of Alaska. Hydrobiologia 471: 101–110.

Auster, P. J. 2005. Are deep-water corals important habitats for fishes? Pages 747–760 *in* A. Freiwald and J. M. Roberts, eds. Cold-water corals and ecosystems. Springer-Verlag, Heidelberg.

_____ and R. W. Langton. 1999. The effects of fishing on fish habitat. Am. Fish. Soc. Symp. 22: 150–187.

Bayer, F. M. 1981. Key to the genera of Octocorallia exclusive of Pennatulacea (Coelenterata: Anthozoa), with diagnoses of new taxa. Proc. Biol. Soc. Wash. 94: 902–947.

Breeze, H. and D. G. Fenton. 2007. Designing management measures to protect cold-water corals off Nova Scotia, Canada. Pages 123–133 *in* R. Y. George and S. D. Cairns, eds. Conservation and adaptive management of seamount and deep-sea coral ecosystems. Rosenstiel School of Marine and Atmospheric Science, University of Miami. Miami. 324 p.

_____, D. S. Davis, M. Butler, and V. Kostlyev. 1997. Distribution and status of deep sea corals off Nova Scotia. Marine Issues Committee Special Publication1, Ecology Action Centre, Halifax.

Buhl-Mortensen, L. and P. B. Mortensen. 2005. Distribution and diversity of species associated with deep-sea gorgonian corals off Atlantic, Canada. Pages 771–805 *in* A. Freiwald and J. M. Roberts, eds. Cold-water corals and ecosystems. Springer-Verlag, Heidelberg.

Cairns, S. D. and R. E. Chapman. 2001. Biogeographic affinities of the North Atlantic deep-water Scleractinia. Pages 30–57 *in* J. H. M. Willison, J. Hall, S. E. Gass, E. L. R. Kenchington, M. Butler, and P. Doherty, eds. Proc. First Int. Symp. on Deep-Sea Corals. Ecology Action Centre, Halifax.

Collins, J. W. 1884. On the occurrence of corals on the Grand Banks. Bull. U.S. Fish. Comm. 237.

Deichmann, E. 1936. The Alcyonaria of the western part of the Atlantic Ocean. Mem. Mus. Comp. Zool. 53: 1–317 + 37 plates.

Drinkwater, K. F. and G. C. Harding. 2001. Effects of the Hudson Strait outflow on the biology of the Labrador Shelf. Can. J. Fish. Aquat. Sci. 58: 171–184.

Edinger, E. N., V. E. Wareham, and R. L. Haedrich. 2007. Patterns of groundfish diversity and abundance in relation to deep-sea coral distributions in Newfoundland and Labrador waters. Pages 101–122 *in* R. Y. George and S. D. Cairns, eds. Conservation and adaptive management of seamount and deep-sea coral ecosystems. Rosenstiel School of Marine and Atmospheric Science, University of Miami. Miami. 324 p.

Fader, G. B. and R. O. Miller. 1986. A reconnaissance study of the surficial and shallow bedrock geology of the southeastern Grand Banks of Newfoundland Geol. Surv. Can. Pap. 86-1B: 591–604.

Fernandes, L., J. Day, A. Lewis, S. Slegers, B. Kerrigan, D. Breen, D. Cameron, B. Jago, J. Hall, D. Lowe, J. Innes, J. Tanzer, V. Chadwick, L. Thompson, K. Gorman, M. Simmons, B. Barnett, K. Sampson, G. De'Ath, B. Mapstone, H. Marsh, H. H. Possingham, I. Ball, T. Ward, K. Dobbs, J. Aumend, D. Slater, and K. Stapleton. 2005. Establishing representative no-take areas in the Great Barrier Reef: large-scale implementation of theory on marine protected areas. Conserv. Biol. 19: 1733–1751.

Fosså, J. H., P. B. Mortensen, and D. M. Furevik. 2002. The deep-water coral *Lophelia pertusa* in Norwegian waters: distribution and fishery impacts. Hydrobiologia 471: 1–12.

Freiwald, A. and J. M. Roberts, eds. 2005. Cold-water corals and ecosystems. Springer-Verlag, Berlin Heidelberg.

Gass, S. E. and J. H. M. Willison. 2005. An assessment of the distribution of deep-sea corals in Atlantic Canada by using both scientific and local forms of knowledge. Pages 223–245 *in* A. Freiwald and J. M. Roberts, eds. Cold-water corals and ecosystems. Springer-Verlag, Heidelberg.

GEBCO. 2003. GEBCO digital atlas: centenary edition of the IHO/IOC general bathymetry of the oceans. National Environment Research Council, Swindon, UK.

Haedrich, R. L. and J. M. Gagnon. 1991. Rock wall fauna in a deep Newfoundland fiord. Cont. Shelf. Res. 11: 1199–1207.

Hall-Spencer, J., V. Allain, and J. H. Fosså. 2002. Trawling damage to Northeast Atlantic ancient coral reefs. Proc. R. Soc. Lond., B. 269: 507–511.

Harding, G. C. 1998. Submarine canyons: deposition centers for detrital organic matter? Pages 105–107 *in* W. G. Harrison and D. G. Fenton, eds. The Gully; A scientific review of its envi-

Harding, G. C. 1998. Submarine canyons: deposition centers for detrital organic matter? Pages 105–107 *in* W. G. Harrison and D. G. Fenton, eds. The Gully; A scientific review of its environment and ecosystem. Dept. Fish Oceans, Can. Stock. Assess. Secr. Res. Doc. 98/83, Dartmouth. 202 p.

Hecker, B., G. Blechschmidt, and P. Gibson. 1980. Final report-Canyon assessment study in the Mid and North Atlantic areas of the U.S. outer continental shelf. US Dept. Interior, Bureau of Land Management, Washington, DC. Contract No BLM-AA551-CT8-49.

Jourdan, E. 1895. Zoanthaires provenant des campagnes du yacht l'Hirondelle (Golfe de Gascogne, Açores, Terra-Neuve). Rés. Camp. Scient. Prince de Monaco. 8: 1–36.

Krieger, K. L. 2001. Coral (*Primnoa*) impacted by fishing gear in the Gulf of Alaska. Pages 106–116 *in* J. H. M. Willison, J. Hall, S. E. Gass, E. L. R. Kenchington, M. Butler, and P. Doherty, eds. Proc. First Int. Symp. on Deep-sea Corals. Ecology Action Centre and Nova Scotia Museum, Halifax,

_____ and B. L. Wing. 2002. Megafauna associations with deepwater corals (*Primnoa* spp.) in the Gulf of Alaska. Hydrobiologia 471: 83–90.

Koslow, J. A., G. W. Boehlert, J. D. M. Gordon, R. L. Haedrich, P. Lorance, and N. Parin. 2000. Continental slope and deep-sea fisheries: implications for a fragile ecosystem. ICES J. Mar. Sci. 57: 548–557.

Kulka, D. W. and J. R. Firth. 1987. Observer Program Training Manual-Newfoundland Region. Canadian Technical Report of Fisheries and Aquatic Sciences No 1355 (Revised). 197 p.

_____ and D. A. Pitcher. 2002. Spatial and temporal patterns in trawling activity in the Canadian Atlantic and Pacific. ICES CM 2001/R:02. 57 p.

Lazier, A., J. E. Smith, M. J. Risk, and H. P. Schwarcz. 1999. The skeletal structure of *Desmophyllum cristagalli*: the use of deep-water corals in sclerochronology. Lethaia 32: 119–130.

Lazier, L. R. N. and D. G. Wright. 1993. Annual velocity variations in the Labrador Current. J. Phys. Oceanogr. 23: 659–678.

Litvin, V. M. and V. D. Rvachev. 1963. The bottom topography and sediments of the Labrador and Newfoundland fishing areas. Pages 100–112 *in* Y. Y. Marti, ed. Translated from Russian. United States Department of the Interior and the National Science Foundation.

MacIsaac, K., C. Bourbonnais, E. Kenchington, D. Gordon, and S. Gass. 2001. Observations on the occurrence and habitat preference of corals in Atlantic Canada. Pages 58–75 *in* J. H. M. Willison, J. Hall, S. E. Gass, E. L. R. Kenchington, M. Butler, and P. Doherty, eds. Proc. First Int. Symp. on Deep-sea Corals. Ecology Action Centre and Nova Scotia Museum, Halifax.

McCallum, B. R. and S. J. Walsh. 1996. Groundfish survey trawls used at the Northwest Atlantic fisheries centre, 1971-present. Doc. 96/50. Serial No. N2726. 18 p.

Miner, R. W. 1950. Field book of seashore life. G. P. Putnam's Son, New York. 215 p.

Mortensen, P. B. and L. Buhl-Mortensen. 2004. Distribution of deep-water gorgonian corals in relation to benthic habitat features in the Northeast Channel (Atlantic Canada). Mar. Biol. 144: 1223–1238.

_____ and _____. 2005a. Morphology and growth of the deep-water gorgonians *Primnoa resedaeformis* and *Paragorgia arborea*. Mar. Biol. 147: 775–788.

_____ and _____. 2005b. Deep-water corals and their habitats in the The Gully, a submarine canyon off Atlantic Canada. Pages 247–277 *in* A. Freiwald and J. M. Roberts, eds. Cold-water corals and ecosystems. Springer-Verlag, Heidelberg.

_____, _____, and D. C. Gordon, Jr. 2006. Distribution of deep-water corals in Atlantic Canada. Pages 1832–1848 *in* Y. Suzuki, T. Nakamori, M. Hidaka, H. Kayanne, B. E. Casareto, K. Nadaoka, H. Yamano, and M. Tsuchiya, eds. Proc. 10th Int. Coral Reef Symp., Okinawa.

_____, _____, D. C. Gordon, Jr., G. B. J. Fader, D. L. McKeown, and D. G. Fenton. 2005. Effects of fisheries on deepwater gorgonian corals in the Northeast Channel, Nova Scotia. Am. Fish. Soc. Symp. 41: 369–382.

Nesis, K. I. 1963. Soviet investigations of the benthos of the Newfoundland-Labrador fishing area. Pages 214–222 *in* Y. Y. Marti, ed. Translated from Russian. United States Department of the Interior and the National Science Foundation.

Opresko, D. M. 2002. Revision of the Antipatharia (Cnidaria: Anthozoa). Part II. Schizopathidae. Zool. Med. Leiden. 76: 411–442.

Piper, D. J. W., G. D. M. Cameron, and M. A. Best (comp.) 1988: Quaternary geology of the continental margin of eastern Canada; Geological Survey of Canada, Map 1711A, scale 1:5,000,000.

Risk, M. J., J. M. Heikoop, M. G. Snow, and R. Beukens. 2002. Lifespans and growth patterns of two deep-sea corals: *Primnoa resedaeformis* and *Desmophyllum cristagalli*. Hydrobiologia 471: 125–131.

Roark, E. B., T. P. Guilderson, S. Flood-Page, R. B. Dunbar, B. L. Ingram, S. J. Fallon, and M. Mc-Culloch. 2005. Radiocarbon-based ages and growth rates of bamboo corals from the Gulf of Alaska, Geophys. Res. Lett., 32, L04606, doi:10.1029/2004GL021919.

Sherwood, O. A., D. B. Scott, and M. J. Risk. 2006. Late Holocene radiocarbon and aspartic acid racemization dating of deep-sea octocorals. Geochimica et Cosmochimica Acta 70: 2806–2814.

Stone, R. P. 2006. Coral habitat in the Aleutian Islands of Alaska: depth distribution, fine-scale species associations, and fisheries interactions. Coral Reefs 25: 229–238.

Tendal, O. S. 1992. The North Atlantic distribution of the octocoral *Paragorgia arborea* (L., 1758) (Cnidaria, Anthozoa). Sarsia 77: 213–217.

_____. 2004. The Bathyal Greenlandic black coral refound: alive and common. Deep-sea newsletter [Serial online]; 33. 28-30. Available from: http://www.le.ac.uk/biology/gat/deepsea/DSN33-final.pdf via the Internet. Accessed 10 Sept. 2005.

Thrush, S. F. and P. K. Dayton. 2002. Disturbance to marine benthic habitats by trawling and dredging: implications for marine biodiversity. Annu. Rev. Ecol. Syst. 2002. 33: 449–473.

Verseveldt, J. 1940. Studies on Octocorallia of the families Briareidae, Paragorgiidae and Anthothelidae. Temminckia 5: 1–142.

Watling, L. and P. J. Auster. 2005. Distribution of deep-water Alcyonacea off the Northeast Coast of the United States. Pages 279–296 *in* A. Freiwald and J. M. Roberts, eds. Cold-water corals and ecosystems. Springer-Verlag, Heidelberg.

_____ and E. A. Norse. 1998. Disturbance of the seabed by mobile fishing gear: a comparison to forest clear cutting. Conserv. Biol. 12: 1180–1197.

Williams, G. C. 1995. Living genera of sea pens (Coelenterata: Octocorallia: Pennatulacea): illustrated key and synopses. Zool. J. Linn. Soc. 113: 93–140.

_____. 1999. Index Pennatulacea: Annotated bibliography and indexes of the sea pens (Coelenterata: Octocorallia) of the world 1469–1999. Proc. Calif. Acad. Sci. 51: 19–103.

ADDRESSES: (V.E.W.) *Environmental Science Program, Memorial University, St. John's, Newfoundland and Labrador, Canada; and Fisheries and Oceans Canada, St. John's, Newfoundland and Labrador, Canada A1C 5X1.* (E.E.) *Geography Dept. and Biology Dept., Memorial University, St. John's, Newfoundland and Labrador, Canada A1B 3X9.* CORRESPONDING AUTHOR: (V.E.W.) *E-mail: <warehamv@dfo-mpo.gc.ca>.*

# Note

# Seabed characteristics and *Lophelia pertusa* distribution patterns at sites in the northern and eastern Gulf of Mexico

WILLIAM W. SCHROEDER

The often patchy, discontinuous distribution of *Lophelia pertusa* (Linnaeus, 1758) was unequivocally established by Wilson (1979a,b) during his investigations of Rockall Bank and other sites in the north-east Atlantic. This work corroborated earlier findings by Joubin (1922) in the Bay of Biscay, Stetson et al. (1962) on Blake Plateau, and Squires (1964) in Wairarapa, New Zealand. Two decades later, Rogers (1999), in his review of the biology of *L. pertusa* and other deep-water reef-forming corals, concluded that factors influencing the distribution of deep-water corals by and large continued to be poorly understood. However, he goes on to state that over small scales topography and hydrographic conditions play important roles in structuring distribution patterns. In their study off Norway, Mortensen et al. (2001) found that although *L. pertusa* reefs were not evenly distributed over the seabed they did occur in geographic and bathymetric patterns that appeared to be regulated by external factors such as: (1) presence of suitable substrate; (2) topography; (3) physical and chemical properties of water masses; and (4) availability of food. Even in the most recent literature uncertainties remain as to exactly which factors play controlling roles in determining distribution patterns (Roberts et al., 2003; Taviani et al., 2005).

In the Gulf of Mexico (GoM), below depths of 300 m, the seafloor is dominated by fine-grained pelagic and hemipelagic sediments (Coleman et al., 1991). Notable exceptions are limestone hardgrounds, lithoherms, and buildups/mounds on the west Florida and Banco de Campeche slopes and in the Florida Straits, and authigenic carbonates/chemoherms (i.e., hydrocarbon-derived and microbially mediated carbonates) in the form of clasts, slabs, hardgrounds, and mound-like buildups (see Roberts and Aharon, 1994) on the Texas-Louisiana, De Soto, and Gulfo de Campeche slopes. These indurated substrates provide suitable surfaces for the colonization and development of sessile megafauna assemblages, which include *L. pertusa* (Schroeder et al., 2005). The purpose of this note is to present initial results of an investigation that examined seabed characteristics and *L. pertusa* distribution patterns at six upper slope sites in the northern and eastern GoM (Table 1, Fig. 1).

## Methods

Landscape descriptions of the seabed surrounding each site are based on published literature (e.g., Reilly et al., 1996; Sager et al., 1999; MacDonald et al., 2003), unpublished reports (e.g., Fu-

George, R. Y. and S. D. Cairns, eds. 2007. Conservation and adaptive management of seamount and deep-sea coral ecosystems. Rosenstiel School of Marine and Atmospheric Science, University of Miami.

Table 1. Landscape descriptions of the seabed surrounding the six study sites. The sites are listed from west to east and named for the Minerals Management Service lease block in which they occur. See Figure 1.

| Study sites | Landscape descriptions |
|---|---|
| Green Canyon (GC) 354 | Plateau rim and scarp |
| Green Canyon (GC) 234 | Grabens and half-grabens within a shallow depression |
| Mississippi Canyon (MC) 885 | Broad, smooth low-relief mound |
| Viosca Knoll (VK) 862 | Low-relief mounds and ridges adjacent to a small submarine canyon |
| Viosca Knoll (VK) 826 | Knoll and depression on a steepening seaward slope |
| Vernon Basin (VB) 640 | Erosional scarp and terrace at the top of the steepening portion of a carbonate ramp |

gro-McClelland, unpubl. data), maps (e.g., Taylor et al., 2000) and available archived geophysical data (e.g., multibeam bathymetry and side-scan sonograph records). Descriptions of site-specific features and *L. pertusa* distribution patterns for each of the study sites are based on photographic documentation (video and digital images), in situ observations obtained during manned submersible and USN submarine NR-1 operations and published literature (e.g., Roberts and Aharon, 1994).

## Results

**Green Canyon 354.**—GC354 is the western most study site. It is located on the seaward edge of the rim of a plateau-like feature adjacent to a steep scarp in the upper central dome and basin subprovince on the Texas-Louisiana slope (Fig. 1). The explored portion of this site, centered at approximately 27°35.85′N, 91°49.50′W, lies on a slope descending from 520–560 m. Scattered across the slope are small chemoherms and authigenic carbonate capped mounds (Fig. 2A). Some of these features are covered with mostly dead *L. pertusa* colonies (Fig. 2B) while others are un-colonized and often have large tubeworm aggregations at their bases. In addition, living and dead *Madrepora oculata* Linnaeus, 1758 colonies have been observed on the lower-slope mounds.

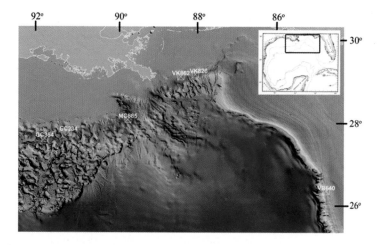

Figure 1. Locations of the six study sites in the northern and eastern GoM.

Figure 2. (A) A 1.9 m high carbonate capped mound at GC354 covered with mostly dead *Lophelia pertusa* colonies; (B) Close-up of mostly dead *L. pertusa* near the top of a mound at GC354; (C) Mostly dead *L. pertusa* colonies on a western rim outcrop at GC234 with adjacent outcrop visible in the background; (D) Small *L. pertusa* and *Callogorgia americana delta* colonies on a low-relief buildup at MC855.

**Green Canyon 234.**—Also located in the upper central dome and basin subprovince on the Texas-Louisiana slope, GC234 is situated within a graben/half-graben complex in a shallow depression (Reilly et al., 1996; Fig. 1). The primary study area is an extensive set of low-relief, linear authigenic carbonate outcrops on the western rim of a half-graben centered around 27°44.8′N, 91°13.4′W, in water depths of 500–530 m. Attached to many of the outcrops are both individual and aggregated colonies of primarily dead *L. pertusa* (Fig. 2C). Down slope to the east assemblages of the gorgonian *Callogorgia americana delta* (Cairns and Bayer, 2002) and tubeworm bushes replace *L. pertusa* as the dominate megafauna colonizing the outcrops and carbonate capped mounds.

**Mississippi Canyon 885.**—MC885, the deepest site investigated, is located on a broad, smooth low-relief mound on Henderson Ridge west of the Mississippi Canyon (Fig. 1). The site is predominantly a small, gently sloping mound of fine sediment with some shell hash centered at about 28°03.9′N, 89°43.0′W, at depths of 626–643 m. It is a region of active brine seepage and includes bacterial mats, mussel beds, and tubeworm aggregations. Also present are scattered clusters of authigenic carbonate ranging from clasts and nodules of various sizes and shapes to low-relief outcrops/buildups. Attached to some of the carbonate are *L. pertusa* (Fig. 2D) and *M. oculata* colonies. The largest coral colonies observed were *M. oculata*; up to 65 cm high. Also present are large colonies of *C. a. delta* (Fig. 2D). It is estimated that up to 50% of the carbonate at this site is not colonized by megafauna.

**Viosca Knoll 862.**—VK862, the shallowest site investigated, is located in a low-relief mound and ridge complex adjacent to a small submarine canyon on the upper De Soto Slope subprovince (Fig. 1). The surveyed portion of the site is a small, low-relief mound composed of large, exposed authigenic carbonate blocks and boulders, some with vertical and/or horizontal dimensions in the order of 5 m, flanked by aprons of smaller material and hardgrounds. It is centered on 29°06.4′N, 88°23.0′W, approximately 2 km northeast of the eastern rim of a submarine canyon. Water depths range between 304–333 m.

Only one significant development of *L. pertusa* was found at this site. An aggregation of at least five large colonies, up to 90 cm in height, and numerous small colonies on top of a 5 m tall boulder at a depth of 309 m near the crest of the south side of the mound (Fig. 3A). This is the shallowest reported site for *L. pertusa* in the GoM (Schroeder et al., 2005). Other smaller colonies, either solitary or in clusters were observed scattered throughout the site with a slight preference for the northern flank of the mound. The dominant taxa at this site, in terms of numbers and biomass are unidentified anemones, and the largest megafauna are antipatharians (Fig. 3B). Large portions of the carbonate were observed to be bare or colonized by only small individuals. There is no evidence of any chemosynthetic fauna at this site. Schroeder et al. (2005) suggest that the adjacent submarine canyon is likely the area where Moore and Bullis (1960) first reported *L. pertusa* (= *Lophelia prolifera*) from the GoM.

**Viosca Knoll 826.**—VK826, first described by Schroeder (2002), has the most extensive *L. pertusa* development found in the GoM to date (Schroeder et al., 2005). It is located on a 90 m tall, isolated knoll on the seaward steepening upper De Soto Slope in the northeastern GoM (Fig. 1). The explored region is centered on 29°09.5′N, 88°01.0′W, in water depths of 430–534 m, and covers most of the crest, south, west and north sides, and portions of the southern and western base/flank regions. The presence of apparently very old living tubeworm aggregations and numerous accumulations of disarticulated lucinid and vesicomyid shells suggest this area is at a senescent phase of chemosynthethic community activity. A second site, the crest and upper portions of a small mound, lies approximately 1 km northeast of the top of the main knoll at 29°10.2′N, 88°00.7′W, in water depths of 458–481 m.

The generally flat crest of the knoll is covered with a combination of broken hardground, low-relief outcrops/buildups, shell pavement, and fine sediment while the adjacent, moderately sloping crest-rim consists of locally hummocky terrain made up of carbonate knolls and ridges, some with abrupt vertical relief, and relatively flat terraces built from carbonate substrate, shell pavement and fine sediment. Individual colonies of *L. pertusa*, as large as 1.5 m in diameter, and colony aggregations up to 1.5 m wide × 1.5 m high × 4 m long occur in this region (Fig. 3C) along with carbonate surfaces that are totally un-colonized (Fig. 3D). The upper regions of the south and north sides of the knoll are a mixture of moderate to steep slopes and terrace-like features composed of carbonate outcrops/buildups, sediment veneered hardgrounds and/or open flats of fine sediment. Extensive assemblages of *L. pertusa* have developed within these areas. Fields of individual colonies, up to 2 m in diameter, and clusters/aggregations of colonies, some as large as 2 m wide × 2 m high × 4 m long occur on both exposed hardground/buildups and sediment veneered hardgrounds (Figs. 3E–F).

The slope on the western side of the knoll is steep and consists of a series of ridges or hummocks and swales of unconsolidated sediment that are oriented along-isobath and have vertical relief of 1–3 m from crest to trough. Carbonate outcrops occasionally oc-

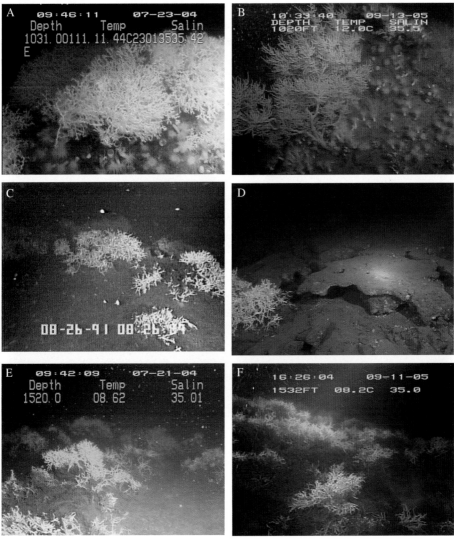

Figure 3. *Lophelia pertusa* colonies on large boulder near the top of the south side of the mound at VK862; (B) Large antipatharian colony and unidentified anemones on a large boulder near the top of the small mound at VK862; (C) Individual and aggregations of *L. pertusa* colonizing carbonate covered knolls and ridges on the southwestern crest-rim at VK826; (D) Mostly un-colonized authigenic carbonate on the southeastern crest-rim at VK826; (E) *L. pertusa* aggregations/thickets on sediment covered hardground on the south side of the knoll at VK826; (F) *L. pertusa* aggregations/thickets on exposed hardground/buildups on north side of the knoll at VK826.

cur on the crests of these features and provide substrate for individual and aggregations of *L. pertusa* colonies. To the immediate east and southeast, the bottom levels off into a relatively flat area consisting mostly of fine, unconsolidated sediment and patches of both exposed and sediment veneered carbonate material some with small *L. pertusa* colonies. The lower portion of the south-southwestern side of the knoll exhibits evidence of slumping. At the base of this zone, at depths of 510–540 m, the bottom flattens out into at least one terrace-like feature that is covered with broken hardground, large boulders and blocks (up to 3–4 m tall × 1–2 m wide) and smaller debris of various sizes and shapes. Most of the carbonate substrate is barren, however, large colonies of *C. a. delta* as well as smaller *L. pertusa* have occasional been observed (Fig. 4A).

*Lophelia pertusa* development at the small mound to the northeast, named "Knobby Knoll" because of its very hummocky crest, is extraordinary. It is so dense that carbonate substrate is visible in only a few locations and thicket production has either covered the numerous knob-like features or has reached a level of initial coppice development (Mullins et al., 1981; Fig. 4B).

**Vernon Basin 640.**—VB640 is the eastern most study site. It is located on an erosional scarp and terrace complex in 450–550 m of water at the top of the steepening portion of a carbonate ramp (Fig. 1). In 1984–1985 Newton et al. (1987) carried out seismic and dredge surveys in this area and described it as a 20 km long linear zone of Late Pleistocene age *L. pertusa* (= *L. prolifera*) buildups. In 2003, during a NOAA-OE expedition, multibeam bathymetric surveys and ROV operations were conducted at a site near the southern end of this zone at 26°20.0′N, 84°45.0′W (Reed et al., 2004). They documented that possibly hundreds of small, low-relief (5–15 m) lithoherms, constructed of rugged black phosphorite-coated limestone, have formed on the terrace in water depths of 450–500 m (Fig. 4C; Reed et al., 2006). Living and dead *L. pertusa* colonies, up to 1 m in diameter and 0.6 m high, were observed on the tops and sides of the two lithoherms they ground-truthed (Fig. 4D; Reed et al., 2004, 2006). However, *L. pertusa* colonization was not observed on either the bases of the lithoherms or on the face or the rim of the erosional scrap.

**Hydrography.**—Near-bottom temperature and salinity values measured where *L. pertusa* was observed at these sites ranged from 12.0 °C and 35.5 in 305 m of water at VK862 to 6.4 °C and 34.9 in 631 m of water at MC885.

## Discussion

Landscapes of the surrounding seabed differed for each of the six sites investigated where *L. pertusa* occur in the northern and eastern GoM. They varied across a spectrum from relatively simple to moderately-complex; MC855, GC354, GC234, VK826, VK862, and VB640. The landscape at the site with the most extensive *L. pertusa* development, VK826, is a 90 m tall, isolated knoll on the seaward steepening upper De Soto Slope in the northeastern GoM. The site with the second largest development, VB640, consists of an erosional scarp and terrace at the top of the steepening portion of a carbonate ramp in the eastern GoM.

The composition of the substrate forming the site-specific features at the five northern GoM sites is authigenic carbonate/chemoherms. Features at the single site in the eastern GoM are constructed of black phosphorite-coated limestone. The types of site-specific features *L. pertusa* has successfully colonized include various sizes and shapes of clasts, nodules, blocks and boulders, hardgrounds and isolated slabs (often fractured and/or sediment veneered), outcrops, buildups, and carbonate-capped mounds and ridges. The extent of areal coverage varies from scattered, generally small solitary features (MC885 and GC354) to more widespread accumulations of larger structures (G234 and VK862) to fairly complex settings (VK826 and VB640). Vertical relief of individual features range from none up to 5 m.

Figure 4. (A) *Callogorgia americana delta* on a large carbonate block at the base of the south side of the knoll at VK826. Small *Lophelia* colonies are often found in this area; (B) A 4.5 m high knob covered with either *Lophelia pertusa* thickets or coppice at VK826-Knobby Knoll. (C) A 12 m high limestone lithoherm at VB640. (Courtesy NOAA-OE and John Reed, HBOI). (D) Upper portion of a lithoherm with living and dead *L. pertusa* at VB640. (Courtesy NOAA-OE and John Reed, HBOI).

In agreement with previous studies, colonization of *L. pertusa* on these hard substrates was observed to be highly variable. Distribution patterns range from scattered, isolated individuals to aggregations of varying densities that in some areas are in the initial phase of thicket building and at one site, VK826-Knobby Knoll, is possibly at a thicket/coppice stage (Mullins et al., 1981). To date, the analysis of the available data has not provided sufficient insight into determining what factors are controlling distribution patterns in the GoM. What can be concluded, considering the percentage of un-colonized substrate at all the sites, except VK826-Knobby Knoll, is that space does not appear to be a limiting factor in the development of *L. pertusa* or other sessile megafauna assemblages in the deep GoM. This suggests that megafauna development and survival at these sites is likely a function of larval recruitment and/or food availability; which is nec mirabile dictu for deep-sea environments.

## Acknowledgments

Thanks are due J. Reed, P. Aharon, S. Brooke, B. Carney, E. Cordis, H. Roberts, B. Phanef, M. Rex, and K. Sulak for providing material and insight that contributed to the preparation of this manuscript. In addition, thanks are due to the CSA/MMS scientific parties and HBOI sub and vessel crews for support during the 2004 and 2005 MMS cruises and to the crews of the US Navy Submarine NR1 during 1993 and 2002 cruises. Funding for this research has been provided by the Minerals Management Service, NOAA-OE and the US Navy.

## Literature Cited

Coleman, J. M., H. H. Roberts, and W. R. Bryant. 1991. Late Quaternary sedimentation, Pages 325–352 in A. Salvador, ed. The geology of North America, Vol. J, The Gulf of Mexico Basin, The Geological Society of America, Boulder.

Joubin, L. 1922. Les coraux de mar profonde nuisibles aux chalutiers. Notes et Memoires. Office Scientifique et Technique des Peches Maritimes, No. 18, 16 p.

MacDonald, I. R., W. W. Sager, and M. B. Peccini. 2003. Gas hydrate and chemosynthetic biota in mounded bathymetry at mid-slope hydrocarbon seeps: northern Gulf of Mexico. Mar. Geol. 198: 133–158.

Moore, D. R. and H. R. Bullis. 1960. A deep-water coral reef in the Gulf of Mexico. Bull. Mar. Sci. 10: 25–128.

Mortensen, P. B., M. Hovland, J. H. Fossa, and D. M. Furevik. 2001. Distribution, abundance and size of Lophelia pertusa coral reefs in mid-Norway in relation to seabed characteristics. J. Mar. Biol. Assoc. U.K. 81: 581–597.

Mullins, H. T., C. R. Newton, K. Heath, and H. M. Vanburen. 1981. Modern deep-water coral mounds north of Little Bahama Bank: Criteria for recognition of deep-water bioherms in the rock record. J. Sediment. Petrol. 51: 999–1013.

Newton, C. R., H. T. Mullins, and A. F. Gardulski. 1987. Coral mounds on the west Florida slope: Unanswered questions regarding the development of deep-water banks. Palaios 2: 59–367.

Reed, J. K., D. Weaver, and S. A. Pomponi. 2006. Habitat and fauna of deep-water Lophelia pertusa coral reefs off the Southeastern USA: Blake Plateau, Straits of Florida, and Gulf of Mexico. Bull. Mar. Sci. 78: 343–375.

_____, A. Wright, and S. Pomponi. 2004. Medicines from the deep sea: exploration of the northeastern Gulf of Mexico. Pages 58–70 in Proc. Amer. Acad. Underwater Sci. 23th Annual Scientific Diving Symp., March 11–13, 2004, Long Beach, California.

Reilly, Jr., J. F., I. R. MacDonald, E. K. Biegert, and J. M. Brooks. 1996. Geologic controls on the distribution of chemosynthetic communities in the Gulf of Mexico. Pages 39–62 in D. Schumacher and M. A. Abrams, eds. Hydrocarbon migration and its near-surface expression. AAPG Memoir 66.

Roberts, H. H. and P. Aharon. 1994. Hydrocarbon-derived carbonate buildups of the northern Gulf of Mexico continental slope: a review of submersible investigations. Geo-Marine Let. 14: 135–148.

Roberts, J. M., D. Long, J. B. Wilson, P. B. Mortensen, and J. D. Gage. 2003. The cold-water coral Lophelia pertusa (Scleractinia) and enigmatic seabed mounds along the north-east Atlantic margin: are they related? Mar. Pollut. Bull. 46: 7–20.

Rogers, A. D. 1999. The biology of Lophelia pertusa (Linnaeus 1758) and other deep-water reef forming corals and impacts from human activities. Int. Rev. Hydrobiol. 844: 315–406.

Sager, W. W., C. S. Lee, I. R. Macdonald, and W. W. Schroeder. 1999. High-frequency near-bottom acoustic reflection signatures of hydrocarbon seeps on the northern Gulf of Mexico continental slope. Geo-Marine Let. 18: 267–276.

Schroeder, W. W. 2002. Observations of Lophelia pertusa and the surficial geology at a deep-water site in the northeastern Gulf of Mexico. Hydrobiology 471: 29–33.

_____, S. Brooke, J. Olson, B. Phaneuf, J. McDonough, and P. Etnoyer. 2005. Occurrence of deep-water Lophelia pertusa & Madrepora oculata in the Gulf of Mexico. Pages 297–307 in A. Freiwald and J. Roberts, eds. Cold-water corals and ecosystems, Springer Verlag, Berlin.

Stetson, T. R., D. F. Squires, and R. M. Pratt. 1962. Coral Banks occurring in deep water on the Blake Plateau. Am. Mus. Novit. 2114: 39.

Squires, D. F. 1964. Fossil coral thickets in Wairarapa, New Zealand. J. Paleontol. 38: 905–915.

Taviani, M., A. Remia, C. Corselli, A. Freiwald, E. Malinverno, F. Mastrototaro, A. Savini, and A. Tursi. 2005. First geo-marine survey of living cold-water Lophelia reefs in the Ionian Sea (Mediterranean basin). Facies 50: 409–417.

Taylor, L. A., T. L. Holcomb, and W. R. Bryant. 2000. Bathymetry of the northern Gulf of Mexico and the Atlantic Ocean east of Florida. Report MGG-16, 17, 18 National Geophysical Data Center, Boulder, Boulder.

Wilson, J. B. 1979a. The distribution of the coral *Lophelia pertusa* (L) (*L. prolifera* (Pallas)) in the north-east Atlantic. J. Mar. Biol. Assoc. UK 59: 149–164.

_____. 1979b. Patch development of the deep-water coral *Lophelia pertusa* (L.) on Rockall Bank. J. Mar. Biol. Assoc. U.K. 59: 165–177.

ADDRESS: (W.W.S.) *Marine Science Program, The University of Alabama, 101 Bienville Boulevard, Dauphin Island, Alabama 36528. E-mail: <wschroeder@disl.org>.*